Empirical Inference

Bernhard Schölkopf • Zhiyuan Luo • Vladimir Vovk
Editors

Empirical Inference

Festschrift in Honor of Vladimir N. Vapnik

Editors
Bernhard Schölkopf
Max Planck Institute for Intelligent Systems
Tübingen
Germany

Zhiyuan Luo
Vladimir Vovk
Department of Computer Science
Royal Holloway, University of London
Egham
Surrey
United Kingdom

ISBN 978-3-642-41135-9   ISBN 978-3-642-41136-6 (eBook)
DOI 10.1007/978-3-642-41136-6
Springer Heidelberg New York Dordrecht London

Library of Congress Control Number: 2013954388

© Springer-Verlag Berlin Heidelberg 2013
This work is subject to copyright. All rights are reserved by the Publisher, whether the whole or part of the material is concerned, specifically the rights of translation, reprinting, reuse of illustrations, recitation, broadcasting, reproduction on microfilms or in any other physical way, and transmission or information storage and retrieval, electronic adaptation, computer software, or by similar or dissimilar methodology now known or hereafter developed. Exempted from this legal reservation are brief excerpts in connection with reviews or scholarly analysis or material supplied specifically for the purpose of being entered and executed on a computer system, for exclusive use by the purchaser of the work. Duplication of this publication or parts thereof is permitted only under the provisions of the Copyright Law of the Publisher's location, in its current version, and permission for use must always be obtained from Springer. Permissions for use may be obtained through RightsLink at the Copyright Clearance Center. Violations are liable to prosecution under the respective Copyright Law.
The use of general descriptive names, registered names, trademarks, service marks, etc. in this publication does not imply, even in the absence of a specific statement, that such names are exempt from the relevant protective laws and regulations and therefore free for general use.
While the advice and information in this book are believed to be true and accurate at the date of publication, neither the authors nor the editors nor the publisher can accept any legal responsibility for any errors or omissions that may be made. The publisher makes no warranty, express or implied, with respect to the material contained herein.

Printed on acid-free paper

Springer is part of Springer Science+Business Media (www.springer.com)

Fig. 1 Some participants in the Empirical Inference Symposium (Photo taken by Robert Williamson)

Fig. 2 Vladimir listening to a talk (Photo by Robert Williamson)

Fig. 3 Vladimir delivering a talk (Photo by Robert Williamson)

Preface

Vladimir Vapnik is a rare example of a scientist for whom the following three statements hold true simultaneously: his work has led to the inception of a whole new field of research, he has lived to see the field blossom and gain in popularity, and he is still as active as ever in his field.

His field, the theory of statistical learning and empirical inference, did not exist when he started his PhD in Moscow in the early 1960s. He was working at the Institute of Control Sciences of the Russian Academy of Sciences under the supervision of Professor Aleksandr Lerner, a cyberneticist and expert in control theory who was later to become a prominent "Refusenik" (a Soviet Jew who was denied permission to emigrate by the Soviet government). Vladimir Vapnik started analyzing learning algorithms and invented the first version of a pattern recognition algorithm termed the "Generalized Portrait", whose successor (the "Support Vector Machine", co-invented by him 30 years later) would become the method of choice in many pattern recognition problems ranging from computer vision to bioinformatics. Following this, he started to collaborate with Aleksey Chervonenkis, also a PhD student in Lerner's laboratory at the time, on the Generalized Portrait and the theory of the empirical risk minimization inductive principle.

Vapnik and Chervonenkis found a stimulating intellectual environment at the Institute of Control Sciences. In 1951 Vadim Trapeznikov was appointed as the institute's director, and it is largely due to his efforts that in the early 1960s it became a hub of new ideas and creativity. It included Mark Aizerman's laboratory, working on the theory of learning systems and having Emmanuel Braverman and Lev Rozonoer among its members, as well as Lerner's laboratory, where Vapnik and Chervonenkis carried out their work that would lead to profound changes in our understanding of machine learning.

The impact of Vapnik and Chervonenkis's work has been considerable in many areas of mathematics and computer science. In theoretical statistics and probability theory, their work is known for extensions of the Glivenko–Cantelli theorem and for uniform laws of large numbers. The latter can be considered as the starting point of an important branch of probability theory, the theory of empirical processes.

The introduction of certain classes of functions (now called Vapnik–Chervonenkis classes) and of the notion now referred to as Vapnik–Chervonenkis, or VC, dimension was a key contribution to this area, relating concepts from analysis (approximation properties of classes of functions) to combinatorial parameters. Also, it was the starting point of what Vapnik called "statistical learning theory", an interdisciplinary field combining mathematical statistics and machine learning. The revolutionary aspect of this theory, when introduced, was that it focused on non-asymptotic estimation properties of non-parametric classes.

The main achievement of Vapnik and Chervonenkis's development of this theory was the introduction and analysis of the inductive principle called "empirical risk minimization." This principle suggests that, given data and a class of functions, we should choose the function that minimizes the error on the data (in other words, the minimizer of the "training error"). Their work on this inductive principle culminated in obtaining necessary and sufficient conditions for its consistency (i.e., the principle asymptotically leading to optimal estimation), related to the validity of a uniform law of large numbers. They also showed how to relate the uniform law to notions of the "capacity" of the function classes used, such as the VC dimension. In machine learning theory, this gave rise to a sizeable community of researchers computing bounds on or estimates of the VC dimension for most popular function classes, such as neural networks or decision trees.

Following the analysis of empirical risk minimization, Vladimir Vapnik developed an improved inductive principle called "structural risk minimization," which underlies a large number of today's learning algorithms, and significantly contributed to the birth of the research area of model selection in statistics.

As recognition of his groundbreaking work in the theory and applications of pattern recognition, Vladimir Vapnik was awarded the Prize of the USSR Council of Ministers, an achievement which is particularly outstanding considering that Vladimir's career in the USSR was hampered significantly due to his being Jewish.

In 1991, after the coup to overthrow Gorbachev had just taken place, Vladimir decided to emigrate to the USA, where he joined the Adaptive Systems Research Department at AT&T Bell Laboratories. There, he and his co-workers developed one of the most successful methods in machine learning, the Support Vector Machine. Based on an ingenious combination of methods from statistical learning (the Generalized Portrait algorithm) and functional analysis (the theory of positive definite kernels), it transformed the field of machine learning. More than just an algorithm, it is a whole approach to learning problems, pioneering the use of functional analysis and convex optimization in machine learning. It has meanwhile set world records on a variety of real-world pattern-recognition benchmark problems. This success has attracted a large number of researchers as well as engineers from various disciplines to the field of statistical learning theory.

Vladimir has continued to develop the theory of support vector machines in unexpected directions. For example, in 2006 he came up with the idea of using "privileged information" in machine learning: information that is available only at the training stage. In a surprisingly wide range of applications such information improves the performance of learning algorithms.

Vladimir's work has found wide recognition throughout the world. In 2003 his groundbreaking work in theoretical and applied statistics and machine learning was recognized with the Alexander Humboldt Research Award. He received the 2005 Gabor Award of the International Neural Network Society. In 2006 he was elected a member of the United States National Academy of Engineering "for insights into the fundamental complexities of learning and for inventing practical and widely applied machine-learning algorithms". In 2008 he received, together with Corinna Cortes, the Paris Kanellakis Award from the Association for Computing Machinery "for the development of Support Vector Machines, a highly effective algorithm for classification and related machine learning problems". In 2010 came the Neural Networks Pioneer Award from the IEEE Computational Intelligence Society. In 2012 Vladimir was honoured with the IEEE Frank Rosenblatt Award "for development of support vector machines and statistical learning theory as a foundation of biologically inspired learning". In the same year he was awarded the Benjamin Franklin Medal in Computer and Cognitive Science by the Franklin Institute "for his fundamental contributions to our understanding of machine learning, which allows computers to classify new data based on statistical models derived from earlier examples, and for his invention of widely used machine learning techniques".

Vladimir has published a number of papers and books that are considered classics. His monographs contain not only a comprehensive account of statistical learning theory and its applications, but they also host a wealth of original and unexplored directions for future research and improvements.

It is remarkable for the development of humankind that we have now arrived at a point where there exist regularities (non-random structures) in the world that are so complex that they cannot be detected by humans, yet they can reliably be learnt by machine learning methods. We believe this will have a transformative effect on our interaction with the world, and we cannot think of an individual whose impact on this transformation has been larger than that of Vladimir Vapnik.

Vladimir's impact on the machine learning community, both through his technical contributions and his philosophy of research as conveyed by numerous keynote talks, has been so large that his standing in the field is that of a living legend. There is no doubt that he will continue to have a profound influence on the field of machine learning, as more and more of his ideas are being put into practice.

Tübingen, Germany and Egham, UK   Bernhard Schölkopf
June 2013   Zhiyuan Luo
                                  Vladimir Vovk

Festschrift

In December 2011 Vladimir Vapnik, one of the founders of statistical learning theory, turned 75 years. To celebrate this event, the Max Planck Institute for Intelligent Systems (Krikamol Muandet, Yevgeny Seldin, and Bernhard Schölkopf) organized a symposium entitled *Empirical Inference*. It was held from 8 to 10 December 2011 in Tübingen, Germany.

The first 2 days consisted of talks given by invited researchers in the fields of statistical learning theory and kernel methods, to which Vladimir has contributed so much. The third day featured additional talks by alumni of the Empirical Inference department at the MPI for Intelligent Systems.

This Festschrift contains both written versions of many of the talks presented at the symposium and chapters written especially for the Festschrift. They reflect the breadth of the research fields deeply affected by Vladimir's extraordinary research contributions, which have served as a constant source of inspiration for other researchers.

Vladimir is active as ever and continues to generate new ideas and provide profound insights. He is an inspiring teacher to his students and a dear friend to many of his colleagues. We wish him happiness and success for many years to come.

Acknowledgements We are grateful to Alexey Chervonenkis, Alex Gammerman, and Chris Watkins for their help and advice. Thanks to Springer, and Ronan Nugent in particular, for their support of this project.

Credits

- The group photograph and two photographs of Vladimir Vapnik in the book frontmatter are used with permission from Robert Williamson.

- The following translation by Lisa Rosenblatt of the 1968 paper by Vladimir Vapnik and Alexey Chervonenkis on pp. 7–12 is used with permission of the American Mathematical Society: V. N. Vapnik, A. Ya. Červonenkis, Uniform Convergence of Frequencies of Occurrence of Events to Their Probabilities, Soviet Mathematics Doklady 9(4), 915–918 (1968). © 1968 by the American Mathematical Society.

- The original Russian text of the 1968 paper by Vladimir Vapnik and Alexey Chervonenkis is used in our translation of several passages with permission of Akademizdattsentr "Nauka" of the Russian Academy of Sciences (Академиздатцентр "Наука" РАН): В. Н. Вапник, А. Я. Червоненкис, О равномерной сходимости частот появления событий к их вероятностям, Доклады Академии Наук СССР 181(4), 781–783 (1968).

- Figures 1.5 and 7.8 and Tables 7.1 and 14.3 from the book *Boosting* by Schapire and Freund (the exact reference is given below) are used in Chap. 5 with permission of MIT Press: Robert E. Schapire and Yoav Freund. Boosting: Foundations and Algorithms. MIT Press, Cambridge, MA, 2012.

- Figure 1 from the following paper is used in Chap. 5 with permission of the Institute of Mathematical Statistics: Robert E. Schapire, Yoav Freund, Peter Bartlett, and Wee Sun Lee: Boosting the margin: A new explanation for the effectiveness of voting methods. Annals of Statistics **26**(5), 1651–1686 (1998).

Contents

Part I History of Statistical Learning Theory

1. **In Hindsight: Doklady Akademii Nauk SSSR, 181(4), 1968** 3
   Léon Bottou

2. **On the Uniform Convergence of the Frequencies of Occurrence of Events to Their Probabilities** 7
   Vladimir N. Vapnik and Alexey Ya. Chervonenkis

3. **Early History of Support Vector Machines** 13
   Alexey Ya. Chervonenkis

Part II Theory and Practice of Statistical Learning Theory

4. **Some Remarks on the Statistical Analysis of SVMs and Related Methods** 25
   Ingo Steinwart

5. **Explaining AdaBoost** 37
   Robert E. Schapire

6. **On the Relations and Differences Between Popper Dimension, Exclusion Dimension and VC-Dimension** 53
   Yevgeny Seldin and Bernhard Schölkopf

7. **On Learnability, Complexity and Stability** 59
   Silvia Villa, Lorenzo Rosasco, and Tomaso Poggio

8. **Loss Functions** 71
   Robert C. Williamson

9. **Statistical Learning Theory in Practice** 81
   Jason Weston

10. **PAC-Bayesian Theory** 95
    David McAllester and Takintayo Akinbiyi

11 **Kernel Ridge Regression** 105
   Vladimir Vovk

12 **Multi-task Learning for Computational Biology:
   Overview and Outlook** 117
   Christian Widmer, Marius Kloft, and Gunnar Rätsch

13 **Semi-supervised Learning in Causal and Anticausal Settings** 129
   Bernhard Schölkopf, Dominik Janzing, Jonas Peters,
   Eleni Sgouritsa, Kun Zhang, and Joris Mooij

14 **Strong Universal Consistent Estimate of the Minimum
   Mean Squared Error** 143
   Luc Devroye, Paola G. Ferrario, László Györfi,
   and Harro Walk

15 **The Median Hypothesis** 161
   Ran Gilad-Bachrach and Chris J.C. Burges

16 **Efficient Transductive Online Learning via Randomized
   Rounding** 177
   Nicolò Cesa-Bianchi and Ohad Shamir

17 **Pivotal Estimation in High-Dimensional Regression
   via Linear Programming** 195
   Eric Gautier and Alexandre B. Tsybakov

18 **On Sparsity Inducing Regularization Methods for
   Machine Learning** 205
   Andreas Argyriou, Luca Baldassarre, Charles A. Micchelli,
   and Massimiliano Pontil

19 **Sharp Oracle Inequalities in Low Rank Estimation** 217
   Vladimir Koltchinskii

20 **On the Consistency of the Bootstrap Approach
   for Support Vector Machines and Related Kernel-Based Methods** ... 231
   Andreas Christmann and Robert Hable

21 **Kernels, Pre-images and Optimization** 245
   John C. Snyder, Sebastian Mika, Kieron Burke,
   and Klaus-Robert Müller

22 **Efficient Learning of Sparse Ranking Functions** 261
   Mark Stevens, Samy Bengio, and Yoram Singer

23 **Direct Approximation of Divergences Between Probability
   Distributions** 273
   Masashi Sugiyama

Index 285

List of Contributors

Tayo Akinbiyi Statistics Department, University of Chicago, Chicago, IL, USA

Andreas Argyriou École Centrale de Paris, Grande Voie des Vignes, Chatenay-Malabry, France

Luca Baldassarre Laboratory for Information and Inference Systems, EPFL, Lausanne, Switzerland

Samy Bengio Google Research, Mountain View, CA, USA

Léon Bottou Microsoft Research, Redmond, WA, USA

Chris J.C. Burges Microsoft Research, Redmond, WA, USA

Kieron Burke Departments of Chemistry, and Physics, University of California, Irvine, CA, USA

Nicolò Cesa-Bianchi Dipartimento di Informatica, Università degli Studi di Milano, Milano, Italy

Alexey Ya. Chervonenkis Institute of Control Sciences, Laboratory 38, Moscow, Russia

Andreas Christmann Department of Mathematics, Lehrstuhl für Stochastik, University of Bayreuth, Bayreuth, Germany

Luc Devroye School of Computer Science, McGill University, Montreal, Canada

Paola G. Ferrario Department of Mathematics, University of Stuttgart, Stuttgart, Germany

Eric Gautier CREST-ENSAE, Malakoff, France

Ran Gilad-Bachrach Microsoft Research, Redmond, WA, USA

László Györfi Department of Computer Science and Information Theory, Budapest University of Technology and Economics, Budapest, Hungary

Robert Hable Department of Mathematics, Lehrstuhl für Stochastik, University of Bayreuth, Bayreuth, Germany

Dominik Janzing Max Planck Institute for Intelligent Systems, Tübingen, Germany

Marius Kloft Computational Biology Center, Memorial Sloan-Kettering Cancer Center, New York, NY, USA

Courant Institute of Mathematical Sciences, New York University, New York, NY, USA

Vladimir Koltchinskii School of Mathematics, Georgia Institute of Technology, Atlanta, GA, USA

David McAllester Toyota Technological Institute at Chicago, Chicago, IL, USA

Charles A. Micchelli Department of Mathematics and Statistics, University at Albany, Albany, NY, USA

Sebastian Mika idalab GmbH, Berlin, Germany

Joris Mooij Institute for Computing & Information Sciences, Radboud University, Nijmegen, Netherlands

Klaus-Robert Müller Machine Learning Group, Technical University of Berlin, Berlin, Germany

Department of Brain and Cognitive Engineering, Korea University, Anam-dong, Seongbuk-gu, Seoul, Korea

Jonas Peters Max Planck Institute for Intelligent Systems, Tübingen, Germany

Tomaso Poggio Center for Biological and Computational Learning, Massachusetts Institute of Technology, Cambridge, MA, USA

Massimiliano Pontil Department of Computer Science, University College London, London, UK

Gunnar Rätsch Computational Biology Center, Memorial Sloan-Kettering Cancer Center, New York, NY, USA

Lorenzo Rosasco Laboratory for Computational and Statistical Learning, Massachusetts Institute of Technology, Cambridge, MA, USA

Dipartimento di Informatica, Istituto Italiano di Tecnologia, Bioingegneria, Robotica e Ingegneria dei Sistemi, Università di Genova, Genova, Italy

Robert E. Schapire Department of Computer Science, Princeton University, Princeton, NJ, USA

Bernhard Schölkopf Max Planck Institute for Intelligent Systems, Tübingen, Germany

List of Contributors

Yevgeny Seldin Queensland University of Technology, Brisbane, Australia

Eleni Sgouritsa Max Planck Institute for Intelligent Systems, Tübingen, Germany

Ohad Shamir Microsoft Research, Cambridge, MA, USA

Yoram Singer Google Research, Mountain View, CA, USA

John C. Snyder Departments of Chemistry, and Physics, University of California, Irvine, CA, USA

Ingo Steinwart Institute for Stochastics and Applications, University of Stuttgart, Stuttgart, Germany

Mark Stevens Google Research, Mountain View, CA, USA

Masashi Sugiyama Tokyo Institute of Technology, Meguro-ku, Tokyo, Japan

Alexandre B. Tsybakov CREST-ENSAE, Malakoff, France

Vladimir N. Vapnik NEC Laboratories America, Princeton, NJ, USA

Silvia Villa Laboratory for Computational and Statistical Learning, Massachusetts Institute of Technology, Cambridge, MA, USA

Dipartimento di Informatica, Istituto Italiano di Tecnologia, Bioingegneria, Robotica e Ingegneria dei Sistemi, Università di Genova, Genova, Italy

Vladimir Vovk Dept. of Computer Science, Royal Holloway, University of London, United Kingdom

Harro Walk Department of Mathematics, University of Stuttgart, Stuttgart, Germany

Jason Weston Google Inc., New York, NY, USA

Christian Widmer Computational Biology Center, Memorial Sloan-Kettering Cancer Center, New York, NY, USA

Machine Learning Group, Technische Universität Berlin, Berlin, Germany

Robert C. Williamson Australian National University and NICTA, Canberra, Australia

Kun Zhang Max Planck Institute for Intelligent Systems, Tübingen, Germany

Acronyms

AEF	Approximation Error Function
AERM	Asymptotic Empirical Risk Minimization
DFT	Density Functional Theory
ERM	Empirical Risk Minimization
IR	Information Retrieval
KL	Kullback–Leibler (as in Kullback–Leibler divergence)
kPCA	Kernel Principal Component Analysis
KRR	Kernel Ridge Regression
MA	Median Approximation
MAP	Maximum a Posteriori
MKL	Multiple Kernel Learning
MTL	Multi-task Learning
MT-MKL	Multi-task Multiple Kernel Learning
NLGD	Nonlinear Gradient Denoising
PAC	Probably Approximately Correct
PC	Principal Component
PCA	Principal Component Analysis
PE	Pearson (as in Pearson Divergence)
R^2	Randomized Rounding
RBF	Radial Basis Function
RKHS	Reproducing Kernel Hilbert Space
RMSE	Root Mean Square Errors
rPE	Relative Pearson (as in Relative Pearson Divergence)
RR	(Ordinary) Ridge Regression
RRCM	Ridge Regression Confidence Machine
SLT	Statistical Learning Theory
SRM	Structural Risk Minimization
SSL	Semi-supervised Learning
SSR	Semi-supervised Regression
SVM	Support Vector Machine
SVR	Support Vector Regression
VC	Vapnik–Chervonenkis (as in the VC Dimension)

Part I
History of Statistical Learning Theory

Part I of the book contains three chapters describing and witnessing several contributions by Vladimir Vapnik to science. In Chap. 1, Léon Bottou discusses the seminal paper published in 1968 by Vapnik and Chervonenkis and laying the foundations of statistical learning theory. An English translation of this important paper is included as Chap. 2. In Chap. 3, Alexey Chervonenkis, the other main developer of the Vapnik–Chervonenkis theory, presents a first-hand account of the early history of Support Vector Machines and provides valuable insights into the first steps in the development of the SVM in the framework of the "Generalised Portrait" method. Different chapters sometimes use different names for Vapnik and Chervonenkis's home institution at the time; in 1969, the Institute of Automation and Remote Control, where they had worked since the early 1960s, was renamed the Institute of Control Sciences. For a detailed technical exposition of several of the results mentioned in Chap. 3, see Chap. 3 of Vapnik and Chervonenkis's monograph [1].

Reference

1. Vapnik, V.N., Chervonenkis, A.Y.: Теория распознавания образов (Theory of Pattern Recognition). Nauka, Moscow (1974). German translation: Theorie der Zeichenerkennung. Akademie, Berlin, 1979

Chapter 1
In Hindsight: Doklady Akademii Nauk SSSR, 181(4), 1968

Léon Bottou

Abstract This short contribution presents the first paper in which Vapnik and Chervonenkis describe the foundations of Statistical Learning Theory (Vapnik, Chervonenkis (1968) Proc USSR Acad Sci 181(4): 781–783).

This short contribution presents the first paper in which Vapnik and Chervonenkis describe the foundations of Statistical Learning Theory [10]. The original paper was published in the *Doklady*, the Proceedings of the USSR Academy of Sciences, in 1968. An English translation was published the same year in *Soviet Mathematics*, a journal from the American Mathematical Society publishing translations of the mathematical section of the Doklady.[1] The importance of the work of Vapnik and Chervonenkis was noticed immediately. Dudley begins his 1969 review for *Mathematical Reviews* [3] with the simple sentence "*the following very interesting results are announced.*"

This concise paper is historically more interesting than the celebrated 1971 paper [11] because the three-page limit has forced its authors to reveal what they consider essential. Every word in this paper counts. In particular, the introduction explains that a uniform law of large numbers "*is necessary in the construction of learning algorithms.*" The mention of learning algorithms in 1968 seems to be an anachronism. In fact, learning machines were a popular subject in the 1960s at the Institute of Automation and Remote Control in Moscow. The trend possibly started with the work of Aizerman and collaborators on pattern recognition [1] and the work

---

[1] A slightly modified version of this English translation of the 1968 paper follows this brief introduction.

L. Bottou (✉)
Microsoft Research, 1 Microsoft Way, Redmond, WA 98052, USA
e-mail: leon@bottou.org

of Fel'dbaum on dual control [4]. Tsypkin then wrote two monographs [7, 8] that clearly define machine learning as a topic for both research and engineering.

These early works on machine learning are supported by diverse mathematical arguments suggesting that learning takes place. The uniform convergence results introduced in the 1968 paper provide powerful tools to construct such arguments. In fact, in their following works [9, 12], Vapnik and Chervonenkis show how the uniform convergence concept splits such arguments into three clearly defined parts: the approximation properties of the model, the estimation properties of the induction principle, and the computational properties of the learning algorithm. Instead of simply establishing proofs for specific cases, the work of Vapnik and Chervonenkis reveals the structure of the space of all learning algorithms. This is a higher achievement in mathematics.

Whereas the law of large numbers tells us how to estimate the probability of a single event, the uniform law of large numbers explains how to simultaneously estimate the probabilities of an infinite family of events. The passage from the simple law to the uniform law relies on a remarkable combinatorial result (Theorem 1 in the 1968 paper). This result was given without proof, most likely because the paper would have exceeded the three page limit. The independent discovery of this combinatorial result is usually attributed to Vapnik and Chervonenkis [11], Sauer [5], or Shelah [6]. Although this cannot be established with certainty, several details suggest that the 1968 paper and its review by Dudley attracted the attention of eminent mathematicians and diffused into the work of their collaborators.[2] However, Sauer gives a better bound in his 1972 paper than Vapnik and Chervonenkis in their 1971 paper.[3] The combinatorial result of the first theorem directly leads to the best known Vapnik–Chervonenkis theorem, namely, the distribution-independent sufficient condition for uniform convergence. A detailed sketch of the proof supports this second theorem. Although the paper mentions the connection with the Glivenko–Cantelli theorem [2], the paper does not spell out the notion of capacity, now known as the *Vapnik–Chervonenkis dimension*. However, the definition of the growth function is followed by its expression for three simple families of events, including the family of half-spaces associated with linear classifiers.

---

[2]Sauer motivates his work with a single sentence, "*P. Erdös transmitted to me in Nice the following question: is it true that [statement of the theorem]*", without attributing the conjecture to anyone. Sauer kindly replied to my questions with interesting details: "*When I proved that Lemma, I was very young, and have since moved my interest more towards model theoretic type questions. Erdös visited Calgary and told me at that occasion that this question had come up. But I do not remember the context in which he claimed that it did come up. I then produced a proof and submitted it as a paper. I did not know about that question before the visit by Erdös,*" and "*the only thing I can contribute is that I believe Weiss in Israel told me that Shelah had asked Perles to prove such a Lemma, which he did, and subsequently both forgot about it and Shelah then asked Perles again to prove that Lemma.*"

[3]In fact, Sauer gives the optimal bound (Dudley, personal communication).

The third and final theorem states the distribution-dependent necessary and sufficient condition for uniform convergence. The paper provides a minimal proof sketch. The proof takes in fact 7 pages in [11] and 23 pages in [9].

In conclusion, this concise paper deserves recognition because it contains the true beginnings of Statistical Learning Theory. The work is clearly motivated by the design of learning algorithms and its results have provided a new foundation for statistics in the computer age.

References

1. Aizerman, M.A., Braverman, É.M., Rozonoér, L.I.: Theoretical foundations of the potential function method in pattern recognition learning. Autom. Remote Control **25**, 821–837 (1964)
2. Cantelli, F.P.: Sulla determinazione empirica della legi di probabilita. Giornale dell'Istituto Italiano degli Attuari **4**, 421–424 (1933)
3. Dudley, R.M.: Mathematical Reviews MR0231431 (37#6986) (1969)
4. Fel'dbaum, A.A.: Optimal Control Systems. Nauka, Moscow (1963). English translation: Academic, New York, 1965
5. Sauer, N.: On the density of families of sets. J. Comb. Theory A **13**, 145–147 (1972)
6. Shelah, S.: A combinatorial problem: stability and order for models and theories in infinitary languages. Pac. J. Math. **41**, 247–261 (1972)
7. Tsypkin, Y.: Adaptation and Learning in Automatic Systems. Nauka, Moscow (1969). English translation: Academic, New York, 1971
8. Tsypkin, Y.: Foundations of the Theory of Learning Systems. Nauka, Moscow (1970). English translation: Academic, New York, 1973
9. Vapnik, V.N.: Estimation of Dependences Based on Empirical Data. Information Science and Statistics. Springer, Berlin/New York (1982)
10. Vapnik, V.N., Chervonenkis, A.Y.: On the uniform convergence of relative frequencies of events to their probabilities. Proc. USSR Acad. Sci. **181**(4), 781–783 (1968). English translation: Soviet Math. Dokl. **9**, 915–918 (1968)
11. Vapnik, V.N., Chervonenkis, A.Y.: On the uniform convergence of relative frequencies of events to their probabilities. Theory Probab. Appl. **16**(2), 264–281 (1971)
12. Vapnik, V.N., Chervonenkis, A.Y.: Theory of pattern recognition. Nauka, Moscow (1974). German translation: Akademie-Verlag, Berlin, 1979

Chapter 2
On the Uniform Convergence of the Frequencies of Occurrence of Events to Their Probabilities

Vladimir N. Vapnik and Alexey Ya. Chervonenkis

Abstract This chapter is a translation of Vapnik and Chervonenkis's pathbreaking note

> В. Н. Вапник, А. Я. Червоненкис, О равномерной сходимости частот появления событий к их вероятностям, Доклады Академии Наук СССР 181(4), 781–783 (1968)

essentially following the excellent translation

> V. N. Vapnik, A. Ya. Červonenkis, Uniform Convergence of Frequencies of Occurrence of Events to Their Probabilities, Soviet Mathematics Doklady 9(4), 915–918 (1968)

by Lisa Rosenblatt (the editors only corrected a few minor mistakes and in some places made the translation follow more closely the Russian original).

(Presented by Academician V. A. Trapeznikov, 6 October 1967)

2.1 Introduction

According to the classical theorem of Bernoulli, the frequency of occurrence of an event A converges (in probability, in a sequence of independent trials to the probability of this event). In some applications, however, it is necessary to draw conclusions about the probabilities of the events of an entire class S from one and the same sample. (In particular, this is necessary in the construction of learning algorithms.) Here it is important to find out whether the frequencies converge to the probabilities uniformly over the entire class of events S. More precisely, it is important to find out whether the probability that the maximal deviation of frequency from the corresponding probability over the class S exceeds a given small number approaches zero in an unbounded number of trials. It turns out that even in the simplest examples such uniform convergence may not take place. Therefore we would like to have a criterion by which we can decide whether there is such convergence or not. In this note we consider sufficient conditions for such uniform convergence which do not depend on the properties of the distribution but are related only to the internal properties of the class S, and we give a bound on the rate of convergence also not depending on the distribution, and finally we point out necessary and sufficient conditions for the uniform convergence of the frequencies to the probabilities over the class of events S.

2.2 Statement of the Problem

Let X be a set of elementary events on which a probability measure μ is defined. Let S be a collection of random events, i.e., of subsets of the space X measurable relative to the measure μ (the system S belongs to a Borel system but does not necessarily coincide with it).

Let $X^{(l)}$ denote the space of samples from X of length l. On the space $X^{(l)}$ we define the probability product measure by the condition $P(Y_1 \cdot Y_2 \cdot \ldots \cdot Y_l) = P(Y_1) \cdot P(Y_2) \cdot \ldots \cdot P(Y_l)$, where Y_i are measurable subsets of X. This formalises the fact that sampling is repeated, i.e., the elements are chosen independently with a fixed distribution.

For every sample $x_1, \ldots, x_l$ and an event A we can define the frequency $v_A^l = v_A(x_1, \ldots, x_l)$ of occurrence of the event A as equal to the ratio of the number n_A of those elements of the sample which belong to A to the overall length l of the sample:

$$v_A(x_1, \ldots, x_l) = n_A/l.$$

Bernoulli's theorem asserts that

$$\lim_{l \to \infty} P(|v_A^l - P_A| > \varepsilon) = 0.$$

We, however, will be interested in the maximal deviation of the frequency from the probability

$$\pi^{(l)} = \sup_{A \in S} |v_A^l - P_A|$$

over the class. The quantity $\pi^{(l)}$ is a point function on the space $X^{(l)}$.

We will assume that this function is measurable relative to the measure on $X^{(l)}$, i.e., that $\pi^{(l)}$ is a random variable. If $\pi^{(l)}$ approaches 0 in probability with unbounded increase of the sample size l, then we will say that the frequencies of the events $A \in S$ converge in probability to the probabilities of these events uniformly over the class S.

The theorems below are concerned with estimating the probability of the event

$$\pi^{(l)} \xrightarrow[l\to\infty]{} 0$$

and finding out conditions when

$$P\left(\pi^{(l)} \xrightarrow[l\to\infty]{} 0\right) = 1.$$

2.3 Some Additional Definitions

Let $X_r = x_1, \ldots, x_r$ be a finite sample of elements from X. Every set A from S determines a subsample $X_r^A = x_{i_1}, \ldots, x_{i_k}$ on this sample consisting of those elements of the sample X_r which are in A. We will say that the set A induces the subsample X_r^A on the sample X_r.

Denote the set of all distinct subsamples induced by the sets from S on the sample X_r by $S(x_1, \ldots, x_r)$. The number of distinct subsamples of the sample X_r induced by the sets from S (the number of elements of the set $S(x_1, \ldots, x_r)$) will be called the index of the system S relative to the sample X_r and denoted by $\Delta^S(x_1, \ldots, x_r)$.

Obviously it is always true that

$$\Delta^S(x_1, \ldots, x_r) \leq 2^r.$$

The function $m^S(r) = \max_{x_1, \ldots, x_r} \Delta^S(x_1, \ldots, x_r)$, where the maximum is taken over all samples of length r, is called the growth function of the class S.

Example 2.1. Let X be a straight line and S the set of all rays of the form $x < a$; $m^S(r) = r + 1$.

Example 2.2. X is the segment $[0, 1]$; S consists of all open sets; $m^S(r) = 2^r$.

Example 2.3. Let X be n-dimensional Euclidean space. The set of events S consists of all half-spaces of the form $(x\phi) > c$, where ϕ is a vector and c a constant; $m^S(r) < r^n$ $(r > n)$.

Along with the growth function $m^S(r)$ consider the function

$$M^S(r) = \int_{X^{(r)}} \ln \Delta^S(x_1, \ldots, x_r) d\mu(X^r);$$

$M^S(r)$ is the mathematical expectation of the logarithm of the index $\Delta^S(x_1, \ldots, x_r)$ of the system S.

2.4 A Property of the Growth Function

The main property of the growth function of the class S is established by the following theorem.

Theorem 2.1. *The growth function $m^S(r)$ is either identically equal to 2^r or majorized by the function r^n, where n is the first value of r for which $m^S(n) \neq 2^n$.*

2.5 Sufficient Conditions for Uniform Convergence Not Depending on Properties of the Distribution

Sufficient conditions for the uniform convergence (with probability 1) of the frequencies to the probabilities are established by the following theorem.

Theorem 2.2. *If $m^S(r) \leq r^n$, then*

$$P\left(\pi^{(l)} \xrightarrow[l\to\infty]{} 0\right) = 1.$$

To prove this theorem, we establish the following lemma.

Take a sample $x_1, \ldots, x_l, x_{l+1}, \ldots, x_{2l}$ of length $2l$ and compute the frequencies of occurrence of an event A on the first half-sample $x_1, \ldots, x_l$ and the second half-sample $x_{l+1}, \ldots, x_{2l}$. Denote the corresponding frequencies by v'_A and v''_A and consider $\rho^{(l)}_A = |v'_A - v''_A|$. We will be interested in the maximal deviation of $\rho^{(l)}_A$ over all events of S, i.e., $\rho^{(l)} = \sup_{A \in S} \rho^{(l)}_A$.

Lemma 2.1. *For each ε with $l > 2/\varepsilon^2$ we have the inequality*

$$P\left(\pi^{(l)} > \varepsilon\right) \leq 2P\left(\rho^{(l)} > \varepsilon/2\right).$$

We further establish for the proof of Theorem 2.2 that

$$P\left(\rho^{(l)} > \varepsilon/2\right) < 2m^S(2l)e^{-\varepsilon^2 l/16},$$

whence

$$P\left(\pi^{(l)} > \varepsilon\right) < 4m^S(2l)e^{-\varepsilon^2 l/16}. \qquad (*)$$

In the case where $m^S(r) < r^n$, the inequality (*) implies uniform convergence in probability. By a well-known lemma [1] from probability theory, we also establish convergence with probability 1 under the conditions of the theorem.

According to Theorem 2.2 there is uniform convergence in Examples 2.1 and 2.3 considered in Sect. 2.3. The fact that there is uniform convergence in Example 2.1 coincides with the assertion of Glivenko's theorem.

In many applications it is necessary to know the required sample size in order to assert with probability at least $1 - \eta$ that the maximal deviation of the frequency from the probability over the class of events S does not exceed ε.

In the case where the growth function $m^S(l) \leq l^n$ for the class S, the inequality (*) easily yields

$$l \geq \frac{32n}{\varepsilon^2}\left(\ln\frac{32n}{\varepsilon^2} - \ln\frac{\eta}{4}\right).$$

2.6 Necessary and Sufficient Conditions for the Uniform Convergence of Frequencies to Probabilities

Theorem 2.3. *For the uniform convergence (with probability 1) of the frequencies to the probabilities over the class of events S the condition*

$$\lim_{l\to\infty} \frac{M^S(l)}{l} = 0; \quad (M^S(l) = E(\ln \Delta^S(x_1,\ldots,x_l)))$$

is necessary and sufficient (here we assume the measurability of the function $\Delta^S(x_1,\ldots,x_l)$).

For the proof of Theorem 2.3 we consider a lemma.

Lemma 2.2. *The sequence $M^S(l)/l$ has a limit as $l \to \infty$.*

In the case where this limit is equal to 0, the sufficiency of the condition is proved analogously to Theorem 2.2. For the proof of necessity we first establish that

$$P(\pi^{(l)} > \varepsilon) > \frac{1}{2}P(\rho^{(l)} > 2\varepsilon).$$

We further establish that if $\lim_{l\to\infty} M^S(l)/l = t \neq 0$ then there is a δ such that

$$\lim_{l\to\infty} P\left(\rho^{(l)} > 2\delta\right) = 1,$$

whence $\lim_{l\to\infty} P\left(\pi^{(l)} > \delta\right) \neq 0$.

The theorem is proved.

V.N. Vapnik and A. Ya. Chervonenkis
Institute of Automation and Remote Control
(Technical Cybernetics)

Received
6 October 1967

Reference

1. Гнеденко, Б.В.: Курс теории вероятностей, 3rd edn. Fizmatgiz, Moscow (1961). English translation: Gnedenko, B.V.: A Course in Probability Theory, p. 212. Chelsea, New York, 1962. MR 25 #2622

Chapter 3
Early History of Support Vector Machines

Alexey Ya. Chervonenkis

Abstract Many of the ideas now being developed in the framework of Support Vector Machines were first proposed by V. N. Vapnik and A. Ya. Chervonenkis (Institute of Control Sciences of the Russian Academy of Sciences, Moscow, Russia) in the framework of the "Generalised Portrait Method" for computer learning and pattern recognition. The development of these ideas started in 1962 and they were first published in 1964.

3.1 The "Generalised Portrait" Method for Pattern Recognition (1962)

3.1.1 Initial Heuristic Ideas

Patterns (pictures) were considered as unit vectors $\mathbf{x}$ ($|\mathbf{x}| = 1$) in some Euclidian space. That means they are situated on a unit sphere. A class of patterns is then a subset $\mathbf{S}$ of such a sphere. Of course, this is a heuristic: in reality vectors (patterns) may not be unit vectors.

The "generalised portrait" of a class $\mathbf{S}$ was defined as the unit vector φ that attains

$$\max_{\varphi} \min_{\mathbf{x} \in \mathbf{S}} (\varphi, \mathbf{x}) = c.$$

It means that the vector φ should be closest to the vectors in the set $\mathbf{S}$ that are most distant from φ. And we hope that

$$c > 0,$$

A. Ya. Chervonenkis (✉)
Institute of Control Sciences, Laboratory 38, Profsoyusnaya ulitsa 65, Moscow 117997, Russia
e-mail: chervnks@ipu.ru

i.e., that the set **S** is rather compact and does not cover too large part of the sphere. See [3].

In this way we try to cut off the segment of the unit sphere which contains completely the set **S** and has the minimum volume. This is done by the hyperplane

$$(\varphi, \mathbf{x}) = c,$$

and the segment is determined by the inequality $(\varphi, \mathbf{x}) \geq c$.

For all vectors in the set **S** it is true that the scalar product

$$(\varphi, \mathbf{x}) \geq c,$$

and there are some vectors of **S** for which

$$(\varphi, \mathbf{x}) = c.$$

They were called support vectors, or marginal vectors. All marginal vectors of the set **S** are just on the edge of the segment; all others are within it.

3.1.2 Reducing to a Convex (Quadratic) Programming Problem

Instead of maximising the scalar product of marginal vectors with the vector φ of a fixed norm, we may minimise some collinear vector ψ under fixed restrictions. Thus our problem is equivalent to the following: find a vector ψ with minimum norm (or its square (ψ, ψ)) under the restrictions

$$(\psi, \mathbf{x}) \geq 1,$$

for all $\mathbf{x} \in S$. Then by the normalisation of this vector to norm 1,

$$\varphi = \psi / |\psi|,$$

we get the required vector φ (of course, it is true only if $c > 0$). So we have reduced the problem to a quadratic programming problem.

3.1.3 Generalisation of the Idea

When we tried to apply this idea to practical pattern recognition learning (using a training sequence as the set **S**, i.e., using only patterns of the class we want to recognise), we found out that it is impossible in most cases to achieve good results without presenting examples of the opposite class or classes.

In reality, we want the scalar products $(\varphi, \mathbf{x})$ for all vectors $\mathbf{x}$ of the opposite class or classes to be less than those for the worst (marginal) vectors of our initial class.

Now we return to the initial idea, denoting by $\mathbf{S}_0$ our initial class of patterns and by $\mathbf{S}_1$ the opposite one. We want to find such a unit vector φ which delivers

$$\max_{\varphi} \min_{\mathbf{x} \in \mathbf{S}_0}(\varphi, \mathbf{x}) = c$$

under the condition that

$$(\varphi, \mathbf{x}) \leq kc \quad \text{for all vectors } \mathbf{x} \text{ of the class } \mathbf{S}_1,$$

where the constant $0 < k < 1$ determines the margin between the two classes.

Again, we want to find the least volume segment of the sphere which contains all vectors of class $\mathbf{S}_0$ and does not include any vector of class $\mathbf{S}_1$ (with some margin, determined by k). We call this vector φ the "generalised portrait" of the class $\mathbf{S}_0$ against the class $\mathbf{S}_1$. See [1, 2].

And again the problem may be reduced to a quadratic programming problem: find such a vector ψ that gives

$$\min(\psi, \psi)$$

under the constraints

$$(\psi, \mathbf{x}) \geq 1, \qquad \text{for all } \mathbf{x} \in \mathbf{S}_0, \qquad \text{and}$$
$$(\psi, \mathbf{y}) \leq k, \qquad \text{for all } \mathbf{y} \in \mathbf{S}_1,$$

with the constant $k < 1$ determining the margin (there is no need to assume $k > 0$).

Then the original vector φ can be found as $\varphi = \psi/|\psi|$. Those vectors of both classes for which the inequalities turn to equalities are called marginal (or support) vectors. If the system of inequalities is consistent, then the problem has a unique solution.

3.2 The Kuhn–Tucker Theorem and Its Application to the "Generalised Portrait" Method

As long as we had a convex programming problem, we could apply to it the well known Kuhn–Tucker theorem, which says: if you minimise a convex function $F(\mathbf{x})$ under the constraints

$$V_i(\mathbf{x}) \leq 0 \quad (i = 1, \ldots, n) \quad \text{with convex functions } V_i(\mathbf{x}),$$

then the necessary and sufficient condition for **x** to deliver a minimum is that there are such nonnegative values a_i that

$$\text{grad } F(\mathbf{x}) = -\sum a_i \text{ grad } V_i(\mathbf{x}),$$

and $a_i V_i(\mathbf{x}) = 0$, for all $i = 1, \ldots, n$. The last expression means that the coefficients a_i may differ from 0 only for those constraints which are reached.

Applying this result to our problem with $F(\mathbf{x}) = \frac{1}{2}(\boldsymbol{\psi}, \boldsymbol{\psi})$, we get for the optimal $\boldsymbol{\psi}$:

$$\boldsymbol{\psi} = \sum_{\mathbf{x}_i \in S_0} a_i \mathbf{x}_i - \sum_{\mathbf{y}_i \in S_1} b_i \mathbf{y}_i,$$

where the coefficients a_i and b_i are nonnegative, and only those which correspond to marginal (support) vectors are nonzero. From this formula it follows also that, for the optimal $\boldsymbol{\psi}$,

$$(\boldsymbol{\psi}, \boldsymbol{\psi}) = \sum_{\mathbf{x}_i \in S_0} a_i (\boldsymbol{\psi}, \mathbf{x}_i) - \sum_{\mathbf{y}_i \in S_1} b_i (\boldsymbol{\psi}, \mathbf{y}_i) = \sum_{\mathbf{x}_i \in S_0} a_i - k \sum_{\mathbf{y}_i \in S_1} b_i.$$

Now it becomes evident that we can remove anyhow those vectors which are not marginal (until they become marginal), or just exclude them from the training sequence, and it will not affect the "generalised portrait" position. In particular, all of them would be recognised correctly.

These results were also obtained in 1962 (first published in 1964 [1, 2]), though we did not use the Kuhn–Tucker theorem, but just proved the results directly.

3.3 Searching for Coefficients Instead of Coordinates

Now we can see that it is possible, instead of searching for the "generalised portrait" in coordinate form, to look for it in the form of a decomposition into the vectors of the training sequence. For brevity, I shall illustrate this fact for the case of a single class generalised portrait, but it can be easily extended to the case of two classes. So we are looking for

$$\boldsymbol{\psi} = \sum_i a_i \mathbf{x}_i.$$

Now the functional $(\boldsymbol{\psi}, \boldsymbol{\psi})$ can be presented in the form

$$(\boldsymbol{\psi}, \boldsymbol{\psi}) = \left(\boldsymbol{\psi}, \sum_i a_i \mathbf{x}_i\right) = \sum_{(i,j)} a_i a_j (\mathbf{x}_i, \mathbf{x}_j),$$

3 Early History of Support Vector Machines   17

and the constraints in the form

$$(\pmb{\psi}, \mathbf{x}_j) = \left(\sum_i a_i \mathbf{x}_i, \mathbf{x}_j\right) = \sum_i a_i (\mathbf{x}_i, \mathbf{x}_j) \geq 1.$$

So in this form we do not use representations either of the initial vectors or of the $\pmb{\psi}$ vector in the coordinate form, but use only the scalar products of the initial vectors and the coefficients of the decomposition of $\pmb{\psi}$ into these vectors.

This idea was used from the very beginning, as soon as we first used analogue computers for computation, and it was more convenient to present data in this form. Later, when using digital computers, we presented data mostly in the coordinate form. But now V. Vapnik has returned to this form using kernel functions, and this is the main difference between the generalised portrait method and the support vector machine method.

3.4 Optimum (Largest Margin) Hyperplane

If we look over different values of the constant k for the two-class generalised portrait, we will get different margins between the classes. For a generalised portrait φ we determine the width of the margin as

$$d = \min_{\mathbf{x} \in S_0}(\varphi, \mathbf{x}) - \max_{\mathbf{y} \in S_1}(\varphi, \mathbf{y}) = \min_{\mathbf{x},\mathbf{y}}(\varphi, (\mathbf{x} - \mathbf{y})), \qquad (3.1)$$

i.e., the distance between the projections of the classes onto the normal vector to the separating hyperplane.

There has been a desire to find the optimal separating hyperplane, the one that gives the maximum margin width. There are several ways to reduce this problem to generalised portrait search. One is to construct the set of vectors $\mathbf{z}$ of the form

$$\mathbf{z} = \mathbf{x} - \mathbf{y} \text{ for all } \mathbf{x} \in S_0 \text{ and } \mathbf{y} \in S_1;$$

then it is evident that the generalised portrait $\varphi \mathbf{z}$ of this set is the normal vector to the optimum hyperplane. However, this requires looking through all pairs of $\mathbf{x} \in S_0$ and $\mathbf{y} \in S_1$. Notice that marginal (support) vectors $\mathbf{z}$ in this case would be those formed by marginal pairs, i.e., those xs for which

$$(\varphi \mathbf{z}, \mathbf{x}) = \min_{\mathbf{x} \in S_0}(\varphi \mathbf{z}, \mathbf{x})$$

and those ys for which

$$(\varphi \mathbf{z}, \mathbf{y}) = \max_{\mathbf{y} \in S_1}(\varphi \mathbf{z}, \mathbf{y}).$$

Another way is to look through all possible values of the constant k and find the value which delivers the maximum to d in (3.1). The vector φz will correspond to some particular value of the constant k (though maybe the roles of the classes should be swapped).

Indeed, let us define the constant k as

$$k = \max_{y \in S_1}(\varphi z, y) / \min_{x \in S_0}(\varphi z, x).$$

It is always possible for separable classes, perhaps by swapping their roles, to make the values $\min_{x \in S_0}(\varphi z, x) > 0$ and $k < 1$. Now

$$\varphi z = \sum a_i z_i = \sum a_i (x_i - y_i) = \sum a_i x_i - \sum a_i y_i,$$

where $a_i \geq 0$ and z_i are marginal differences formed by pairs of marginal vectors x_i and y_i. Some vectors can be repeated in the latter sums. But in this case the vector ψz satisfies the Kuhn–Tucker sufficient conditions, and thus coincides with the generalised portrait of the classes for this particular value of the constant k.

Moreover, one can see that in this case the sums of coefficients before xs and before ys are equal. And it gives us a third way to find the optimum hyperplane. It is just to look for such a value of k that ensures this condition. So, in some sense, searching for the optimum hyperplane is a particular case of the generalised portrait method.

3.5 Lagrangian Dual Function

Instead of solving the initial quadratic programming problem, it is for many reasons more convenient to maximise the Lagrangian dual function

$$W(a, b) = \sum a_i - k \sum b_i - \frac{1}{2}(\psi, \psi),$$

where

$$\psi = \sum a_i x_i - \sum b_i y_i$$

under the constraints

$$a_i \geq 0,$$
$$b_i \geq 0.$$

In this case we have simpler constraints, but a more complicated quadratic function as compared with the initial problem. From the computational point of view this form is preferable.

To find the normal vector to the optimal hyperplane, it suffices to put $k = -1$ here and add the additional constraint

$$\sum a_i = \sum b_i.$$

3.6 Generalisation Properties

Generalisation properties of the generalised portrait method (and then of the optimal hyperplane and the support vector machine) become obvious if we apply the "Jackknife" method in this case. This method takes off one object from the training sequence, constructs a decision rule using some procedure, and then tests the omitted object by this rule. Then it does the same with other objects of the training sequence one by one and calculates the frequency of correct answers and errors.

Theoretically it will be an unbiased estimation of what you will have for a test sequence (though I am not aware of a good theoretical evaluation of its variance).

In our case, if we remove from the training sequence a vector which is not marginal (support vector), then the position of the generalised portrait will not change, and then the omitted vector will be recognised correctly. So the frequency of errors by the "Jackknife" testing method cannot exceed the share of support vectors within the total training sequence. Of course if you already have some errors in the training sequence, their frequency should be added to this estimate. Even more can be said. If the system of support vectors is linearly independent, there exists a unique decomposition of the generalised portrait into the support vectors. But if not, we can find some decompositions where some coefficients become equal to 0 while others are positive. We call "informative" those support vectors which never have 0 coefficients in such decompositions. It can be easily seen that the number of errors made by the "Jackknife" method can never exceed the number of informative vectors. In turn, the number of informative vectors never exceeds the dimensionality of the parameter space. Due to the fact that the "Jackknife" estimation is unbiased, we obtain a theoretical result: for linearly separable classes and the generalised portrait method, the error expectation for an exam is at most

$$\mathbb{E}\,\text{error} \leq n/(l+1), \tag{3.2}$$

where n is the space dimension and l is the length of the training sequence. In practice, it often happens that the number of marginal (support) vectors is much less than the dimension of the parameter space.

It is interesting to mention that, as our experience shows, the number of support vectors for the optimal hyperplane often turns out to be larger than that for the generalised portrait for other constants k. And, though V. Vapnik now proposes using only the optimal hyperplane, it might be reasonable to return to the idea of the generalised portrait with an arbitrary constant k, looking for a value for it that provides the fewest number of support vectors. It seems that the result will depend on the nature of the problem. If both classes are approximately equal in size the optimum hyperplane is preferable. In the case when we separate a rather narrow class from a rather wide one (for instance, the letter "a" from all the others) the generalised portrait with constant k near 1 is preferable.

The ratio of the number of support vectors to the total length of the training sequence was used by us first as heuristic evaluation. Its theoretical significance was realised only in 1966, and then it led to the notions of the growth function and VC dimension to conditions for uniform convergence, etc.

3.7 Support Vector Machines

As already mentioned, the method of support vector machines differs from the generalised portrait method mainly in the fact that no coordinate representation is used there. Everything is done using only decompositions into support vectors and scalar products presented in kernel form. It allows us to deal with very high dimensional spaces. Still, generalization properties of the method are preserved due to the rather small number of support vectors. Of course, this number depends on the real geometry of classes and on how far we are ready to go in making errors on the training sequence.

References

1. Vapnik, V.N., Chervonenkis, A.Y.: Об одном классе алгоритмов обучения распознаванию образов (On a class of algorithms of learning pattern recognition). Avtomatika i Telemekhanika **25**(6) (1964). The journal is translated into English as *Automation and Remote Control*
2. Vapnik, V.N., Chervonenkis, A.Y.: Об одном классе персептронов (On a class of perceptrons). Avtomatika i Telemekhanika **25**(1) (1964). The journal is translated into English as *Automation and Remote Control*
3. Vapnik, V.N., Lerner, A.Y.: Узнавание образов при помощи обобщенных портретов (Pattern recognition using generalized portraits). Avtomatika i Telemekhanika **24**(6), 774–780 (1968). The journal is translated into English as *Automation and Remote Control*

Part II
Theory and Practice of Statistical Learning Theory

Part II contains 20 technical contributions from some of the leading researchers in the key areas of machine learning, kernel methods, and statistical learning theory influenced by Vladimir Vapnik's work. Some of the papers are accessible surveys of these key research areas, some are original research contributions, and some contain elements of both.

Vladimir Vapnik's Support Vector Machines (SVMs) have become a very popular subject of research in many communities, and in Chap. 4 Ingo Steinwart reviews several recent developments, mainly concentrating on the cases of binary classification and least squares regression. The principal areas that he discusses are the universal consistency and the learning rates of SVMs and related methods.

Another very popular subject of research is boosting, or creating highly accurate prediction rules by combining a large number of relatively weak rules. In Chap. 5, Robert Schapire, one of the founders of the field, reviews several approaches that have been proposed for explaining and understanding the effectiveness of AdaBoost, the most widely used boosting algorithm. He carefully discusses the underlying assumptions, strengths, and weaknesses of various explanations.

At an intuitive level, Karl Popper's notion of falsifiability of scientific theories is clearly related to the notion of overfitting in statistical learning theory. However, at a technical level the situation is subtle. In Chap. 6, Yevgeny Seldin and Bernhard Schölkopf explore the relationship between these notions. Formally, they compare different definitions of dimension of a function class, namely Popper dimension, VC dimension, and exclusion dimension.

In Chap. 7, Silvia Villa, Lorenzo Rosasco, and Tomaso Poggio tackle a key question of statistical learning theory: which function spaces are learnable? They review both older results, based on the notion of complexity, and more recent results, based on the notion of stability.

Loss functions are used, explicitly or implicitly, in all machine learning problems. The three canonical loss functions are 0-1 loss, squared loss, and log loss (corresponding to Vladimir Vapnik's three main learning problems). In Chap. 8, Robert Williamson summarises recent developments in the theory of loss functions.

In particular, he argues that there are many other interesting classes of loss functions and discusses computational and statistical implications of the choice of the loss function.

In Chap. 9, Jason Weston reviews several well-known learning algorithms and discusses their strengths and limitations from the practical point of view. In particular, he covers linear models, embedding models, nearest neighbours, neural networks, and SVMs and other kernel methods, paying special attention to their performance on real-world large-scale datasets.

In Chap. 10, David McAllester and Takintayo Akinbiyi review the basics of PAC-Bayesian theory, an approach to machine learning that blends Bayesian learning and the VC-style uniform convergence approach. They discuss the fundamental equations of the theory with an emphasis on Catoni's basic inequality and Catoni's localisation methods.

In Chap. 11, Vladimir Vovk presents the method of Kernel Ridge Regression (KRR) as a simple special case of Support Vector Regression. The original motivation for KRR was Bayesian, but looking at it as a special case of SVR motivates looking for non-Bayesian performance guarantees for KRR. The chapter discusses two very different kinds of such guarantees.

Multi-task learning is a recent development in machine learning that aims at solving a set of learning problems at the same time. Chapter 12 by Christian Widmer, Marius Kloft, and Gunnar Rätsch focuses on an approach to multi-task learning based on the assumption that closely related tasks yield similar parameters in the learning model. The chapter gives an overview of the area of multi-task learning, presents an example of a successful application in computational biology, and gives practical guidelines for assessing how promising multi-task learning is for a given dataset.

Semi-supervised learning is a class of machine learning techniques that make use of both labelled and unlabelled data for training. In Chap. 13, Bernhard Schölkopf, Dominik Janzing, Jonas Peters, Eleni Sgouritsa, Kun Zhang, and Joris Mooij connect the problem of semi-supervised learning with causal inference. They show that semi-supervised learning can help in an anticausal setting, but not in a causal setting. These results are validated by experiments on public datasets.

In Chap. 14, Luc Devroye, Paola Ferrario, László Györfi, and Harro Walk are concerned with estimating the minimum mean squared error in the problem of regression. They propose a non-recursive estimator and a recursive estimator, prove their strong universal consistency, and bound the rate of convergence of the non-recursive estimator under mild conditions.

Chapter 15 by Ran Gilad-Bachrach and Chris Burges addresses the problem of hypothesis selection, i.e., selecting the best hypothesis from a given hypothesis class. The authors adopt the PAC-Bayesian framework, which is reviewed in Chap. 10, and limit themselves to the binary classification setting. They propose a depth function for classifiers to measure their proximity to the Bayesian optimal classifier and show that deeper classifiers have stronger generalisation bounds. The deepest classifier, which they call the median hypothesis, is, therefore, a good hypothesis to select.

In the framework of online learning, labels are predicted successively, each being revealed before the next one is predicted, and many online learning algorithms combine computational efficiency, lack of distributional assumptions, and strong theoretical guarantees. In Chap. 16, Nicolò Cesa-Bianchi and Ohad Shamir propose an efficient online learning algorithm adapted to transductive settings and present its applications to collaborative filtering and to linking batch learning and transductive online learning.

In Chap. 17, Eric Gautier and Alexandre Tsybakov propose a new estimation method in a high-dimensional linear regression model under the scenario of sparsity, i.e., assuming that only few coefficients are non-zero. Their method, which they call self-tuned Dantzig estimator, is based on linear programming, which makes it computationally efficient. They obtain upper bounds for estimation and prediction errors under weak assumptions on the model and the distribution of the errors; it turns out that their method achieves the same rate as that achieved in a much more restrictive situation.

Regularization is often used in machine learning to prevent overfitting, and an important class of regularizers are those inducing sparsity. In Chap. 18, Andreas Argyriou, Luca Baldassarre, Charles Micchelli, and Massimiliano Pontil discuss a general class of such regularizers: compositions of a non-differentiable convex function with a linear function. They propose a general approach to solving such regularization problems and apply it to SVMs.

The structural risk minimization principle, first proposed by Vladimir Vapnik and then studied by Vapnik and Chervonenkis in the 1970s, has been developed into penalized empirical risk minimization with more general complexity penalties. In Chap. 19, Vladimir Koltchinskii discusses the penalized empirical risk minimization in problems of estimation of large matrices of a relatively small rank with nuclear norm used as a complexity penalty. In particular, he proves sharp low-rank oracle inequalities for such problems.

Chapter 20 by Andreas Christmann and Robert Hable is another chapter devoted to SVMs; the authors, however, understand SVMs in a broad sense, allowing general convex loss functions and general kernels. They consider the problem of approximating the finite sample distributions of SVMs predictions. Such approximations allow various kinds of statistical inferences based on SVMs, such as prediction intervals. The main result of the chapter says that bootstrap approximations are consistent under mild assumptions.

Chapter 21 by John Snyder, Sebastian Mika, Kieron Burke, and Klaus-Robert Müller is devoted to kernel-based methods. They start from a brief review of kernel-based methods in general and of kernel PCA in particular. The main novel contribution of the chapter is in showing how kernel-based methods can be used for property optimization. In conclusion, the authors apply their techniques to problems in quantum chemistry and physics.

In Chap. 22, Mark Stevens, Samy Bengio, and Yoram Singer consider problems in which we need to rank a small number of positive examples over a vast number of negative examples. An appropriate loss function for such problems is the "domination loss", whose definition the authors extend and generalize. They describe and analyze several efficient algorithms for learning a ranking function using the domination loss. The effectiveness of these algorithms is demonstrated in experiments on benchmark datasets.

In Chap. 23, Masashi Sugiyama studies the problem of approximating the divergence between two probability distributions given samples drawn from them. The naive approach of first estimating the probability distributions is inefficient (and violates Vladimir Vapnik's general principle), and the author is interested in direct approximations for four divergence measures: Kullback–Leibler divergence, Pearson divergence, relative Pearson divergence, and L^2-distance. He discusses recent advances in this direction and their applications in machine learning.

Chapter 4
Some Remarks on the Statistical Analysis of SVMs and Related Methods

Ingo Steinwart

Abstract Since their invention by Vladimir Vapnik and his co-workers in the early 1990s, support vector machines (SVMs) have attracted a lot of research activities from various communities. While at the beginning this research mostly focused on generalization bounds, the last decade witnessed a shift towards consistency, oracle inequalities, and learning rates. We discuss some of these developments in view of binary classification and least squares regression.

4.1 Introduction

Given a data set $D := ((x_1, y_1), \ldots, (x_n, y_n))$ sampled from some unknown distribution P on $X \times Y$, the goal of supervised statistical learning is to find an $f_D : X \to \mathbb{R}$ whose L-risk

$$\mathcal{R}_{L,P}(f_D) := \int_{X \times Y} L(x, y, f_D(x)) \, dP(x, y)$$

is small. Here, $L : X \times Y \times \mathbb{R} \to [0, \infty)$ is a loss describing our learning goal. Probably the two best-known examples of such losses are the binary classification loss and the least squares loss. However, other choices, e.g., for quantile regression, weighted classification, and classification with reject option, are important, too. To formalize the concept of "learning", we also need the Bayes risk

$$\mathcal{R}_{L,P}^* := \inf \{ \mathcal{R}_{L,P}(f) \mid f : X \to \mathbb{R} \}.$$

If this infimum is attained we denote a function that achieves $\mathcal{R}_{L,P}^*$ by $f_{L,P}^*$.

I. Steinwart (✉)
Institute for Stochastics and Applications, University of Stuttgart, Pfaffenwaldring 57, D-70569 Stuttgart, Germany
e-mail: ingo.steinwart@mathematik.uni-stuttgart.de

Now, a learning method $\mathcal{L}$ assigns to every finite data set D a function f_D. Such an $\mathcal{L}$ learns in the sense of L-risk consistency for P if

$$\lim_{n\to\infty} P^n\Big(D \in (X \times Y)^n : \mathcal{R}_{L,P}(f_D) \leq \mathcal{R}^*_{L,P} + \varepsilon\Big) = 1 \qquad (4.1)$$

for all $\varepsilon > 0$. Moreover, $\mathcal{L}$ is called universally L-risk consistent if it is L-risk consistent for all distributions P on $X \times Y$.

Recall that the first results on universally consistent learning methods were shown by Stone [33] in a seminal paper. Since then, various learning methods have been shown to be universally consistent. We refer to the books [11] and [15] for binary classification and least squares regression, respectively.

Clearly, consistency does not specify the speed of convergence in (4.1). To address this we fix a sequence $(\varepsilon_n) \subset (0, 1]$ converging to 0. Then, we say that $\mathcal{L}$ learns with rate (ε_n) if there exists a family $(c_\tau)_{\tau\in(0,1]}$ such that for all $n \geq 1$ and all $\tau \in (0, 1]$, we have

$$P^n\Big(D \in (X \times Y)^n : \mathcal{R}_{L,P}(f_D) \leq \mathcal{R}^*_{L,P} + c_\tau \varepsilon_n\Big) \geq 1 - \tau.$$

In addition, we say that $\mathcal{L}$ learns with expected rate (ε_n) if $\mathbb{E}_{D\sim P^n}\mathcal{R}_{L,P}(f_D) \preceq \varepsilon_n$. Here, $a_n \preceq b_n$ means that there exists a constant $c \geq 0$ with $a_n \leq cb_n$ for all $n \geq 1$. Analogously, we sometimes write $a_n \sim b_n$ if $a_n \preceq b_n$ and $b_n \preceq a_n$.

Unlike consistency, learning rates usually require assumptions on P by the no-free-lunch theorem of Devroye; see [10] and [11, Theorem 7.2]. In Sect. 4.4 we will discuss such assumptions and the resulting rates for SVMs.

To recall the definition of SVMs and related methods, we fix a reproducing kernel Hilbert space (RKHS) H, a loss L that is convex in its third argument, and a $\lambda > 0$. Then, the optimization problem

$$f_{D,\lambda} \in \arg\min_{f \in H} \lambda \|f\|_H^2 + \mathcal{R}_{L,D}(f), \qquad (4.2)$$

where $\mathcal{R}_{L,D}(f)$ is the empirical risk of f, that is

$$\mathcal{R}_{L,D}(f) = \frac{1}{n} \sum_{i=1}^{n} L(x_i, y_i, f(x_i)),$$

has a unique solution $f_{D,\lambda} \in H$; see [28, Lemma 5.1 and Theorem 5.2].

Let us briefly make some historical remarks: In 1992 V. Vapnik and co-workers [6] presented the first SVM, namely the hard-margin SVM, which combined the generalized portrait algorithm from [37] with a kernel embedding inspired by [1]. A few years later, C. Cortes and V. Vapnik [8] proposed the first soft-margin SVMs, which are instances of (4.2) for which L is the (squared) hinge loss. Almost at the same time, the ϵ-insensitive loss for regression was proposed in [12, 36, 38]. However, approaches of the form (4.2) are actually significantly older. In 1971, for example, G. Kimeldorf and G. Wahba [16] showed a form of the representer theorem for the Sobolev space case $H = W^m([0,1]^d)$ with $m > d/2$ and the least squares loss L. Until the end of the 1980s a substantial amount of further research dealt with this and similar cases; see e.g., [24, 39]. Inspired by this work, [23] presented an approach called regularization network to the learning community in 1990, which basically considers (4.2) for the least squares loss.

Ideally, a learning method is automatic, i.e., no parameters need to be set by the user. In the SVM case, this means that λ and possible kernel parameters such as the width $\gamma > 0$ of the Gaussian kernel

$$k_\gamma(x, x') := \exp(-\gamma^{-2}\|x - x'\|), \qquad x, x' \in \mathbb{R}^d,$$

are set automatically. In practice, such parameters are usually determined by cross-validation. Let us briefly describe a simplified version of this; see [28, Definition 6.28]. To this end, we split D into two (almost) equally sized parts D_1 and D_2. In addition, let Λ be a finite set of candidates for λ and, if necessary, Γ be a finite set of candidates for the kernel parameter. Then, for all combinations $(\lambda, \gamma) \in \Lambda \times \Gamma$, the optimization (4.2) is solved for the data set D_1, and the resulting *clipped* SVM solution—see (4.4)—is validated on D_2, i.e., its empirical D_2-error is computed. Finally, the SVM solution with the smallest D_2-error is taken as the decision function f_D.

In the following, we try to give a brief survey on what is known about consistency and learning rates for SVMs. To this end, we first recall some key concepts related to their analysis in Sect. 4.2. We then consider consistency and learning rates in Sects. 4.3 and 4.4, respectively. Due to limited space, these discussions are restricted to binary classification and least squares regression. However, most of the results we discuss are actually derived from generic oracle inequalities and thus they can be naturally extended to other losses. Here, differences usually only occur if assumptions on P are made to guarantee, e.g., variance bounds or approximation properties. For an example we refer to quantile regression with the pinball loss in [13, 29].

4.2 Mathematical Prerequisites

In the following, let $(X, \mathcal{A})$ be a measurable space, $Y \subset \mathbb{R}$ be a closed subset, and P be a distribution on $X \times Y$ whose marginal distribution on X is denoted by P_X. In addition, we always assume that H is a separable reproducing kernel Hilbert space (RKHS) of a bounded measurable kernel k on X with $\|k\|_\infty \leq 1$. Finally, if not stated otherwise, L denotes a loss that satisfies $\mathcal{R}_{L,P}(0) < \infty$.

The goal of this section is to recall some concepts that describe interactions between P, L, and H, which are relevant for the analysis of SVMs.

Let us begin by recalling that the "inclusion" operator $I_k : H \to L_2(P_X)$ that maps an $f \in H$ to its equivalence $L_2(P_X)$-class $[f]_\sim$ is a Hilbert-Schmidt operator; see [28, Theorem 4.27]. Moreover, the usual integral operator $T_k : L_2(P_X) \to L_2(P_X)$ with respect to k is well defined and given by $T_k = I_k \circ I_k^*$, where I_k^* denotes the adjoint operator of I_k. In particular, T_k is self-adjoint, positive and nuclear—see again [28, Theorem 4.27]—and thus, the classical spectral theorem can be applied. This yields an at most countable family $(\mu_i)_{i \in I} \subset (0, \infty)$ of non-zero eigenvalues (with geometric multiplicities) of T_k, which, in the case of infinite I, converges to 0. As usual, we assume without loss of generality that $I \subset \mathbb{N}$ and $\mu_1 \geq \mu_2 \geq \cdots > 0$. Some of the results we will review later make explicit assumptions on the decay of the eigenvalues, while other results make assumptions on the behavior of covering numbers or entropy numbers. Since the latter two are essentially the same concepts, let us only recall the latter. To this end, we first consider a compact metric space (M, d). Then, for $n \geq 1$, the nth entropy number of an $A \subset M$ is defined by

$$\varepsilon_n(A, d) := \inf\left\{\varepsilon > 0 : \exists t_1, \ldots, t_n \in M \text{ such that } A \subset \bigcup_{i=1}^n B(t_i, \varepsilon)\right\}$$

where $B(t, \varepsilon)$ denotes the closed ball with centre t and radius ε. Moreover, if E and F are Banach spaces and $T : E \to F$ is a bounded linear operator, then the nth (dyadic) entropy number of T is defined by $e_n(T) := \varepsilon_{2^{n-1}}(TB_E, \|\cdot\|_F)$, where B_E denotes the closed unit ball of E. In the Hilbert space case, eigenvalue and entropy number decays are closely related. For example, [32, Theorem 15] shows that

$$\mu_i(T_k) \preceq i^{-1/p} \quad \Longleftrightarrow \quad e_i(I_k : H \to L_2(P_X)) \preceq i^{-1/2p}. \qquad (4.3)$$

Moreover, the latter is implied by $e_i(\text{id} : H \to \ell_\infty(X)) \preceq i^{-1/2p}$. Assumptions on the eigenvalue or entropy number decay are used to estimate the stochastic error of (4.2). To derive consistency and learning rates, however, we also need to bound the approximation error. For example, for consistency, we obviously need zero approximation error, that is $\mathcal{R}_{L,P,H}^* = \mathcal{R}_{L,P}^*$, where $\mathcal{R}_{L,P,H}^* := \inf_{f \in H} \mathcal{R}_{L,P}(f)$ denotes the smallest possible L-risk in H. If H is universal (cf. [26] and [22]), that is, X is a compact metric space and H is dense in $C(X)$, this equality can be guaranteed; see [28, Corollary 5.29]. For specific losses, however, weaker assumptions on H are sufficient. For example, if L is the least squares loss, the equality $\mathcal{R}_{L,P,H}^* = \mathcal{R}_{L,P}^*$ holds if and only if H is dense in $L_2(P_X)$. For many Lipschitz continuous losses including the hinge loss, the ϵ-insensitive loss, and the pinball loss, an analogous characterization holds in terms of $L_1(P_X)$-denseness; see [28, Corollary 5.37]. Finally, recall that for fixed $\gamma > 0$, the RKHS H_γ of the Gaussian kernel k_γ is dense in $L_p(P_X)$ for all $p \in [1, \infty)$; see [28, Theorem 4.63]. Once we have fixed an H with $\mathcal{R}_{L,P,H}^* = \mathcal{R}_{L,P}^*$, we need to consider the approximation error function (AEF)

$$A(\lambda) := \inf_{f \in H} \lambda \|f\|_H^2 + \mathcal{R}_{L,P}(f) - \mathcal{R}_{L,P}^*, \qquad \lambda \geq 0.$$

It can be shown that $\lim_{\lambda \to 0} A(\lambda) = 0$; see [28, Lemma 5.15]. In general, the speed of convergence cannot be faster than $O(\lambda)$, and this rate is achieved if and only if there exists an $f \in H$ with $\mathcal{R}_{L,P}(f) = \mathcal{R}_{L,P}^*$; see [28, Corollary 5.18]. For the least squares loss, the behavior of the AEF can be described by interpolation spaces $[E, F]_{\theta, r}$ of the real method; see [4, 5]. Namely, [25] shows that $f_{L,P}^* \in [L_2(P_X), H]_{\beta,\infty}$ if and only if $A(\lambda) \in O(\lambda^\beta)$. Here we note that the latter condition is often imposed to derive learning rates. Other authors, however, assume $f_{L,P}^* \in \mathrm{ran}\, T_k^{\beta/2} = [L_2(P_X), [H]_\sim]_{\beta,2}$, where $\mathrm{ran}\, T_k^{\beta/2}$ denotes the image of the $\beta/2$-fractional power of T_k, and the equality of this image to the interpolation space has been recently shown in [31, Theorem 4.6]. To compare these conditions we note that, we always have the continuous embeddings $[L_2(\nu), [H]_\sim]_{\beta-\varepsilon,\infty} \hookrightarrow [L_2(\nu), [H]_\sim]_{\beta,2} \hookrightarrow [L_2(\nu), [H]_\sim]_{\beta,\infty}$ for all $\varepsilon > 0$.

Finally, one often knows in advance that it suffices to look for decision functions of the form $f_D : X \to [-M, M]$ for some $M > 0$. In particular, this is the case if the loss is clippable at M, that is, for all $x \in X$, $y \in Y$, and $t \in \mathbb{R}$, we have

$$L(x, y, \widehat{t}\,) \leq L(x, y, t), \tag{4.4}$$

where $\widehat{t} := \max\{-M, \min\{M, t\}\}$. Note that for convex L this is satisfied if and only if $L(x, y, \cdot) : \mathbb{R} \to [0, \infty)$ has a global minimum that is contained in $[-M, M]$ for all $(x, y) \in X \times Y$; see [28, Lemma 2.23]. The latter is satisfied for many commonly used losses, and for such losses it is beneficial to clip the SVM decision function.

4.3 Universal Consistency

In this section we discuss several results concerning the universal consistency of learning methods of the form (4.2) for binary classification and least squares regression. Due to space constraints we restrict our considerations to a priori chosen parameters. However, Theorems 4.1 and 4.2 below and the results discussed for regression can also be formulated for data splitting approaches; cf. [28, Theorems 7.24 and 8.26].

4.3.1 Binary Classification

Let us first note that the binary classification loss, which defines the actual learning goal, is not even continuous, and hence cannot be used in the SVM optimization problem (4.2). This issue is resolved by using a surrogate loss L such as the (squared) hinge loss or the least squares loss. For these losses, the first consistency results can be found in [27] and [40]. To recall these results, we assume that $X \subset \mathbb{R}^d$ is compact and H is universal. Then [27] establishes universal *classification consistency* if (a) we use the hinge loss, (b) we have $\varepsilon_i(X, d_k) \preceq i^{-1/\alpha}$ for some $\alpha > 0$, where d_k is the kernel metric in the sense of [28, Eq. (4.20)], and (c) we use a sequence of regularization parameters (λ_n) with $\lambda_n \to 0$ and $n\lambda_n^\alpha \to \infty$. In addition, for the Gaussian kernel k_γ with fixed but arbitrary width γ we can choose $\alpha := d$. By completely different methods, [40] shows universal classification consistency for a variety of losses including the (squared) hinge loss and the least squares loss if $\lambda_n \to 0$ and $n\lambda_n \to \infty$. A key idea in both articles is to compare the excess L-risk $\mathcal{R}_{L,P}(f) - \mathcal{R}_{L,P}^*$ of arbitrary f to the excess classification risk of f. Namely, in [27] an asymptotic relationship is shown, while [40] goes one step further by establishing inequalities between these excess risks. This idea was picked up in [2], which showed for convex margin-based losses, i.e., for losses L of the form $L(y, t) = \varphi(yt)$, that we have an asymptotic relationship or an inequality between these excess risks if and only if φ is differentiable at 0 with $\varphi'(0) < 0$. For such losses we have the following consistency result:

Theorem 4.1. *Let L be as above and $\varphi(t) \in O(t^q)$ for some $q \geq 1$ and $t \to \infty$. Moreover, let H be dense in $L_q(P_X)$ and $(\lambda_n) \subset (0, \infty)$ with $\lambda_n \to 0$. Then the clipped SVM is classification-consistent for P if one of these conditions is satisfied:*

1. $n\lambda_n / \ln n \to \infty$ and $n\lambda_n^{q/2} \to \infty$.
2. $n\lambda_n^{q/2} \to \infty$ and $n\lambda_n^p \to \infty$ for some $p \in (0, 1)$ with $\mu_i(T_k) \preceq i^{-1/p}$.

If X is compact and H is universal, then all assumptions involving q can be dropped.

Proof. The first assertion follows from [28, Lemma 5.15 and Theorem 5.31] together with a simple generalization of [28, Theorem 7.22]. The second result can be shown analogously by employing [28, Theorem 7.23] together with [28, Corollary 7.31] and (4.3). Now assume that X is compact and H is universal. We fix an $\varepsilon \in (0, 2]$ and pick an $f : X \to \mathbb{R}$ with $\mathcal{R}_{L,P}(f) \leq \mathcal{R}_{L,P}^* + \varepsilon$. Since L is clippable, say at M, we may assume that f maps into $[-M, M]$. By [3, Theorem 29.14] we then find a $g \in C(X)$ with $\|f - g\|_{L_1(P_X)} \leq \varepsilon$. Again, we can assume that $\|g\|_\infty \leq M$. Since H is universal, there also exists an $h_\varepsilon \in H$ with $\|h_\varepsilon - g\|_H \leq \varepsilon$. Here we note that we can additionally assume that the resulting function $\varepsilon \mapsto \|h_\varepsilon\|_H$ is decreasing. Our construction yields $\|h_\varepsilon\|_\infty \leq 2 + M$ and $\|f - h_\varepsilon\|_{L_1(P_X)} \leq 2\varepsilon$. Since L is locally Lipschitz—see [28, Lemma 2.25]—we find $\mathcal{R}_{L,P}(h_\varepsilon) - \mathcal{R}_{L,P}(f) \leq 2c_L \varepsilon$ by [28, Lemma 2.19], where $c_L \geq 1$ is a constant only depending on L. This gives $\mathcal{R}_{L,P}(h_\varepsilon) - \mathcal{R}_{L,P}^* \leq 3c_L \varepsilon$. For $\lambda \in (0, 1]$ we now define $\varepsilon_\lambda := 2\inf\{\varepsilon \in (0, 1] : \|h_\varepsilon\|_H^2 \leq \lambda^{-1/2}\}$. We then obtain $\|h_{\varepsilon_\lambda}\|_\infty \leq 2 + M$ and

$$\lambda \|h_\varepsilon\|_H^2 + \mathcal{R}_{L,P}(h_\varepsilon) - \mathcal{R}_{L,P}^* \leq \lambda^{1/2} + 3c_L \varepsilon_\lambda \to 0 \qquad \lambda \to 0.$$

Choosing $f_0 := h_{\varepsilon_\lambda}$ in (the proofs) of [28, Theorems 7.22 and 7.23] gives the assertions. $\square$

The result above yields universal classification consistency, if, e.g., $X = \mathbb{R}^d$ and $H = H_\gamma$ with *fixed* kernel width γ. For Gaussian kernels, it is, however, common practice to vary γ with the sample size, too. The following result covers this case:

Theorem 4.2. *Let L be convex, clippable, and margin-based with $\varphi'(0) < 0$. Furthermore, let $(\lambda_n) \subset (0, 1]$ and $\gamma_n \subset (0, 1]$ satisfy $\lambda_n \gamma_n^{-d} \to 0$. Then the clipped SVM is universally classification-consistent if one of the following conditions holds:*

1. *$X = \mathbb{R}^d$, $\varphi(t) \in O(t^q)$ for some $q \geq 1$ and $t \to \infty$, $n\lambda_n / \ln n \to \infty$ and $n\lambda_n^{q/2} \to \infty$.*
2. *$X \subset \mathbb{R}^d$ is compact and $\lambda_n^\varepsilon \gamma_n^d n \to \infty$ for some $\varepsilon > 0$.*

Proof. Using $\| \text{id} : H_1 \to H_\gamma \| \leq \gamma^{-d/2}$—see [28, Proposition 4.46]—it is easy to check that the AEFs A_γ and A_1 of the Gaussian RKHSs H_γ and H_1 satisfy $A_\gamma(\lambda) \leq A_1(\lambda \gamma^{-d})$. Then the first assertion follows as for Theorem 4.1. The second assertion can be shown using the arguments for compact X in the proof of Theorem 4.1.

4.3.2 Least Squares Regression

We already noted in the introduction that least squares regression methods of the form (4.2) had already been around when SVMs were proposed. Despite their earlier appearance, the first[1] universal consistency result in our sense seems to be shown relatively late by [17]. Under the moment condition $\mathcal{R}_{L,P}(0) = \mathbb{E}_{(x,y)\sim P} y^2 < \infty$, the authors obtain consistency for $H = W^m([0, 1]^d)$ if $\lambda_n \to 0$, $n\lambda_n \to \infty$, and the decision functions f_{D,λ_n} are clipped at $\ln n$. In [15, Theorem 20.4] the condition $n\lambda_n \to \infty$ was relaxed to $n\lambda_n^p/(\ln n)^7 \to \infty$ with $p := d/(2m)$, and it seems plausible that their proof allows us to remove the logarithmic factor at least partially if Y is bounded and a more aggressive clipping is applied. In any case, for bounded Y the general theory tells us that the logarithmic factors can be removed. Indeed, for bounded Y, it is easy to check that the conditions ensuring consistency in Theorems 4.1 and 4.2 also ensure consistency for least squares regression if we set $q = 2$. In the case of Theorem 4.1, for example, we obtain consistency for generic H if $\lambda_n \to 0$ and $n\lambda_n / \ln n \to \infty$, and the latter can be replaced by $n\lambda_n^p \to \infty$ for some $p \in (0, 1)$ if X is compact, H is universal, and $\mu_i(T_k) \preceq i^{-1/p}$. Note that this covers the case $H = W^m([0, 1]^d)$ for $p := d/(2m)$ by the well-known estimate $e_i(I_k : W^m([0, 1]^d) \to \ell_\infty([0, 1]^d)) \preceq i^{-\frac{m}{d}}$; see, e.g., [14, p. 118].

---

[1] In [15] the authors actually give some credit to the 1987 paper [35] for the case $d = 1$.

4.4 Learning Rates

In this section we discuss some known learning rates for SVMs for binary classification and least squares regression.

4.4.1 Binary Classification

Probably the earliest established learning rates for SVMs with (squared) hinge loss can be found in [26]. To formulate this result we define $\eta(x) := P(Y = 1|x)$, $x \in X$, as well as $X_- := \{\eta < 1/2\}$ and $X_+ := \{\eta > 1/2\}$. We say that P has zero noise if $|2\eta - 1| = 1$ holds P_X–almost surely, and has strictly separated classes if $d(X_-, X_+) > 0$ for a version of η and a metric d on X. Now assume that (X, d) is compact, H is universal and $\lambda_n = n^{-1}$. Then [26] shows that $P^n(D : \mathcal{R}_{L,P}(f)_{D,\lambda_n} = 0) \geq 1 - e^{-cn}$ for all $n \geq n_0$, where L is the classification loss and c and n_0 depend on P and H.

In [19] exponentially fast expected rates under similar but weaker conditions were shown. There the authors assume that (X, d) is compact, η has a Lipschitz continuous version and that P has Tsybakov's noise exponent $q = \infty$; see below. Note that together these assumptions imply that P has strictly separated classes. For universal kernels and the logistic loss for classification, they then show that there are constants $c_1, c_2 > 0$ with $\mathbb{E}_{D \sim P^n} \mathcal{R}_{L,P}(f)_{D,\lambda_n} - \mathcal{R}^*_{L,P} \leq \exp(-c_1 n \lambda_n)$ if $\lambda_n \leq c_2$ and $n \lambda_n^{1+p} \to \infty$. Here, $p \in (0, 1)$ is a constant such that $\sup_\nu e_i(I_k : H \to L_2(\nu)) \leq c i^{-1/(2p)}$ for all $i \geq 1$, where the supremum is taken over all distributions ν on X.

For both results discussed so far, it seems fair to say that (a) the assumptions on P are very strong and that (b) similar rates can also be achieved without much effort for classical histogram rules. In the case of the hinge loss and Gaussian kernels with varying widths, more realistic assumptions on P have been proposed in [30], which, to some extent, generalize the assumptions above. To briefly describe them, we define the distance to the decision boundary by $\Delta(x) := d(x, X_+)$ if $x \in X_-$, $\Delta(x) := d(x, X_-)$ if $x \in X_+$, and $\Delta(x) = 0$ otherwise. Then P is said to have margin noise exponent $\beta \in (0, \infty]$ if $\mathbb{E}_{P_X} \mathbf{1}_{\{\Delta < t\}} |2\eta - 1| \leq (ct)^\beta$ for a constant $c \geq 1$ and all $t \geq 0$. A detailed discussion of this assumption can be found in [28, Sect. 8.2], so we only mention that β is large if there is not much mass and/or a lot of noise in the area $\{\Delta < t\}$ around the decision boundary. In addition, we need Tsybakov's noise condition [34] that bounds the total amount of noise by $P_X(|2\eta - 1| < t) \leq (ct)^q$ for constants $c > 0$ and $q \in [0, \infty]$, and all $t \geq 0$. Then [28, Theorem 8.26] shows that the data splitting approach with polynomially growing n^{-1}-nets Λ_n and $n^{-1/d}$-nets Γ_n of $(0, 1]$ learns with rate $n^{-\frac{\beta(q+1)}{\beta(q+2)+d(q+1)}+\varepsilon}$ for all $\varepsilon > 0$. Note that depending on β and q the exponent in the rate varies between 0 and 1; in particular, rates up to n^{-1} are possible in all dimensions d provided that β and q are large enough.

Finally, let us briefly discuss some rates for generic H and the hinge loss (the least squares case will be considered at the end of our discussions on least squares regression). To this end, we assume that P satisfies Tsybakov's noise condition for some $q \in [0, \infty]$, as well as $\mu_i(T_k) \preceq i^{-1/p}$ and $A(\lambda) \in O(\lambda^\beta)$ for some $p \in (0, 1)$ and $\beta \in (0, 1]$. Then we usually have to expect $\beta < 1$, since for $\beta = 1$ the Bayes decision function, which is a step function, must be contained in H, and for commonly used H this is impossible. In addition, Tsybakov's noise condition gives a variance bound, which in turn can be used, e.g., in [28, Theorem 7.24]. The resulting learning rate is $n^{-\min\{\frac{2\beta}{\beta+1}, \frac{\beta(q+1)}{\beta(q+2-p)+p(q+1)}\}}$ for the data splitting approach if (Λ_n) is a sequence of polynomially growing n^{-2}-nets of $(0, 1]$.

4.4.2 Least Squares Regression

Similarly to the case of consistency, the first learning rates were established for the space $H = W^m([0, 1]^d)$. Indeed, based on some techniques from empirical processes pioneered by S. van de Geer, [18] showed expected rates of the form $(\ln n)^2 n^{-\frac{2s}{2s+d}}$ for a structural risk minimization procedure to choose the parameters m and λ. Here $s > d/2$ describes the unknown smoothness of the regression function in the sense of $f^*_{L,P} \in W^s([0, 1]^d)$. The procedure is thus adaptive to the unknown smoothness s, and in addition, no assumptions except supp $P_X \subset [0, 1]^d$ are necessary. Let us now turn to the generic case. Here, beginning with [9], various investigations have been made, so we only focus on the ones that established (nearly) optimal rates. To the best of our knowledge, the first result in this direction was established in [7] under the assumptions $\mu_i(T_k) \sim i^{-1/p}$ and $f^*_{L,P} \in \operatorname{ran} T^{\beta/2}$ for some $p \in (0, 1)$ and $\beta \in [1, 2]$. Note that $\beta \geq 1$ implies that $f^*_{L,P} \in H$. Then, modulo some logarithmic factor in the case $\beta = 1$, the authors establish the rate

$$n^{-\frac{\beta}{\beta+p}}, \tag{4.5}$$

and they also show that this rate is optimal. Especially remarkable is the fact that the authors are able to deal with values $\beta > 1$, since for such values the classical approach that splits the analysis into a stochastic part and the AEF fails due to the fact that the AEF does not converge faster than linearly. To avoid this issue, the authors split quite differently with the help of spectral methods.

From a practical point of view, however, the case $\beta < 1$ is the more realistic one. For this case, the first essentially optimal rate was proved in [21] for a variant of (4.2) in which the exponent 2 in the regularization term is replaced by the smaller exponent $2p/(1 + p)$, where $p \in (0, 1)$ is chosen such that $\mu_i(T_k) \preceq i^{-1/p}$. Provided that the eigenvectors of T_k are *uniformly* bounded and $f^*_{L,P} \in [L_2(P_X), H]_{\beta, \infty}$ for some $\beta \in (0, 1]$, [21] then establishes (4.5) modulo some logarithmic factors. A closer look at this assumption on the eigenvectors shows that it is solely used to establish the interpolation inequality $\|f\|_\infty \leq c\|f\|_H^p \|f\|_{L_2(P_X)}^{1-p}$ for all $f \in H$, where $c > 0$ is some constant. Interestingly, this inequality is equivalent to the continuous embedding $[L_2(P_X), H]_{p,1} \hookrightarrow L_\infty(P_X)$. Now, [32] shows that by combining the interpolation inequality with [28, Theorem 7.23] the original algorithm (4.2) also learns with rate (4.5) and the additional logarithmic factors are superfluous. Moreover, if the eigenvalue assumption is two-sided, i.e., $\mu_i(T_k) \sim i^{-1/p}$, then (4.5) is also optimal for all $\beta \in (p, 1]$.

In the Sobolev space case $H = W^m([0, 1]^d)$ and $f_{L,P}^* \in W^s([0, 1]^d)$ for some $m > d/2$ and $s \in (0, m]$, these generic results imply the above-mentioned rates $n^{-\frac{2s}{2s+d}}$ if P_X is (essentially) the uniform distribution; see [32]. Moreover, [13] has recently shown that up to some arbitrarily small $\varepsilon > 0$ in the exponent, the rates can also be achieved by Gaussian RKHSs H_γ if γ varies with the sample size. Note that the latter seems to be somewhat necessary, since for fixed γ and $f_{L,P}^* \notin C^\infty$, the AEF can only have logarithmic decay; see [25]. Finally, the rates of [13, 32] can also be achieved by the data splitting approach.

Let us finally return to binary classification with the least squares loss. To this end, we assume $\eta \in [L_2(P_X), H]_{\beta, \infty}$ and that Tsybakov's noise assumption is satisfied for some $q \in [0, \infty]$. Note that the latter implies a stronger calibration inequality between the excess least squares and the excess classification risk; see [2] and [28, Theorem 8.29]. Considering [32], we then obtain the rate $n^{-\frac{\beta q}{(\beta+p)(q+1)}}$, which at first glance seems to be fine, since for large β and q the exponent reaches 1. However, it may be the case that large values for β and q exclude each other. To illustrate this (see [20] for a similar observation), let us consider the Sobolev case $\eta \in W^s([0, 1]^d)$ in which the rates in [32] become $n^{-\frac{2sq}{(2s+d)(q+1)}}$. To get rates close to n^{-1}, we need large s, say $s > 1 + d/2$. Then $\eta \in C^1$ by Sobolev's embedding theorem, which in turn excludes $q > 1$ by some geometric considerations, and hence rates arbitrarily close to n^{-1} are impossible. Finally, the same observation can be made for [13].

References

1. Aizerman, M., Braverman, E., Rozonoer, L.: Theoretical foundations of the potential function method in pattern recognition learning. Autom. Remote Control **25**, 821–837 (1964)
2. Bartlett, P., Jordan, M., McAuliffe, J.: Convexity, classification, and risk bounds. J. Am. Stat. Assoc. **101**, 138–156 (2006)
3. Bauer, H.: Measure and Integration Theory. De Gruyter, Berlin (2001)
4. Bennett, C., Sharpley, R.: Interpolation of Operators. Academic, Boston (1988)
5. Bergh, J., Löfström, J.: Interpolation Spaces: An Introduction. Springer, New York (1976)
6. Boser, B., Guyon, I., Vapnik, V.: A training algorithm for optimal margin classifiers. In: Proceedings of the 5th Annual ACM Workshop on Computational Learning Theory, Pittsburgh, pp. 144–152 (1992)
7. Caponnetto, A., De Vito, E.: Optimal rates for regularized least squares algorithm. Found. Comput. Math. **7**, 331–368 (2007)
8. Cortes, C., Vapnik, V.: Support vector networks. Mach. Learn. **20**, 273–297 (1995)

9. Cucker, F., Smale, S.: On the mathematical foundations of learning. Bull. Am. Math. Soc. **39**, 1–49 (2002)
10. Devroye, L.: Any discrimination rule can have an arbitrarily bad probability of error for finite sample size. IEEE Trans. Pattern Anal. Mach. Intell. **4**, 154–157 (1982)
11. Devroye, L., Györfi, L., Lugosi, G.: A Probabilistic Theory of Pattern Recognition. Springer, New York (1996)
12. Drucker, H., Burges, C., Kaufman, L., Smola, A., Vapnik, V.: Support vector regression machines. In: Advances in Neural Information Processing Systems, Denver, vol. 9, pp. 155–161. MIT, Cambridge (1997)
13. Eberts, M., Steinwart, I.: Optimal regression rates for SVMs using Gaussian kernels. Electron. J. Stat. **7**, 1–42 (2013)
14. Edmunds, D., Triebel, H.: Function Spaces, Entropy Numbers, Differential Operators. Cambridge University Press, Cambridge (1996)
15. Györfi, L., Kohler, M., Krzyżak, A., Walk, H.: A Distribution-Free Theory of Nonparametric Regression. Springer, New York (2002)
16. Kimeldorf, G., Wahba, G.: Some results on Tchebycheffian spline functions. J. Math. Anal. Appl. **33**, 82–95 (1971)
17. Kohler, M., Krzyżak, A.: Nonparametric regression estimation using penalized least squares. IEEE Trans. Inform. Theory **47**, 3054–3058 (2001)
18. Kohler, M., Krzyżak, A., Schäfer, D.: Application of structural risk minimization to multivariate smoothing spline regression estimates. Bernoulli **4**, 475–489 (2002)
19. Koltchinskii, V., Beznosova, O.: Exponential convergence rates in classification. In: Proceedings of the 18th Annual Conference on Learning Theory, Bertinoro, pp. 295–307 (2005)
20. Loustau, S.: Aggregation of SVM classifiers using Sobolev spaces. J. Mach. Learn. Res. **9**, 1559–1582 (2008)
21. Mendelson, S., Neeman, J.: Regularization in kernel learning. Ann. Stat. **38**, 526–565 (2010)
22. Micchelli, C., Xu, Y., Zhang, H.: Universal kernels. J. Mach. Learn. Res. **7**, 2651–2667 (2006)
23. Poggio, T., Girosi, F.: A theory of networks for approximation and learning. Proc. IEEE **78**, 1481–1497 (1990)
24. Silverman, B.: Some aspects of the spline smoothing approach to nonparametric regression. J. Royal Stat. Soc. B Stat. Methodol. **47**, 1–52 (1985)
25. Smale, S., Zhou, D.: Estimating the approximation error in learning theory. Anal. Appl. **1**, 17–41 (2003)
26. Steinwart, I.: On the influence of the kernel on the consistency of support vector machines. J. Mach. Learn. Res. **2**, 67–93 (2001)
27. Steinwart, I.: Support vector machines are universally consistent. J. Complexity **18**, 768–791 (2002)
28. Steinwart, I., Christmann, A.: Support Vector Machines. Springer, New York (2008)
29. Steinwart, I., Christmann, A.: Estimating conditional quantiles with the help of the pinball loss. Bernoulli **17**, 211–225 (2011)
30. Steinwart, I., Scovel, C.: Fast rates for support vector machines using Gaussian kernels. Ann. Stat. **35**, 575–607 (2007)
31. Steinwart, I., Scovel, C.: Mercer's theorem on general domains: on the interaction between measures, kernels, and RKHSs. Constr. Approx. **35**, 363–417 (2012)
32. Steinwart, I., Hush, D., Scovel, C.: Optimal rates for regularized least squares regression. In: Proceedings of the 22nd Annual Conference on Learning Theory, Montreal, pp. 79–93 (2009)
33. Stone, C.: Consistent nonparametric regression. Ann. Stat. **5**, 595–645 (1977)
34. Tsybakov, A.: Optimal aggregation of classifiers in statistical learning. Ann. Stat. **32**, 135–166 (2004)
35. van de Geer, S.: A new approach to least squares estimation, with applications. Ann. Stat. **15**, 587–602 (1987)
36. Vapnik, V.: The Nature of Statistical Learning Theory. Springer, New York (1995)

37. Vapnik, V., Lerner, A.: Pattern recognition using generalized portrait method. Autom. Remote Control **24**, 774–780 (1963)
38. Vapnik, V., Golowich, S., Smola, A.: Support vector method for function approximation, regression estimation, and signal processing. In: Advances in Neural Information Processing Systems, Denver, vol. 9, pp. 81–287. MIT, Cambridge (1997)
39. Wahba, G.: Spline Models for Observational Data. Series in Applied Mathematics, vol. 59. SIAM, Philadelphia (1990)
40. Zhang, T.: Statistical behaviour and consistency of classification methods based on convex risk minimization. Ann. Stat. **32**, 56–134 (2004)

Chapter 5
Explaining AdaBoost

Robert E. Schapire

Abstract Boosting is an approach to machine learning based on the idea of creating a highly accurate prediction rule by combining many relatively weak and inaccurate rules. The AdaBoost algorithm of Freund and Schapire was the first practical boosting algorithm, and remains one of the most widely used and studied, with applications in numerous fields. This chapter aims to review some of the many perspectives and analyses of AdaBoost that have been applied to explain or understand it as a learning method, with comparisons of both the strengths and weaknesses of the various approaches.

5.1 Introduction

Boosting is an approach to machine learning based on the idea of creating a highly accurate prediction rule by combining many relatively weak and inaccurate rules. The AdaBoost algorithm of Freund and Schapire [10] was the first practical boosting algorithm, and remains one of the most widely used and studied, with applications in numerous fields. Over the years, a great variety of attempts have been made to "explain" AdaBoost as a learning algorithm, that is, to understand why it works, how it works, and when it works (or fails). It is by understanding the nature of learning at its foundation—both generally and with regard to particular algorithms and phenomena—that the field is able to move forward. Indeed, this has been the lesson of Vapnik's lifework.

This chapter aims to review some of the numerous perspectives and analyses of AdaBoost that have been applied to explain or understand it as a learning method, with comparisons of both the strengths and weaknesses of the various approaches.

R.E. Schapire (✉)
Department of Computer Science, Princeton University, 35 Olden Street, Princeton, NJ 08540, USA
e-mail: schapire@cs.princeton.edu

Given: $(x_1, y_1), \ldots, (x_m, y_m)$ where $x_i \in \mathcal{X}$, $y_i \in \{-1, +1\}$.
Initialize: $D_1(i) = 1/m$ for $i = 1, \ldots, m$.
For $t = 1, \ldots, T$:
- Train weak learner using distribution D_t.
- Get weak hypothesis $h_t : \mathcal{X} \to \{-1, +1\}$.
- Aim: select h_t with low weighted error:

$$\epsilon_t = \Pr_{i \sim D_t}[h_t(x_i) \neq y_i].$$

- Choose $\alpha_t = \frac{1}{2} \ln\left(\frac{1-\epsilon_t}{\epsilon_t}\right)$.
- Update, for $i = 1, \ldots, m$:

$$D_{t+1}(i) = \frac{D_t(i) \exp(-\alpha_t y_i h_t(x_i))}{Z_t}$$

where Z_t is a normalization factor (chosen so that D_{t+1} will be a distribution).

Output the final hypothesis:

$$H(x) = \text{sign}\left(\sum_{t=1}^{T} \alpha_t h_t(x)\right).$$

Fig. 5.1 The boosting algorithm AdaBoost

For brevity, the presentation is at a high level with few technical details. A much more in-depth exposition of most of the topics of this chapter, including more complete references to the relevant literature, can be found in the recent book by Schapire and Freund [30].

Pseudocode for AdaBoost is shown in Fig. 5.1. Here we are given m labeled training examples $(x_1, y_1), \ldots, (x_m, y_m)$, where the x_i's are in some domain $\mathcal{X}$ and the labels $y_i \in \{-1, +1\}$. On each round $t = 1, \ldots, T$, a distribution D_t is computed as in the figure over the m training examples, and a given *weak learner* or *weak learning algorithm* is applied to find a *weak hypothesis* $h_t : \mathcal{X} \to \{-1, +1\}$, where the aim of the weak learner is to find a weak hypothesis with low weighted error ϵ_t relative to D_t. The *final* or *combined hypothesis* H computes the sign of a weighted combination of weak hypotheses

$$F(x) = \sum_{t=1}^{T} \alpha_t h_t(x). \tag{5.1}$$

This is equivalent to saying that H is computed as a weighted majority vote of the weak hypotheses h_t where each hypothesis is assigned weight α_t. (In this chapter, we use the terms "hypothesis" and "classifier" interchangeably.)

5.2 Direct Application of VC Theory

We begin by considering how the general theory of Vapnik and Chervonenkis can be applied directly to AdaBoost.

Intuitively, for a learned classifier to be effective and accurate in its predictions, it should meet three conditions: (1) it should have been trained on "enough" training examples; (2) it should provide a good fit to those training examples (usually meaning that it should have low training error); and (3) it should be "simple." This last condition, our expectation that simpler rules are better, is often referred to as *Occam's razor*.

In formalizing these conditions, Vapnik and Chervonenkis [34, 35] established a foundation for understanding the fundamental nature of learning, laying the groundwork for the design of effective and principled learning algorithms. Specifically, they derived upper bounds on the generalization error of a classifier that could be stated in terms of the three conditions above, and along the way provided workable definitions (such as the VC-dimension) of the slippery and mysterious notion of simplicity.

To understand AdaBoost, the very general and encompassing VC theory is the most sensible starting point. All analyses of learning methods depend in some way on assumptions, since otherwise, learning is quite impossible. From the very beginning, much of the work studying boosting has been based on the assumption that each of the weak hypotheses has accuracy just a little bit better than random guessing; for two-class problems, this means they should each have error below $1/2$, that is, each ϵ_t should be at most $1/2 - \gamma$ for some $\gamma > 0$. This assumption, called the *weak learning condition*, is intrinsic to the mathematical definition of a boosting algorithm, which, given this assumption and sufficient data, can provably produce a final hypothesis with arbitrarily small generalization error.

Given the weak learning condition, it is possible to prove that the training error of AdaBoost's final hypothesis decreases to zero very rapidly; in fact, in just $O(\log m)$ rounds (ignoring all other parameters of the problem), the final hypothesis will perfectly fit the training set [10]. Furthermore, we can measure the complexity (that is, lack of simplicity) of the final hypothesis using the VC-dimension, which can be computed using combinatorial arguments [2, 10]. Having analyzed both the complexity and training fit of the final hypothesis, one can immediately apply the VC theory to obtain a bound on its generalization error.

Such an analysis predicts the kind of behavior depicted on the left of Fig. 5.2, which shows the error (both training and test) of the final hypothesis as a function of the number of rounds of boosting. As noted above, we expect training error to drop very quickly, but at the same time, the VC-dimension of the final hypothesis is increasing roughly linearly with the number of rounds T. Thus, with improved fit to the training set, the test error drops at first, but then rises again as a result of the final hypothesis becoming overly complex. This is classic overfitting behavior. Indeed, overfitting can happen with AdaBoost as seen on the right side of the figure, which shows training and test errors on an actual benchmark dataset. However, as we will

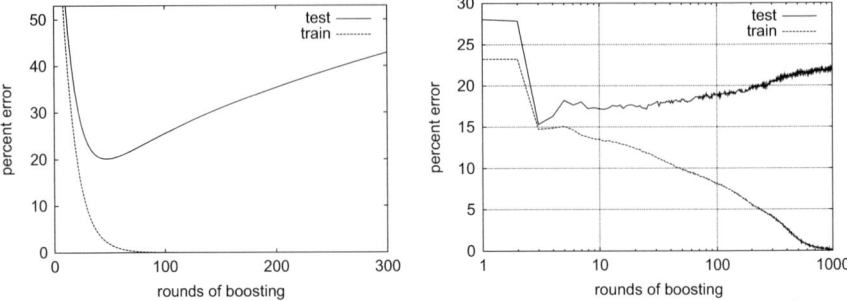

Fig. 5.2 *Left:* A plot of the theoretical training and test percent errors for AdaBoost, as predicted by the arguments of Sect. 5.2. *Right:* The training and test percent error rates obtained using boosting on the Cleveland heart-disease benchmark dataset. (Reprinted from [30] with permission of MIT Press)

see shortly, AdaBoost often does not overfit, apparently in direct contradiction of what is predicted by VC theory.

Summarizing this first approach to understanding AdaBoost, a direct application of VC theory shows that AdaBoost can work if provided with enough data and simple weak classifiers which satisfy the weak learning condition, and if run for enough but not too many rounds. The theory captures the cases in which AdaBoost does overfit, but also predicts (incorrectly) that AdaBoost will *always* overfit.

As in all of the approaches to be discussed in this chapter, the numerical bounds on generalization error that can be obtained using this technique are horrendously loose.

5.3 The Margins Explanation

Another actual typical run on a different benchmark dataset is shown on the left of Fig. 5.3. In this case, boosting was used in combination with the decision-tree learning algorithm C4.5 [26] as the weak learner. A single decision tree produced by C4.5 on this dataset has a test error rate of 13.8 %. In this example, boosting very quickly drives down the training error; in fact, after only five rounds, the training error is zero so that all training examples are correctly classified. (Note that there is no reason why AdaBoost cannot proceed beyond this point.)

The test performance of boosting on this dataset is extremely good, far better than that of a single decision tree. And surprisingly, unlike in the earlier example, the test error on this dataset never increases, even after 1,000 trees have been combined, by which point, the combined classifier involves more than two million decision nodes. Even after the training error hits zero, the test error continues to drop, from 8.4 % on round 5 down to 3.1 % on round 1,000. This pronounced lack of overfitting seems to flatly contradict the intuition and theory discussed in Sect. 5.2 which says

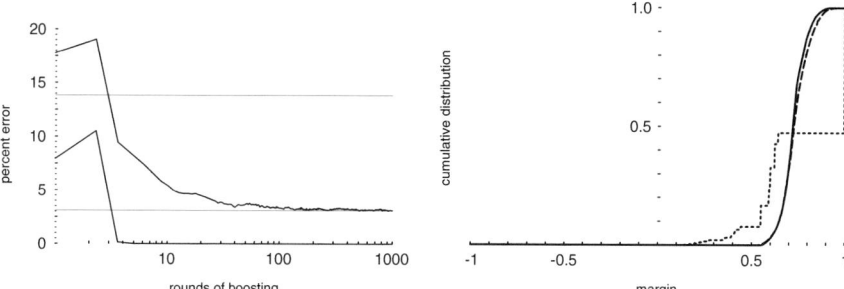

Fig. 5.3 *Left:* The training and test percent error rates obtained using boosting on an OCR dataset with C4.5 as the weak learner. The *top* and *bottom curves* are test and training error, respectively. The *top horizontal line* shows the test error rate using just C4.5. The *bottom line* shows the final test error rate of AdaBoost after 1,000 rounds. *Right:* The margin distribution graph for this same case showing the cumulative distribution of margins of the training instances after 5, 100 and 1,000 iterations, indicated by *short-dashed, long-dashed (mostly hidden)* and *solid curves*, respectively. (Both figures are reprinted from [32] with permission of the Institute of Mathematical Statistics)

that simpler is better. Surely, a combination of five trees is much simpler than a combination of 1,000 trees (about 200 times simpler, in terms of raw size), and both perform equally well on the training set (perfectly, in fact). So how can it be that the far larger and more complex combined classifier performs so much better on the test set?

Such resistance to overfitting is typical of boosting, although, as we saw earlier, boosting certainly *can* overfit. This resistance is one of the properties that make it such an attractive learning algorithm. But how can we understand this behavior?

The *margins explanation* of Schapire et al. [32] was proposed as a way out of this seeming paradox. Briefly, the main idea is the following. The description above of AdaBoost's performance on the training set only took into account the training error, which is zero already after only five rounds. However, training error only tells part of the story in that it only reports the number of examples that are correctly or incorrectly classified. Instead, to understand AdaBoost, we also need to consider how *confident* are the predictions made by the algorithm. According to this explanation, although the training error—that is, whether or not the predictions are correct—is not changing after round 5, the confidence in those predictions is increasing dramatically with additional rounds of boosting. And it is this increase in confidence which accounts for the better generalization performance.

To measure confidence, we use a quantity called the *margin*. Recall that the combined classifier H is simply a weighted majority vote—that is, the result of a small-scale "election"—of the predictions of the weak classifiers. In a real-world election, confidence in the outcome is measured by the margin of victory, the difference between the fraction of votes received by the winner and the fraction of votes received by the loser. In the same way, we can define margin in our setting as the difference between the weighted fraction of the weak classifiers predicting the correct label and the weighted fraction predicting the incorrect label. When this vote is very close, so that the predicted label $H(x)$ is based on a narrow majority, the margin will be small in magnitude and, intuitively, we will have little confidence in the prediction. On the other hand, when the prediction $H(x)$ is based on a clear and substantial majority of the weak classifiers, the margin will be correspondingly large, lending greater confidence in the predicted label. Thus, the magnitude of the margin is a reasonable measure of confidence. Furthermore, the margin will be positive if and only if the overall prediction $H(x)$ is correct.

We can visualize the effect AdaBoost has on the margins of the training examples by plotting their distribution. In particular, we can create a plot showing, for each $\theta \in [-1, +1]$, the fraction of training examples with margin at most θ. For such a cumulative distribution curve, the bulk of the distribution lies where the curve rises the most steeply. Fig. 5.3, on the right, shows such a *margin distribution graph* for the same dataset as above, showing the margin distribution after 5, 100 and 1,000 rounds of boosting. Whereas nothing at all is happening to the training error, these curves expose dramatic changes happening to the margin distribution. For instance, after five rounds, although the training error is zero (so that no example has negative margin), a rather substantial fraction of the training examples (7.7%) has margin below 0.5. By round 100, all of these examples have been swept to the right so that not a single example has margin below 0.5, and nearly all have margin above 0.6.

Thus, this example is indicative of the powerful effect AdaBoost has on the margins, aggressively pushing up those examples with small or negative margin. Moreover, comparing the two sides of Fig. 5.3, we see that this overall increase in the margins appears to be correlated with better performance on the test set.

AdaBoost can be analyzed theoretically along exactly these lines. It is possible to prove first a bound on the generalization error of AdaBoost—or any other voting method—that depends only on the margins of the training examples, and *not* on the number of rounds of boosting. Such a bound predicts that AdaBoost will not overfit regardless of how long it is run, provided that large margins can be achieved (and provided, of course, that the weak classifiers are not too complex relative to the size of the training set).

The second part of such an analysis is to prove that, as observed empirically in Fig. 5.3, AdaBoost generally tends to increase the margins of all training examples, and moreover, the higher the accuracy of the weak hypotheses, the larger will be the margins.

All this suggests that perhaps a more effective learning algorithm could be designed by explicitly attempting to maximize the margins. This was attempted by Breiman [4] (among others) who created an algorithm called arc-gv for maximizing the smallest margin of any training example. Although this algorithm did indeed produce larger margins, its test performance turned out to be slightly *worse* than that of AdaBoost, apparently contradicting the margins theory. In a follow-up study, Reyzin and Schapire [28] suggested two possible explanations. First, more aggressive margin maximization seems to produce more complex weak hypotheses, which tends to raise the potential for overfitting, confounding the experiments. And second, in some cases, arc-gv produces a higher *minimum* margin, but a distribution of margins that is lower overall.

In summary, according to the margins explanation, AdaBoost will succeed without overfitting if the weak-hypothesis accuracies are substantially better than random (since this will lead to large margins), and if provided with enough data relative to the complexity of the weak hypotheses. This is really the only known theory that explains the cases in which overfitting is not observed. On the other hand, attempted extensions of AdaBoost based on direct maximization of margins have not been entirely successful, though work in this area is ongoing (see, for instance, [22, 36]).

5.4 Loss Minimization

Many, perhaps even most, learning and statistical methods that are in common use can be viewed as procedures for minimizing a *loss function* (also called a cost or objective function) that in some way measures how well a model fits the observed data. A classic example is least squares regression in which a sum of squared errors is minimized. AdaBoost, though not originally designed for this purpose, also turns out to minimize a particular loss function. Viewing the algorithm in this light can be helpful for a number of reasons. First, such an understanding can help to clarify the goal of the algorithm and can be useful in proving convergence properties. And second, by decoupling the algorithm from its objective, we may be able to derive better or faster algorithms for the same objective, or alternatively, we might be able to generalize AdaBoost for new challenges.

AdaBoost can be understood as a procedure for greedily minimizing what has come to be called the *exponential loss*, namely,

$$\frac{1}{m} \sum_{i=1}^{m} \exp(-y_i F(x_i))$$

where $F(x)$ is as given in Eq. (5.1). In other words, it can be shown that the choices of α_t and h_t on each round happen to be the same as would be chosen so as to cause the greatest decrease in this loss. This connection was first observed by Breiman [4] and later expanded upon by others [7, 12, 23, 25, 27, 31].

Why does this loss make sense? Intuitively, minimizing exponential loss strongly favors the choice of a function F for which the sign of $F(x_i)$ is likely to agree with the correct label y_i; since the final hypothesis H is computed as the sign of F, this is exactly the behavior we seek in attempting to minimize the number of mistaken classifications. Another argument that is sometimes made is that the real goal of minimizing classification errors requires the optimization of an objective that is not continuous, differentiable or easily minimized, but which can be approximated by a smooth and convex "surrogate" objective function such as the exponential loss. The exponential loss is also related to the loss used for logistic regression [12].

As a procedure for minimizing this loss, AdaBoost can be viewed as a form of coordinate descent (in which each step is made greedily along one of the coordinate directions), as noted by Breiman [4]. Alternatively, AdaBoost can be viewed as a form of functional gradient descent, as observed by Mason et al. [23] and Friedman [11]. This understanding has led to the immediate generalization of boosting to a wide range of other learning problems and loss functions, such as regression.

From this perspective, it might seem tempting to conclude that AdaBoost's effectiveness as a learning algorithm is derived from the choice of loss function that it apparently aims to minimize, in other words, that AdaBoost works *only because* it minimizes exponential loss. If this were true, then it would follow plausibly that a still better algorithm could be designed using more powerful and sophisticated approaches to optimization than AdaBoost's comparatively meek approach.

However, it is critical to keep in mind that minimization of exponential loss by itself is *not* sufficient to guarantee low generalization error. On the contrary, it is very much possible to minimize the exponential loss (using a procedure other than AdaBoost), while suffering quite substantial generalization error (relative, say, to AdaBoost). To demonstrate this point, consider the following experiment from Schapire and Freund [30], which is similar in spirit to the work of Mease and Wyner [24, 37]. Data for this experiment was generated synthetically with each instance **x** a 10,000-dimensional $\{-1, +1\}$-valued vector, that is, a point in $\{-1, +1\}^{10,000}$. Each of the 1,000 training and 10,000 test examples were generated uniformly at random from this space. The label y associated with an instance **x** was defined to be the majority vote of three designated coordinates of **x**. The weak hypotheses used were associated with coordinates so that each was of the form $h(\mathbf{x}) = x_j$ for all **x**, and for some coordinate j. (The negatives of these were also included.)

Three different algorithms were tested. The first was ordinary AdaBoost using an exhaustive weak learner that, on each round, finds the minimum weighted error weak hypothesis. We refer to this as *exhaustive AdaBoost*. The second algorithm was *gradient descent* on the exponential loss function (which can be written in a parametric form so that ordinary gradient descent can be applied). The third algorithm was actually the same as AdaBoost except that the weak learner does not actively search for the best weak hypothesis, but rather selects one uniformly at random from the space of possible weak hypotheses; we refer to this method as *random AdaBoost*.

All three algorithms are guaranteed to minimize the exponential loss, but that does *not* mean that they will necessarily perform the same on actual data in terms of classification accuracy. It is true that the exponential loss is convex, and therefore can have no local minima. But it is possible, and even typical, for the minimum either to be non-unique, or to not exist at all at any finite setting of the parameters. Therefore, different algorithms for the same (convex) loss can yield very different hypotheses.

The results of these experiments are shown in Table 5.1. Regarding speed (measured by number of rounds), the table shows that gradient descent is extremely

5 Explaining AdaBoost

Table 5.1 Results of the experiment described in Sect. 5.4. The numbers in brackets show the number of rounds required for each algorithm to reach specified values of the exponential loss. The unbracketed numbers show the percent test error achieved by each algorithm at the point in its run at which the exponential loss first dropped below the specified values. All results are averaged over ten random repetitions of the experiment. (Reprinted from [30] with permission of MIT Press)

Exp. loss	% Test error [# rounds]					
	Exhaustive AdaBoost		Gradient descent		Random AdaBoost	
10^{-10}	0.0	[94]	40.7	[5]	44.0	[24,464]
10^{-20}	0.0	[190]	40.8	[9]	41.6	[47,534]
10^{-40}	0.0	[382]	40.8	[21]	40.9	[94,479]
10^{-100}	0.0	[956]	40.8	[70]	40.3	[234,654]

fast at minimizing exponential loss, while random AdaBoost is unbearably slow, though eventually effective. Exhaustive AdaBoost is somewhere in between. As for accuracy, the table shows that both gradient descent and random AdaBoost performed very poorly on this data, with test errors never dropping significantly below 40 %. In contrast, exhaustive AdaBoost quickly achieved and maintained perfect test accuracy after the third round.

Of course, this artificial example is not meant to show that exhaustive AdaBoost is always a better algorithm than the other two methods. Rather, the point is that AdaBoost's strong performance as a classification algorithm cannot be credited—at least not exclusively—to its effect on the exponential loss. If this were the case, then any algorithm achieving equally low exponential loss should have equally low generalization error. But this is far from what we see in this example, where exhaustive AdaBoost's very low exponential loss is matched by the competitors, but their test errors are not even close. Clearly, some other factor beyond its exponential loss must be at work to explain exhaustive AdaBoost's comparatively strong performance.

So to summarize, minimization of exponential loss is a fundamental property of AdaBoost, and one that opens the door for a range of practical generalizations of the algorithm. However, it is important to keep in mind that this perspective is rather limited in terms of what it can tell us about AdaBoost's accuracy as a learning algorithm. The example above demonstrates that any understanding of AdaBoost's generalization capabilities must in some way take into account the particular dynamics of the algorithm—not just the objective function, but what procedure is actually being used to minimize it.

5.5 Regularization

Without question, AdaBoost minimizes exponential loss. And yet, as was just seen, other algorithms for minimizing this same loss can perform far worse. If the choice of loss function cannot explain how AdaBoost avoids the poor performance of these other algorithms, then how does it do it?

In general, when minimizing a loss function, it has become quite popular and standard to *regularize*, that is, to modify or constrain the optimization problem in a way that attempts to avoid overfitting by limiting complexity or encouraging smoothness. In our context, we have seen that AdaBoost constructs a linear combination F of weak hypotheses (as in Eq. (5.1)), and does so in a way that minimizes exponential loss over all such linear combinations. To regularize, we might instead choose our objective to be the minimization of this same loss, but subject to the constraint that the weak-hypothesis weights appearing in F, when viewed collectively as a vector, have ℓ_1-norm bounded by some preset parameter $B > 0$. There are many other ways of regularizing (for instance, using a different norm), but this particular form based on the ℓ_1-norm, sometimes called the "lasso," has the especially favorable property that it seems to encourage sparsity, that is, a solution with relatively few nonzero weights [33].

AdaBoost certainly does not explicitly regularize—there is nothing about the algorithm that overtly limits the weights on the weak hypotheses. Nevertheless, is it possible that it is somehow applying some kind of implicit form of regularization? In fact, it turns out that a simple variant of AdaBoost, when stopped after any number of rounds, can often be viewed as providing an approximate solution to ℓ_1-regularized minimization of exponential loss. To see this, consider an experiment in which we compute the solution to this regularized optimization problem for all possible values of the preset bound B. As B varies, these weight vectors trace out a path or trajectory, which can be plotted in the unrealistic but illustrative case where the space of possible weak hypotheses is very small. This is shown on the left of Fig. 5.4 on benchmark data using just six possible weak hypotheses. Each curve corresponds to one of the six weak hypotheses and plots its weight at the regularized solution as a function of B. Thus, the figure depicts the entire trajectory.

For comparison, consider a variant of AdaBoost in which α_t, rather than being set as in Fig. 5.1, is chosen on each round to be equal to a fixed small constant $\alpha > 0$. As above, we can plot the trajectory of the weights on the weak hypotheses which define the combined classifier as a function of the number of iterations T, multiplied by the constant α so that the resulting scale αT is equal to the cumulative sum of weight updates after T iterations. This is shown, for the same dataset, on the right of Fig. 5.4 (using $\alpha = 10^{-6}$).

Remarkably, the two plots are practically indistinguishable. This shows that, at least in this case, a variant of AdaBoost, when run for T rounds, computes essentially the same solution vectors as when using ℓ_1-regularization with B set to αT. Thus, early stopping—that is, halting boosting after a limited number of rounds—is in this sense apparently equivalent to regularization. This correspondence was first observed by Hastie et al. [13], and explored further by Rosset et al. [29]. Later, Zhao and Yu [40] showed theoretically that the correspondence will hold generally under certain but not all conditions.

All this suggests a plausible explanation for how AdaBoost works: Regularization is a general technique that protects against overfitting by constraining, smoothing, and/or promoting sparsity. As just discussed, AdaBoost with early

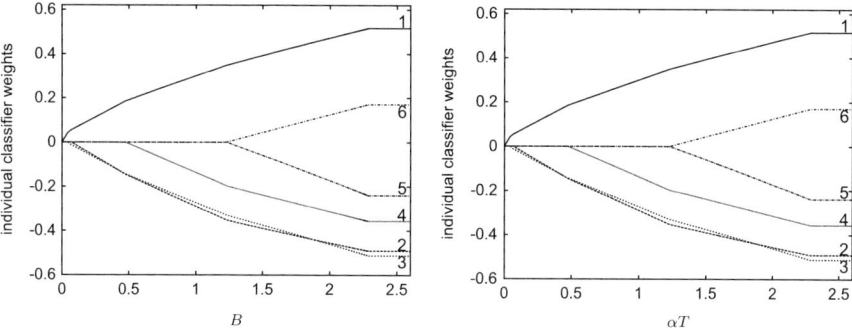

Fig. 5.4 The trajectories of the weight vectors computed on a benchmark dataset using only six possible weak hypotheses. Trajectories are plotted for ℓ_1-regularized exponential loss as the parameter B varies (*left*), and for a variant of AdaBoost in which $\alpha_t = \alpha = 10^{-6}$ on every round (*right*). Each figure includes one curve for each of the six weak hypotheses, showing its associated weight as a function of the total weight added. (Reprinted from [30] with permission of MIT Press)

stopping is related to ℓ_1-regularization. Therefore, AdaBoost avoids overfitting through implicit regularization.

However, there are important deficiencies in this argument. First of all, strictly speaking, it does not apply to AdaBoost, but only to a variant of AdaBoost in which the weights on each round are set to a small fixed constant. And second, this argument only makes sense if we stop AdaBoost after a relatively small number of rounds since it is through early stopping, according to this view, that regularization is actually applied.

What happens if AdaBoost is run for a large number of rounds, as in the cases described in Sect. 5.3 where overfitting was apparently absent? According to this view, making the number of rounds T large corresponds to choosing a regularization parameter B that is also large. Thus, when T is very large, the purported regularization must be extremely weak, and in the limit must become so vanishingly weak as to apparently have no constraining influence at all on the optimization problem that it is meant to constrain. When this happens, how can it be having any effect at all?

In fact, Rosset et al. [29] proved that if the regularization is relaxed to the limit so that $B \to \infty$, then the resulting (anemically regularized) solutions turn out asymptotically to maximize the margins of the training examples. This means that we can prove something about how well such solutions will perform on new data, but only as a result of their margin-maximizing properties and by applying the margins theory. It is not the regularization that is explaining good performance here since it has been weakened to the point of essentially disappearing altogether.

So to summarize, we have seen a perspective in which boosting with early stopping can be related to ℓ_1-regularization. However, this view does not apply to

AdaBoost, but only to a variant. And furthermore, for a large number of rounds, we can only explain good performance, according to this view, by again appealing to the margins theory rather than as a direct result of implicit regularization.

5.6 Inherently Unpredictable Data

As discussed in Sect. 5.3, the margins theory shows that, if given "enough" data, and if the weak learning condition holds, then the generalization error can be made arbitrarily close to zero so that the resulting classifier is essentially perfect. This obviously seems like a good thing. But it should also make us suspicious since, even under the most ideal circumstances, it is usually impossible on real data to get perfect accuracy due to intrinsic noise or uncertainty. In other words, the *Bayes error*, the minimum possible error of any classifier, is usually strictly positive.

So on the one hand, the margins theory tells us that, with enough data, it should be possible to train a perfectly accurate classifier, but on the other hand, the data itself usually makes this impossible. In practice, this is not necessarily a contradiction, even when the weak learning assumption holds. This is because the weak-hypothesis space typically is not fixed, but grows in complexity with the size of the training set; for instance, this happens "automatically" when using decision trees as weak hypotheses since the generated trees will usually be bigger if trained with more data. Nevertheless, it would certainly be desirable to have a theory that more directly handles the case in which the Bayes error is nonzero.

Indeed, it has been proved that AdaBoost's combined classifier has an error rate that converges to the Bayes optimal provided that the algorithm is given enough data, that it is run for enough but not too many rounds, and that the weak hypotheses come from a class of functions that is "sufficiently rich." In this sense, the algorithm is said to be *universally consistent*, a property that was proved by Bartlett and Traskin [1] following the work of many others [3, 5, 14, 19, 21, 38, 39].

This means that AdaBoost can (theoretically) learn optimally even in noisy settings. Furthermore, this theory does not depend on the weak learning condition. However, the theory does not explain why AdaBoost can often work even when run for a very large number of rounds since, like all explanations other than the margins theory, it depends on the algorithm being stopped after a finite and relatively small number of rounds. Furthermore, the assumption of sufficient richness among the weak hypotheses can also be problematic.

Regarding this last point, Long and Servedio [18] presented an example of a learning problem which shows just how far off a universally consistent algorithm like AdaBoost can be from optimal when this assumption does not hold, even when the noise affecting the data is seemingly very mild. In this example, each data point has its label inverted with quite low probability, say 1%. The Bayes optimal classifier has an error rate that is also just 1%, and is obtainable by a classifier of the same form as that used by AdaBoost. Nevertheless, AdaBoost, in this case, will provably produce a classifier whose error exceeds 50%, in other words, at least as bad as random guessing. In fact, this result holds even if the learning algorithm is provided with unlimited training data. And it is really not a result about AdaBoost at all—it is about algorithms based on loss minimization. The same result applies to any method that minimizes exponential loss, as well as most other commonly used convex losses. It also holds even if regularization is applied. For instance, it can be shown that the same result applies to support vector machines, logistic regression, linear regression, lasso, ridge regression, and many more methods.

So this example shows that such consistency results can fail badly if the weak classifiers are not rich enough. It also shows that AdaBoost (and most other loss-based methods) can be very susceptible to noise, even with regularization, at least on artificially constructed datasets. This susceptibility to noise has also been observed in practice, for instance, by Dietterich [6] and Maclin and Opitz [20].

How then should we handle noise and outliers? Certainly, these must be a problem on "real-world" datasets, and yet, AdaBoost often works well anyway. So one approach is simply not to worry about it. Theoretically, various approaches to handling noise in boosting have also been proposed, often using techniques based on "branching programs" [15–17].

Yet another approach is based on an entirely different boosting algorithm called *boost-by-majority*, due to Freund [8]. In a certain sense, this algorithm turns out to be exactly optimally efficient as a boosting algorithm. Furthermore, it does not appear to minimize any convex loss function. Like AdaBoost, the algorithm on each round puts more weight on the harder examples. However, unlike AdaBoost, it has a very interesting behavior in which it can "give up" on the very hard examples. This property might make the algorithm more robust to noise by its eventually ignoring outliers and noise-corrupted examples rather than "spinning its wheels" on them as AdaBoost does. Unfortunately, unlike AdaBoost, the boost-by-majority algorithm is not *adaptive* in the sense that it requires prior knowledge about the number of rounds and the degree to which the weak learning assumption holds. Nevertheless, Freund [9] proposed making it adaptive by passing to a kind of limit in which time is moving continuously rather than in discrete steps.

The resulting algorithm, called *BrownBoost*, is somewhat more challenging to implement, but preliminary experiments suggest that it might be more resistant to noise and outliers. See Table 5.2.

Summarizing, we have seen that, under appropriate conditions, AdaBoost provably converges in its accuracy to the best possible, even in the presence of noise and even without the weak learning condition. On the other hand, AdaBoost's performance can be very poor when the weak hypotheses are insufficiently expressive. Noise can be a real problem for AdaBoost, and various approaches have been proposed for handling it, including a form of boosting which operates in continuous time.

Table 5.2 The results of running AdaBoost and BrownBoost on the "letter" and "satimage" benchmark datasets. After being converted to binary by combining the classes into two arbitrary groups, each dataset was split randomly into training and test sets, and corrupted for training with artificial noise at the given rates. The entries of the table show percent error on *uncorrupted* test examples. All results are averaged over 50 random repetitions of the experiment. (These experiments were conducted by Evan Ettinger, Sunsern Cheamanunkul and Yoav Freund, and were reported in [30])

Dataset	Noise (%)	AdaBoost	BrownBoost
Letter	0	3.7	4.2
	10	10.8	7.0
	20	15.7	10.5
Satimage	0	4.9	5.2
	10	12.1	6.2
	20	21.3	7.4

5.7 Conclusions

This chapter has attempted to bring together several different approaches that have been proposed for understanding AdaBoost. These approaches are reflective of broader trends within machine learning, including the rise of methods based on margin maximization, loss minimization, and regularization. As we have seen, these different approaches are based on varying assumptions, and attempt to capture different aspects of AdaBoost's behavior. As such, one can argue as to which of these is most realistic or explanatory, a perspective that is likely to depend on individual taste and experience. Furthermore, direct experimental comparison of the different approaches is especially difficult due to the looseness of the various bounds and theoretical predictions when applied to actual data.

For the most part, the different perspectives that have been presented do not subsume one another, each having something to say about AdaBoost that is perhaps not captured by the others. But taken together, they form a rich and expansive theory for understanding this one algorithm. Perhaps someday a single overriding theory will emerge that encompasses all of them.

Acknowledgements Support for this research was generously provided by NSF under Award #1016029.

References

1. Bartlett, P.L., Traskin, M.: AdaBoost is consistent. J. Mach. Learn. Res. **8**, 2347–2368 (2007)
2. Baum, E.B., Haussler, D.: What size net gives valid generalization? Neural Comput. **1**(1), 151–160 (1989)
3. Bickel, P.J., Ritov, Y., Zakai, A.: Some theory for generalized boosting algorithms. J. Mach. Learn. Res. **7**, 705–732 (2006)

4. Breiman, L.: Prediction games and arcing classifiers. Neural Comput. **11**(7), 1493–1517 (1999)
5. Breiman, L.: Population theory for boosting ensembles. Ann. Stat. **32**(1), 1–11 (2004)
6. Dietterich, T.G.: An experimental comparison of three methods for constructing ensembles of decision trees: bagging, boosting, and randomization. Mach. Learn. **40**(2), 139–158 (2000)
7. Frean, M., Downs, T.: A simple cost function for boosting. Technical report, Department of Computer Science and Electrical Engineering, University of Queensland (1998)
8. Freund, Y.: Boosting a weak learning algorithm by majority. Inf. Comput. **121**(2), 256–285 (1995)
9. Freund, Y.: An adaptive version of the boost by majority algorithm. Mach. Learn. **43**(3), 293–318 (2001)
10. Freund, Y., Schapire, R.E.: A decision-theoretic generalization of on-line learning and an application to boosting. J. Comput. Syst. Sci. **55**(1), 119–139 (1997)
11. Friedman, J.H.: Greedy function approximation: a gradient boosting machine. Ann. Stat. **29**(5), 1189–1232 (2001)
12. Friedman, J., Hastie, T., Tibshirani, R.: Additive logistic regression: a statistical view of boosting. Ann. Stat. **28**(2), 337–407 (2000)
13. Hastie, T., Tibshirani, R., Friedman, J.: The Elements of Statistical Learning: Data Mining, Inference, and Prediction, 2nd edn. Springer, New York (2009)
14. Jiang, W.: Process consistency for AdaBoost. Ann. Stat. **32**(1), 13–29 (2004)
15. Kalai, A.T., Servedio, R.A.: Boosting in the presence of noise. J. Comput. Syst. Sci. **71**(3), 266–290 (2005)
16. Long, P.M., Servedio, R.A.: Adaptive martingale boosting. In: 18th Annual Conference on Learning Theory, Bertinoro (2005)
17. Long, P.M., Servedio, R.A.: Adaptive martingale boosting. Advances in Neural Information Processing Systems 21, Vancouver (2009)
18. Long, P.M., Servedio, R.A.: Random classification noise defeats all convex potential boosters. Mach. Learn. **78**, 287–304 (2010)
19. Lugosi, G., Vayatis, N.: On the Bayes-risk consistency of regularized boosting methods. Ann. Stat. **32**(1), 30–55 (2004)
20. Maclin, R., Opitz, D.: An empirical evaluation of bagging and boosting. In: Proceedings of the 14th National Conference on Artificial Intelligence, Providence, pp. 546–551 (1997)
21. Mannor, S., Meir, R., Zhang, T.: Greedy algorithms for classification—consistency, convergence rates, and adaptivity. J. Mach. Learn. Res. **4**, 713–742 (2003)
22. Mason, L., Bartlett, P., Baxter, J.: Direct optimization of margins improves generalization in combined classifiers. Advances in Neural Information Processing Systems 12. MIT Press, Cambridge (2000)
23. Mason, L., Baxter, J., Bartlett, P., Frean, M.: Functional gradient techniques for combining hypotheses. Advances in Large Margin Classifiers. MIT Press, Cambridge (2000)
24. Mease, D., Wyner, A.: Evidence contrary to the statistical view of boosting. J. Mach. Learn. Res. **9**, 131–156 (2008)
25. Onoda, T., Rätsch, G., Müller, K.R.: An asymptotic analysis of AdaBoost in the binary classification case. In: Proceedings of the 8th International Conference on Artificial Neural Networks, Skövde, pp. 195–200 (1998)
26. Quinlan, J.R.: C4.5: Programs for Machine Learning. Morgan Kaufmann, San Mateo (1993)
27. Rätsch, G., Onoda, T., Müller, K.R.: Soft margins for AdaBoost. Mach. Learn. **42**(3), 287–320 (2001)
28. Reyzin, L., Schapire, R.E.: How boosting the margin can also boost classifier complexity. In: Proceedings of the 23rd International Conference on Machine Learning, Pittsburgh (2006)
29. Rosset, S., Zhu, J., Hastie, T.: Boosting as a regularized path to a maximum margin classifier. J. Mach. Learn. Res. **5**, 941–973 (2004)
30. Schapire, R.E., Freund, Y.: Boosting: Foundations and Algorithms. MIT Press, Cambridge (2012)
31. Schapire, R.E., Singer, Y.: Improved boosting algorithms using confidence-rated predictions. Mach. Learn. **37**(3), 297–336 (1999)

32. Schapire, R.E., Freund, Y., Bartlett, P., Lee, W.S.: Boosting the margin: a new explanation for the effectiveness of voting methods. Ann. Stat. **26**(5), 1651–1686 (1998)
33. Tibshirani, R.: Regression shrinkage and selection via the Lasso. J. Royal Stat. Soc. B (Methodol.) **58**(1), 267–288 (1996)
34. Vapnik, V.N., Chervonenkis, A.Y.: On the uniform convergence of relative frequencies of events to their probabilities. Theory Prob. Appl. **16**(2), 264–280 (1971)
35. Vapnik, V.N., Chervonenkis, A.Y.: Theory of Pattern Recognition. Nauka, Moscow (1974). (In Russian)
36. Wang, L., Sugiyama, M., Jing, Z., Yang, C., Zhou, Z.H., Feng, J.: A refined margin analysis for boosting algorithms via equilibrium margin. J. Mach. Learn. Res. **12**, 1835–1863 (2011)
37. Wyner, A.J.: On boosting and the exponential loss. In: Proceedings of the 9th International Workshop on Artificial Intelligence and Statistics, Key West (2003)
38. Zhang, T.: Statistical behavior and consistency of classification methods based on convex risk minimization. Ann. Stat. **32**(1), 56–134 (2004)
39. Zhang, T., Yu, B.: Boosting with early stopping: convergence and consistency. Ann. Stat. **33**(4), 1538–1579 (2005)
40. Zhao, P., Yu, B.: Stagewise Lasso. J. Mach. Learn. Res. **8**, 2701–2726 (2007)

Chapter 6
On the Relations and Differences Between Popper Dimension, Exclusion Dimension and VC-Dimension

Yevgeny Seldin and Bernhard Schölkopf

Abstract A high-level relation between Karl Popper's ideas on "falsifiability of scientific theories" and the notion of "overfitting" in statistical learning theory can be easily traced. However, it was pointed out that at the level of technical details the two concepts are significantly different. One possible explanation that we suggest is that the process of falsification is an active process, whereas statistical learning theory is mainly concerned with supervised learning, which is a passive process of learning from examples arriving from a stationary distribution. We show that concepts that are closer (although still distant) to Karl Popper's definitions of falsifiability can be found in the domain of learning using membership queries, and derive relations between Popper's dimension, exclusion dimension, and the VC-dimension.

6.1 Introduction

There is a clear relation between Karl Popper's notion of unfalsifiability of a scientific theory [4] and the notion of overfitting in statistical learning theory [8–10]. However, when we go down to Karl Popper's definition of complexity of a scientific theory and to the definition of complexity of a function class in statistical learning theory, the VC-dimension, we find significant dissimilarities [2, 6]. Corfield et al. showed that Karl Popper's definition of complexity of a function class would not

---

This work was primarily done when YS was with the Max Planck Institute for Intelligent Systems.

Y. Seldin (✉)
Queensland University of Technology, Brisbane, QLD 4001, Australia
e-mail: yevgeny.seldin@gmail.com

B. Schölkopf
Max Planck Institute for Intelligent Systems, Spemannstrasse, 72076 Tübingen, Germany
e-mail: bs@tuebingen.mpg.de

work in supervised learning and wondered whether he failed to provide an accurate definition or was concerned with a different setting.

We argue that the process of falsification of a scientific theory is fundamentally different from the process of learning in supervised learning and, therefore, the parallels and applications to Popper's dimension are to be sought in other fields, probably in learning using membership queries. In supervised learning we are passively learning from examples generated from a stationary distribution and we evaluate the hypothesis on the same distribution. In most guarantees provided by statistical learning theory it is assumed that the test environment is similar to the training environment. In the process of falsification of a scientific theory we deliberately look for a test environment where the hypothesis will fail. The test (or falsification) environment is very likely to be different from the training environment. In this sense the process is more similar to learning using membership queries.

Below we compare Popper's dimension and VC dimension with exclusion dimension [1], used in learning a finite concept class over a finite domain using membership queries.

6.2 Setting

For the sake of comparison of the dimensions we consider a class of functions $\mathcal{F}$ from an arbitrary domain $\mathcal{X}$ to $\mathcal{Y} = \{0, 1\}$. The *VC-dimension* [7] of $\mathcal{F}$ is defined as "the maximal number of points that can be shattered by $\mathcal{F}$", whereas a set of points is said to be *shattered* by $\mathcal{F}$ if "all possible labellings of the points can be implemented by $\mathcal{F}$" [5]. The *Popper dimension* of $\mathcal{F}$ [2] is defined as "the minimal number of points in $\mathcal{X}$ that cannot be shattered by $\mathcal{F}$". (Such sets of points can potentially falsify $\mathcal{F}$.) Finally, the exclusion dimension of $\mathcal{F}$ is defined as "the maximum over $g \notin \mathcal{F}$ of the minimal size of any specifying set for g with respect to $\mathcal{F}$", where a *specifying set* is "a set S, such that at most one concept $f \in \mathcal{F}$ agrees with g on S" [1]. The exclusion dimension can be seen as the minimal number of examples (minus 1) that are required in order to prove that $g \notin \mathcal{F}$ in the worst case (over g).

In the next section we translate the definitions of the three dimensions into logical quantifiers and in the section after that bring some examples that help us figure out relations between the dimensions.

6.3 Definitions with Logical Quantifiers

Below we define the dimensions using logical quantifiers. The colon symbol ":" is used as an abbreviation of "such that".

VC-dimension:

$$VC(\mathcal{F}) = \max\{n : \exists \{x_1, \ldots, x_n\} : \forall \{y_1, \ldots, y_n\} \exists f \in \mathcal{F} : \forall i, f(x_i) = y_i\} \tag{6.1}$$

Popper dimension:

$$PD(\mathcal{F}) = \min\{n : \exists\{x_1,\ldots,x_n\} \exists\{y_1,\ldots,y_n\} : \forall f \in \mathcal{F} \exists i : f(x_i) \neq y_i\} \quad (6.2)$$

Exclusion dimension:

$$XD(\mathcal{F}) = \left(\max_{g \notin \mathcal{F}} \min\{n : \exists\{x_1,\ldots,x_n\} : \forall f \in \mathcal{F} \exists i : f(x_i) \neq g(x_i)\}\right) - 1. \quad (6.3)$$

6.4 Relations

It is easy to observe the following relations between the dimensions:

Lemma 6.1. *For any function class $\mathcal{F}$:*

$$PD(\mathcal{F}) \leq VC(\mathcal{F}) + 1 \quad (6.4)$$

$$PD(\mathcal{F}) \leq XD(\mathcal{F}) + 1. \quad (6.5)$$

Proof. Follows from the definitions of the dimensions (6.1)–(6.3). □

Next we use examples inspired by the analysis of the teaching dimension by Goldman and Kearns [3] in order to show that no bounds in the other direction can be obtained, that bounds (6.4) and (6.5) are tight in certain situations, and that the relation between the exclusion and VC dimensions can be arbitrary.

Lemma 6.2. *For any $n \geq 2$ there exists a function class $\mathcal{F}$ for which $XD(\mathcal{F}) = |\mathcal{F}| - 1 = n - 1$, $VC(\mathcal{F}) = 1$, and $PD(\mathcal{F}) = 2$.*

Proof. Consider the function class $\mathcal{F}$ in Fig. 6.1. Each $f_i \in \mathcal{F}$ equals 0 on x_i and 1 elsewhere. Clearly, $VC(\mathcal{F}) = 1$ and $PD(\mathcal{F}) = 2$. For calculation of the exclusion dimension consider a function g which equals 1 everywhere. In order to prove that $g \notin \mathcal{F}$ we have to reveal the value of g for all x_i; hence $XD(\mathcal{F}) = n - 1 = |\mathcal{F}| - 1$. □

Lemma 6.3. *For any $n \geq 0$ there exists a function class $\mathcal{F}$ for which $PD(\mathcal{F}) = 2$ and $XD(\mathcal{F}) = VC(\mathcal{F}) = \log_2 |\mathcal{F}| = n$.*

Proof. Consider the function class $\mathcal{F}$ in Fig. 6.2. The functions in $\mathcal{F}$ implement all possible labellings of $x_1,\ldots,x_n$, whereas for each $f_i \in \mathcal{F}$ we have that $f_i(x_{n+j}) = 1$ for $i = j$ and 0 otherwise. It is easy to see that $VC(\mathcal{F}) = n = \log_2 |\mathcal{F}|$ and $PD(\mathcal{F}) = 2$ (because we cannot implement a function f such that $f(x_{n+1}) = f(x_{n+2}) = 1$). For the same class $XD(\mathcal{F}) = n = VC(\mathcal{F})$. The "hard" example g that proves this bound is $g(x_{n+j}) = 0$ for all $1 \leq j \leq 2^n$, and $g(x_j)$ is arbitrary for $1 \leq j \leq n$. Indeed, in order to prove that $g \notin \mathcal{F}$ we have to reveal the

Fig. 6.1 A function class $\mathcal{F}$ for which $XD(\mathcal{F}) = |\mathcal{F}| - 1$, $VC(\mathcal{F}) = 1$, and $PD(\mathcal{F}) = 2$

	x_1	x_2	x_3	$\ldots$	x_n
f_1	$-$	$+$	$+$	$\ldots$	$+$
f_2	$+$	$-$	$+$	$\ldots$	$+$
f_3	$+$	$+$	$-$	$\ldots$	$+$
$\vdots$				$\ddots$	
f_n	$+$	$+$	$+$	$\ldots$	$-$

Fig. 6.2 A function class $\mathcal{F}$ for which $PD(\mathcal{F}) = 2$ and $XD(\mathcal{F}) = VC(\mathcal{F}) = \log_2 |\mathcal{F}|$

	x_1	x_2	x_3	$\ldots$	x_n	x_{n+1}	x_{n+2}	x_{n+3}	x_{n+4}	$\ldots$	x_{n+2^n}
f_1	$-$	$-$	$-$	$\ldots$	$-$	$+$	$-$	$-$	$-$	$\ldots$	$-$
f_2	$+$	$-$	$-$	$\ldots$	$-$	$-$	$+$	$-$	$-$	$\ldots$	$-$
f_3	$-$	$+$	$-$	$\ldots$	$-$	$-$	$-$	$+$	$-$	$\ldots$	$-$
f_4	$+$	$+$	$-$	$\ldots$	$-$	$-$	$-$	$-$	$+$	$\ldots$	$-$
$\vdots$				$\ddots$							
f_{2^n}	$+$	$+$	$+$	$\ldots$	$+$	$-$	$-$	$-$	$-$	$\ldots$	$+$

Fig. 6.3 A function class $\mathcal{F}$ for which $PD(\mathcal{F}) = 2$, $XD(\mathcal{F}) = 1$, and $VC(\mathcal{F}) = \log_2 \frac{|\mathcal{F}|}{2}$

	x_1	x_2	x_3	$\ldots$	x_n	x_{n+1}	x_{n+2}	x_{n+3}	x_{n+4}	$\ldots$	x_{n+2^n}
f_1	$-$	$-$	$-$	$\ldots$	$-$	$+$	$-$	$-$	$-$	$\ldots$	$-$
f_2	$+$	$-$	$-$	$\ldots$	$-$	$-$	$+$	$-$	$-$	$\ldots$	$-$
f_3	$-$	$+$	$-$	$\ldots$	$-$	$-$	$-$	$+$	$-$	$\ldots$	$-$
f_4	$+$	$+$	$-$	$\ldots$	$-$	$-$	$-$	$-$	$+$	$\ldots$	$-$
$\vdots$				$\ddots$							
f_{2^n}	$+$	$+$	$+$	$\ldots$	$+$	$-$	$-$	$-$	$-$	$\ldots$	$+$
f_{2^n+1}	$-$	$-$	$-$	$\ldots$	$-$	$-$	$-$	$-$	$-$	$\ldots$	$-$
f_{2^n+2}	$+$	$-$	$-$	$\ldots$	$-$	$-$	$-$	$-$	$-$	$\ldots$	$-$
f_{2^n+3}	$-$	$+$	$-$	$\ldots$	$-$	$-$	$-$	$-$	$-$	$\ldots$	$-$
f_{2^n+4}	$+$	$+$	$-$	$\ldots$	$-$	$-$	$-$	$-$	$-$	$\ldots$	$-$
$\vdots$				$\ddots$							
$f_{2^n+2^n}$	$+$	$+$	$+$	$\ldots$	$+$	$-$	$-$	$-$	$-$	$\ldots$	$-$

values of g for all $x_1, \ldots, x_n$ and then to check that the corresponding "check bit" in $x_{n+1}, \ldots, x_{n+2^n}$ is "off". $\square$

Lemma 6.4. *For any $n \geq 0$ there exists a function class $\mathcal{F}$ for which $PD(\mathcal{F}) = 2$ and $XD(\mathcal{F}) = 1$, and $VC(\mathcal{F}) = \log_2 \frac{|\mathcal{F}|}{2} = n$.*

Proof. Consider the function class $\mathcal{F}$ in Fig. 6.3. It is the same as the function class in Fig. 6.2, but augmented with the "hard" examples considered in the proof of Lemma 6.3, i.e., in addition to the functions that $\mathcal{F}$ included in the previous lemma, $\mathcal{F}$ further includes functions that provide arbitrary classification of $x_1, \ldots, x_n$ and are equal to 0 for $x_{n+1}, \ldots, x_{n+2^n}$. As previously, $VC(\mathcal{F}) = n = \log_2 \frac{|\mathcal{F}|}{2}$ and $PD(\mathcal{F}) = 2$. However, this time $XD(\mathcal{F}) = 1 = PD(\mathcal{F}) - 1$. This is because in this example $g \notin \mathcal{F}$ either has to be 1 on two samples x_{n+j_1} and x_{n+j_2} for $1 \leq j_1, j_2 \leq 2^n$ (and then we can prove that $g \notin \mathcal{F}$ by revealing its value at x_{n+j_1} and x_{n+j_2}), or g is 1 for some x_{n+j} for $1 \leq j \leq 2^n$, but there is some x_i for $1 \leq i \leq n$ such that $g(x_i) \neq f_j(x_i)$ (and then once again by revealing the value of g at x_{n+j} and x_i we prove that $g \notin \mathcal{F}$). $\square$

6.5 Discussion

We have shown that both VC and exclusion dimension bound Popper dimension from above; however, the relation between the exclusion dimension and the VC dimension can be arbitrary.

Popper dimension can be regarded as the minimal number of examples required to falsify $\mathcal{F}$ in the best case (when the simplest nature phenomenon that is not explained by $\mathcal{F}$ is most distant from all $f \in \mathcal{F}$), whereas exclusion dimension can be regarded as the minimal number of examples required to falsify $\mathcal{F}$ in the worst case (when the simplest nature phenomenon not explained by $\mathcal{F}$ is very similar to some $f \in \mathcal{F}$).

Acknowledgements We would like to thank Vladimir Vovk for his careful reading of and comments on this manuscript.

References

1. Angluin, D.: Queries revisited. Theor. Comput. Sci. **313**(2), 175–194 (2004)
2. Corfield, D., Schölkopf, B., Vapnik, V.N.: Falsification and statistical learning theory: comparing the Popper and Vapnik-Chervonenkis dimensions. J. Gen. Philos. Sci. **40**, 51–58 (2009)
3. Goldman, S.A., Kearns, M.J.: On the complexity of teaching. J. Comput. Syst. Sci. **50**(1), 20–31 (1995)
4. Popper, K.: Logik der Forschung. Mohr Siebeck, Vienna (1934). English translation: The Logic of Scientific Discovery, 1959
5. Schölkopf, B., Smola, A.: Learning with Kernels. Support Vector Machines, Regularization, Optimization and Beyond. MIT Press, Cambridge (2002)
6. Vapnik, V.N.: The Nature of Statistical Learning Theory. Springer, New York (1995)
7. Vapnik, V.N.: Statistical Learning Theory. Wiley, New York (1998)
8. Vapnik, V.N., Chervonenkis, A.Y.: On the uniform convergence of relative frequencies of events to their probabilities. Proc. USSR Acad. Sci. **181**(4), 781–783 (1968). English translation: Soviet Math. Dokl. **9**, 915–918, 1968
9. Vapnik, V.N., Chervonenkis, A.Y.: On the uniform convergence of relative frequencies of events to their probabilities. Theory Prob. Appl. **16**(2), 264–281 (1971)
10. Vapnik, V.N., Chervonenkis, A.Y.: Theory of pattern recognition. Nauka, Moscow (1974) (in Russian). German translation: W.N. Wapnik, A.Ya. Tschervonenkis (1979), Theorie der Zeichenerkennug, Akademia, Berlin

Chapter 7
On Learnability, Complexity and Stability

Silvia Villa, Lorenzo Rosasco, and Tomaso Poggio

Abstract We consider the fundamental question of learnability of a hypothesis class in the supervised learning setting and in the general learning setting introduced by Vladimir Vapnik. We survey classic results characterizing learnability in terms of suitable notions of complexity, as well as more recent results that establish the connection between learnability and stability of a learning algorithm.

7.1 Introduction

A key question in statistical learning is which hypothesis (function) spaces are learnable. Roughly speaking, a hypothesis space is learnable if there is a consistent learning algorithm, i.e., one returning an optimal solution as the sample size goes to infinity. Classic results for supervised learning characterize learnability of a function class in terms of its complexity (combinatorial dimension) [1–3, 9, 16, 17]. Indeed, minimization of the empirical risk on a function class having finite complexity can be shown to be consistent. A key aspect in this approach is the connection with empirical process theory results showing that finite combinatorial dimensions characterize function classes for which a uniform law of large numbers holds, namely uniform Glivenko–Cantelli classes [7].

More recently, the concept of stability has emerged as an alternative and effective method to design consistent learning algorithms [4]. Stability refers broadly to continuity properties of a learning algorithm to its input and it is known to play a crucial role in regularization theory [8]. Surprisingly, for certain classes of loss functions, a suitable notion of stability of ERM can be shown to characterize learnability of a function class [10–12].

In this chapter, after recalling some basic concepts (Sect. 7.2), we review results characterizing learnability in terms of complexity and stability in supervised learning (Sect. 7.3) and in the so called general learning (Sect. 7.4). We conclude with some remarks and open questions.

7.2 Supervised Learning, Consistency and Learnability

In this section, we introduce basic concepts in Statistical Learning Theory (SLT). First, we describe the supervised learning setting, and then define the notions of consistency of a learning algorithm and learnability of a hypothesis class.

Consider a probability space $(\mathcal{Z}, \rho)$, where $\mathcal{Z} = \mathcal{X} \times \mathcal{Y}$, with $\mathcal{X}$ a measurable space and $\mathcal{Y}$ a closed subset of $\mathbb{R}$. A loss function is a measurable map $\ell : \mathbb{R} \times \mathcal{Y} \to [0, +\infty)$. We are interested in the problem of minimizing the expected risk,

$$\inf_{\mathcal{F}} \mathcal{E}_\rho, \qquad \mathcal{E}_\rho(f) = \int_{\mathcal{X} \times \mathcal{Y}} \ell(f(x), y) \, d\rho(x, y), \tag{7.1}$$

where $\mathcal{F} \subset \mathcal{Y}^{\mathcal{X}}$ is the set of measurable functions from $\mathcal{X}$ to $\mathcal{Y}$ (endowed with the product topology and the corresponding Borel σ-algebra). The probability distribution ρ is assumed to be fixed but known only through a training set, i.e., a set of pairs $\mathbf{z}_n = ((x_1, y_1), \ldots, (x_n, y_n)) \in \mathcal{Z}^n$ sampled identically and independently according to ρ. Roughly speaking, the problem of supervised learning is that of approximatively solving Problem (7.1) given a training set $\mathbf{z}_n$.

Example 7.1 (Regression and Classification). In (bounded) regression, $\mathcal{Y}$ is a bounded interval in $\mathbb{R}$, while in binary classification $\mathcal{Y} = \{0, 1\}$. Examples of loss functions are the square loss $\ell(t, y) = (t - y)^2$ in regression and the misclassification loss $\ell(t, y) = \mathbb{1}_{\{t \neq y\}}$ in classification. See [16] for a more exhaustive list of loss functions.

In the next section, the notion of approximation considered in SLT is defined rigorously. We first introduce the concepts of hypothesis space and learning algorithm.

Definition 7.1. A *hypothesis space* is a set of functions $\mathcal{H} \subseteq \mathcal{F}$. We say that $\mathcal{H}$ is *universal* if $\inf_{\mathcal{F}} \mathcal{E}_\rho = \inf_{\mathcal{H}} \mathcal{E}_\rho$, for all distributions ρ on $\mathcal{Z}$.

7 On Learnability, Complexity and Stability

Definition 7.2. A *learning algorithm* A on $\mathcal{H}$ is a map,

$$A : \bigcup_{n \in \mathbb{N}} \mathcal{Z}^n \to \mathcal{H}, \quad \mathbf{z}_n \mapsto A_{\mathbf{z}_n} = A(\mathbf{z}_n),$$

such that, for all $n \geq 1$, $A_{|\mathcal{Z}^n}$ is measurable with respect to the completion of the product σ-algebra on $\mathcal{Z}^n$.

Empirical Risk Minimization (ERM) is arguably the most popular example of a learning algorithm in SLT.

Example 7.2. Given a training set $\mathbf{z}_n$ the empirical risk $\mathcal{E}_{\mathbf{z}_n} : \mathcal{F} \to \mathbb{R}$ is defined as

$$\mathcal{E}_{\mathbf{z}_n}(f) = \frac{1}{n} \sum_{i=1}^{n} \ell(f(x_i), y_i).$$

Given a hypothesis space $\mathcal{H}$, ERM on $\mathcal{H}$ is defined by minimization of the empirical risk on $\mathcal{H}$.

We add one remark.

Remark 7.1 (ERM and Asymptotic ERM). In general some care is needed while defining ERM since a (measurable) minimizer might not be ensured to exist. When $\mathcal{Y} = \{0, 1\}$ and ℓ is the misclassification loss function, it is easy to see that a minimizer exists (possibly non-unique). In this case measurability is studied, for example, as in Lemma 6.17 of [15]. When considering more general loss functions or regression problems one might need to consider learning algorithms defined by suitable (measurable) almost-minimizers of the empirical risk (see, e.g., Definition 7.10).

7.2.1 Consistency and Learnability

Aside from computational considerations, the following definition formalizes in what sense a learning algorithm approximatively solves Problem (7.1).

Definition 7.3. We say that a learning algorithm A on $\mathcal{H}$ is *uniformly consistent*[1] if

$$\forall \epsilon > 0, \quad \lim_{n \to +\infty} \sup_{\rho} \rho^n \left(\{ \mathbf{z}_n : \mathcal{E}_\rho(A_{\mathbf{z}_n}) - \inf_{\mathcal{H}} \mathcal{E}_\rho > \epsilon \} \right) = 0,$$

and *universally uniformly consistent* if $\mathcal{H}$ is universal.

The next definition shifts the focus from a learning algorithm on $\mathcal{H}$, to $\mathcal{H}$ itself.

---

[1] Consistency can be defined with respect to other convergence notions for random variables. If the loss function is bounded, convergence in probability is equivalent to convergence in expectation.

Definition 7.4. We say that a space $\mathcal{H}$ is *uniformly learnable* if there exists a uniformly consistent learning algorithm on $\mathcal{H}$. If $\mathcal{H}$ is also universal we say that it is *universally uniformly learnable*.

Note that, in the above definition, the term "uniform" refers to the distribution for which consistency holds, whereas "universal" refers to the possibility of solving Problem (7.1) without a bias due to the choice of $\mathcal{H}$. The requirement of uniform learnability implies the existence of a learning rate for A [15] or equivalently a bound on the sample complexity [2]. The following classical result, sometimes called the "no free lunch" theorem, shows that uniform universal learnability of a hypothesis space is too much to hope for.

Theorem 7.1. *Let* $\mathcal{Y} = \{0, 1\}$, *and* $\mathcal{X}$ *be such that there exists a measure* μ *on* $\mathcal{X}$ *having an atom-free distribution. Let* ℓ *be the misclassification loss. If* $\mathcal{H}$ *is universal, then* $\mathcal{H}$ *is not uniformly learnable.*

The proof of the above result is based on Theorem 7.1 in [6], which shows that for each learning algorithm A on $\mathcal{H}$ and any fixed n, there exists a measure ρ on $\mathcal{X} \times \mathcal{Y}$ such that the expected value of $\mathcal{E}_\rho(A_{\mathbf{z}_n}) - \inf_\mathcal{H} \mathcal{E}_\rho$ is greater than $1/4$. A general form of the no free lunch theorem, beyond classification, is given in [15] (see Corollary 6.8). In particular, this result shows that the no free lunch theorem holds for convex loss functions as soon as there are two probability distributions ρ_1, ρ_2 such that $\inf_\mathcal{H} \mathcal{E}_{\rho_1} \neq \inf_\mathcal{H} \mathcal{E}_{\rho_2}$ (assuming that minimizers exist). Roughly speaking, if there exist two learning problems with distinct solutions, then $\mathcal{H}$ cannot be universally uniformly learnable (this latter condition becomes more involved when the loss is not convex).

The no free lunch theorem shows that universal uniform consistency is too strong of a requirement. Restrictions on either the class of considered distributions ρ or the hypothesis spaces/algorithms are needed to define a meaningful problem. In the following, we will follow the latter approach, where assumptions on $\mathcal{H}$ (or A), but not [...] on the class of distributions [...] ρ, are made.

7.3 Learnability of a Hypothesis Space

In this section we study uniform learnability by putting appropriate restrictions on the hypothesis space $\mathcal{H}$. We are interested in conditions which are not only sufficient but also necessary. We discuss two series of results. The first is classical and characterizes learnability of a hypothesis space in terms of suitable complexity measures. The second, more recent, is based on the stability (in a suitable sense) of ERM on $\mathcal{H}$.

7.3.1 Complexity and Learnability

Classically, assumptions on $\mathcal{H}$ are imposed in the form of restrictions on its "size" defined in terms of suitable notions of combinatorial dimensions (complexity). The following definition of complexity for a class of binary-valued functions has been introduced in [17].

Definition 7.5. Assume $\mathcal{Y} = \{0, 1\}$. We say that $\mathcal{H}$ *shatters* $S \subseteq \mathcal{X}$ if for each $E \subseteq S$ there exists $f_E \in \mathcal{H}$ such that $f_E(x) = 0$ if $x \in E$, and $f_E(x) = 1$ is $x \in S \setminus E$. The *VC-dimension* of $\mathcal{H}$ is defined as

$$\text{VC}(\mathcal{H}) = \max\{d \in \mathbb{N} : \exists S = \{x_1, \ldots x_d\} \text{ shattered by } \mathcal{H}\}.$$

The VC-dimension turns out to be related to a special class of functions, called uniform Glivenko–Cantelli, for which a uniform form of the law of large numbers holds [7].

Definition 7.6. We say that $\mathcal{H}$ is a *uniform Glivenko–Cantelli (uGC) class* if it has the following property:

$$\forall \epsilon > 0, \quad \lim_{n \to +\infty} \sup_\rho \rho^n\left(\left\{\mathbf{z}_n : \sup_{f \in \mathcal{H}} |\mathcal{E}_\rho(f) - \mathcal{E}_{\mathbf{z}_n}(f)| > \epsilon\right\}\right) = 0.$$

The following theorem completely characterizes learnability in classification.

Theorem 7.2. *Let $\mathcal{Y} = \{0, 1\}$ and ℓ be the misclassification loss. Then the following conditions are equivalent:*

1. *$\mathcal{H}$ is uniformly learnable,*
2. *ERM on $\mathcal{H}$ is uniformly consistent,*
3. *$\mathcal{H}$ is a uGC-class,*
4. *The VC-dimension of $\mathcal{H}$ is finite.*

The proof of the above result can be found, for example, in [2] (see Theorems 4.9, 4.10 and 5.2). The characterization of uGC classes in terms of combinatorial dimensions is a central theme in empirical process theory [7]. The results on binary-valued functions are essentially due to Vapnik and Chervonenkis [17]. The proof that uGC of $\mathcal{H}$ implies its learnability is straightforward. The key step in the above proof is showing that learnability is sufficient for a finite VC-dimension, i.e., $\text{VC}(\mathcal{H}) < \infty$. The proof of this last step crucially depends on the considered loss function.

A similar result holds for bounded regression with the square [1, 2] and absolute loss functions [3, 9]. In these cases, a new notion of complexity needs to be defined since the VC-dimension of real-valued function classes is not defined. Here, we recall the definition of the γ-fat shattering dimension of a class of functions $\mathcal{H}$ originally introduced in [9].

Definition 7.7. Let $\mathcal{H}$ be a set of functions from $\mathcal{X}$ to $\mathbb{R}$ and $\gamma > 0$. Consider $S = \{x_1, \ldots, x_d\} \subset \mathcal{X}$. Then S is γ-*shattered by* $\mathcal{H}$ if there are real numbers $r_1, \ldots, r_d$ such that for each $E \subseteq S$ there is a function $f_E \in \mathcal{H}$ satisfying

$$\begin{cases} f_E(x) \leq r_i - \gamma & \forall x \in S \setminus E \\ f_E(x) \geq r_i + \gamma & \forall x \in E. \end{cases}$$

We say that $(r_1, \ldots, r_d)$ witnesses the shattering. The γ-*fat shattering dimension of* $\mathcal{H}$ is

$$\text{fat}_{\mathcal{H}}(\gamma) = \max\{d \ : \ \exists S = \{x_1, \ldots, x_d\} \subseteq \mathcal{X} \text{ s.t. } S \text{ is } \gamma\text{-shattered by } \mathcal{H}\}.$$

As mentioned above, an analog of Theorem 7.2 can be proved for bounded regression with the square and absolute losses if condition 4 is replaced by $\text{fat}_{\mathcal{H}}(\gamma) < +\infty$ for all $\gamma > 0$. We end noting that it is an open problem to generalize the above results for loss functions other than the square and the absolute loss functions.

7.3.2 Stability and Learnability

In this section we show that learnability of a hypothesis space $\mathcal{H}$ is equivalent to the stability (in a suitable sense) of ERM on $\mathcal{H}$. It is useful to introduce the following notation. For a given loss function ℓ, let $L : \mathcal{F} \times Z \to [0, \infty)$ be defined as $L(f, z) = \ell(f(x), y)$, for $f \in \mathcal{F}$ and $z = (x, y) \in \mathcal{Z}$. Moreover, let $\mathbf{z}_n^i$ be the training $\mathbf{z}_n$ with the ith point removed. With the above notation, the relevant notion of stability is given by the following definition.

Definition 7.8. A learning algorithm A on $\mathcal{H}$ is *uniformly* CV_{loo} *stable* if there exist sequences $(\beta_n, \delta_n)_{n \in \mathbb{N}}$ such that $\beta_n \to 0$, $\delta_n \to 0$ and

$$\sup_{\rho} \rho^n \{|L(A_{\mathbf{z}_n^i}, z_i) - L(A_{\mathbf{z}_n}, z_i)| \leq \beta_n\} \geq 1 - \delta_n,$$

for all $i \in \{1, \ldots, n\}$.

Before illustrating the implications of the above definition to learnability we first make a few comments and historical remarks. We note that, in a broad sense, stability refers to a quantification of the continuity of a map with respect to its input. The key role of stability in learning has long been advocated on the basis of the interpretation of supervised learning as an ill-posed inverse problems [11]. Indeed, the concept of stability is central in the theory of regularization of ill-posed problems [8]. The first quantitative connection between the performance of a symmetric learning algorithm[2] and a notion of stability is derived in the seminal

---

[2] We say that a learning algorithm A is symmetric if it does not depend on the order of the points in $\mathbf{z}_n$.

7 On Learnability, Complexity and Stability

paper [4]. Here a notion of stability, called uniform stability, is shown to be sufficient for consistency. If we let $\mathbf{z}_n^{i,u}$ be the training $\mathbf{z}_n$ with the ith point replaced by u, uniform stability is defined as

$$|L(A_{\mathbf{z}_n^{i,u}}, z) - L(A_{\mathbf{z}_n}, z)| \leq \beta_n, \tag{7.2}$$

for all $\mathbf{z}_n \in \mathcal{Z}^n$, $u, z \in \mathcal{Z}^n$ and $i \in \{1, \ldots, n\}$. A thorough investigation of weaker notions of stability is given in [10]. Here, many different notions of stability are shown to be sufficient for consistency (and learnability) and the question is raised of whether stability (of ERM on $\mathcal{H}$) can be shown to be necessary for learnability of $\mathcal{H}$. In particular, a definition of CV stability for ERM is shown to be necessary and sufficient for learnability in a Probably Approximate Correct (PAC) setting, that is when $\mathcal{Y} = \{0, 1\}$ and for some $h^* \in \mathcal{H}$, $y = h^*(x)$, for all $x \in \mathcal{X}$. Finally, Definition 7.8 of uniform CV_{loo} stability is given and studied in [11]. When compared to uniform stability, we see that: (1) the "replace one" training set $\mathbf{z}_n^{i,u}$ is considered instead of the "leave one out" training set $\mathbf{z}_n^i$; (2) the error is evaluated on the point z_i which is left out, rather than on any possible $z \in \mathcal{Z}$; finally, (3) the condition is assumed to hold for a fraction $1 - \delta_n$ of training sets (which becomes increasingly larger as n increases) rather than uniformly for any training set $\mathbf{z}_n \in \mathcal{Z}^n$.

The importance of CV_{loo} stability is made clear by the following result.

Theorem 7.3. *Let* $\mathcal{Y} = \{0, 1\}$ *and* ℓ *be the misclassification loss function. Then the following conditions are equivalent,*

1. $\mathcal{H}$ *is uniformly learnable,*
2. *ERM on* $\mathcal{H}$ *is* CV_{loo} *stable.*

The proof of the above result is given in [11] and is based on essentially two steps. The first is proving that CV_{loo} stability of ERM on $\mathcal{H}$ implies that ERM is uniformly consistent. The second is showing that if $\mathcal{H}$ is a uGC class then ERM on $\mathcal{H}$ is CV_{loo} stable. Theorem 7.3 then follows from Theorem 7.2 (since uniform consistency of ERM on $\mathcal{H}$ and $\mathcal{H}$ being uGC are equivalent).

Both steps in the above proof can be generalized to regression as long as the loss function is assumed to be bounded. The latter assumption holds, for example, if the loss function satisfies a suitable Lipschitz condition and $\mathcal{Y}$ is compact (so that $\mathcal{H}$ is a set of uniformly bounded functions). However, generalizing Theorem 7.3 beyond classification requires the generalization of Theorem 7.2. For the square and absolute loss functions and $\mathcal{Y}$ compact, the characterization of learnability in terms of a γ-fat shattering dimension can be used. It is an open problem whether there is a more direct way to show that learnability is sufficient for stability, independently of Theorem 7.2. Another open problem is how to extend the above results to more general classes of loss functions. We will see a partial answer to this question in Sect. 7.4.

7.4 Learnability in the General Learning Setting

In the previous sections we focused our attention on supervised learning. Here we ask whether the results we discussed extend to the so-called general learning [16].

Let $(\mathcal{Z}, \rho)$ be a probability space and $\mathcal{F}$ a measurable space. A loss function is a map $L : \mathcal{F} \times \mathcal{Z} \to [0, \infty)$, such that $L(f, \cdot)$ is measurable for all $f \in \mathcal{F}$. We are interested in the problem of minimizing the expected risk,

$$\inf_{\mathcal{H}} \mathcal{E}_\rho, \qquad \mathcal{E}_\rho(f) = \int_{\mathcal{Z}} L(f, z) \, d\rho(z),$$

when ρ is fixed but known only through a training set, $\mathbf{z}_n = (z_1, \ldots, z_n) \in \mathcal{Z}^n$, sampled identically and independently according to ρ. Definition 7.2 of a learning algorithm on $\mathcal{H}$ applies as is to this setting and ERM on $\mathcal{H}$ is defined by the minimization of the empirical risk

$$\mathcal{E}_{\mathbf{z}_n}(f) = \frac{1}{n} \sum_{i=1}^{n} L(f, z_i).$$

While general learning is close to supervised learning, there are important differences. The data space $\mathcal{Z}$ has no natural decomposition, and in general $\mathcal{F}$ is not a space of functions. Indeed, $\mathcal{F}$ and $\mathcal{Z}$ are related only via the loss function L. For our discussion it is important to note that the distinction between $\mathcal{F}$ and the hypothesis space $\mathcal{H}$ becomes blurred. In supervised learning $\mathcal{F}$ is the largest set of functions for which Problem (7.1) is well defined (measurable functions in $\mathcal{Y}^{\mathcal{X}}$). The choice of a hypothesis space corresponds intuitively to a more "manageable" function space. In general, learning the choice of $\mathcal{F}$ is more arbitrary, and as a consequence the definition of universal hypothesis space is less clear. The setting is too general for an analogue of the no free lunch theorem to hold. Given these premises, in what follows we will simply identify $\mathcal{F} = \mathcal{H}$ and consider the question of learnability, noting that the definition of uniform learnability extends naturally to general learning. We present two sets of ideas. The first, due to Vapnik, focuses on a more restrictive notion of consistency of ERM. The second investigates the characterization of uniform learnability in terms of stability.

7.4.1 Vapnik's Approach and Non-trivial Consistency

The extension of the classical results characterizing learnability in terms of complexity measure is tricky. Since $\mathcal{H}$ is not a function space, the definitions of VC or V_γ dimensions do not make sense. A possibility is to consider the class $L \circ \mathcal{H} := \{z \in \mathcal{Z} \mapsto L(f, z) \text{ for some } f \in \mathcal{H}\}$ and the corresponding VC dimension (if L is binary-valued) or V_γ dimension (if L is real-valued). Classical results about the equivalence between the uGC property and finite complexity apply to the class $L \circ \mathcal{H}$. Moreover, uniform learnability can be easily proved if $L \circ \mathcal{H}$ is a uGC class. On the contrary, the reverse implication does not hold in the general learning setting. A counterexample is given in [16] (Sect. 3.1), showing that it is possible to design hypothesis classes with infinite VC (or V_γ) dimension, which are uniformly learnable with ERM. The construction is as follows. Consider an arbitrary set $\mathcal{H}$ and a loss L for which the class $L \circ \mathcal{H}$ has infinite VC (or V_γ) dimension. Define a new space $\widetilde{\mathcal{H}} := \mathcal{H} \cup \{\tilde{h}\}$ by adding to $\mathcal{H}$ an element $\tilde{h}$ such that $L(\tilde{h}, z) \leq L(h, z)$ for all $z \in \mathcal{Z}$ and $h \in \mathcal{H}$.[3] The space $L \circ \widetilde{\mathcal{H}}$ has infinite VC, or V_γ, dimension and is trivially learnable by ERM, which is constant and coincides with $\tilde{h}$ for each probability measure ρ. The previous counterexample proves that learnability, and in particular learnability via ERM, does not imply finite VC or V_γ dimension. To avoid these cases of "trivial consistency" and to restore the equivalence between learnability and finite dimension, the following stronger notion of consistency for ERM has been introduced by Vapnik [16].

Definition 7.9. ERM on $\mathcal{H}$ is *strictly uniformly consistent* if and only if

$$\forall \epsilon > 0, \quad \lim_{n \to \infty} \sup_\rho \rho^n (\inf_{\mathcal{H}_c} \mathcal{E}_{\mathbf{z}_n}(f) - \inf_{\mathcal{H}_c} \mathcal{E}_\rho(f) > \epsilon) = 0,$$

where $\mathcal{H}_c = \{f \in \mathcal{H} : \mathcal{E}_\rho(f) \geq c\}$.

The following result characterizes strictly uniform consistency in terms of the uGC property of the class $L \circ \mathcal{H}$ (see Theorem 3.1 and its corollary in [16]).

Theorem 7.4. *Let $B > 0$ and assume $L(f, z) \leq B$ for all $f \in \mathcal{H}$ and $z \in \mathcal{Z}$. Then the following conditions are equivalent,*

1. *ERM on $\mathcal{H}$ is strictly consistent,*
2. *$L \circ \mathcal{H}$ is a uniform one-sided Glivenko–Cantelli class.*

The definition of a one-sided Glivenko–Cantelli class simply corresponds to omitting the absolute value in Definition 7.6.

7.4.2 Stability and Learnability for General Learning

In this section we discuss ideas from [14] extending the stability approach to general learning. The following definitions are relevant.

Definition 7.10. A *uniform Asymptotic ERM (AERM) algorithm* A on $\mathcal{H}$ is a learning algorithm such that

---

[3]Note that this construction is not possible in classification or in regression with the square loss.

$$\forall \epsilon > 0, \quad \lim_{n \to \infty} \sup_{\rho} \rho^n(\{\mathbf{z}_n : \mathcal{E}_{\mathbf{z}_n}(A_{\mathbf{z}_n}) - \inf_{\mathcal{H}} \mathcal{E}_{\mathbf{z}_n} > \epsilon\}) = 0.$$

Definition 7.11. A learning algorithm A on $\mathcal{H}$ is *uniformly replace one (RO) stable* if there exists a sequence $\beta_n \to 0$ such that

$$\frac{1}{n} \sum_{i=1}^n |L(A_{\mathbf{z}_n^{i,u}}, z) - L(A_{\mathbf{z}_n}, z)| \leq \beta_n$$

for all $\mathbf{z}_n \in \mathcal{Z}^n$, $u, z \in \mathcal{Z}^n$ and $i \in \{1, \ldots, n\}$.

Note that the above definition is close to that of uniform stability (7.2), although the latter turns out to be a stronger condition. The importance of the above definitions is made clear by the following result.

Theorem 7.5. *Let $B > 0$ and assume $L(f, z) \leq B$ for all $f \in \mathcal{H}$ and $z \in \mathcal{Z}$. Then the following conditions are equivalent:*

1. *$\mathcal{H}$ is uniformly learnable,*
2. *There exists an AERM on $\mathcal{H}$ which is uniform RO stable.*

As mentioned in Remark 7.1, Theorem 7.3 holds not only for exact minimizers of the empirical risk, but also for AERMs. In this view, there is a subtle difference between Theorem 7.3 and Theorem 7.5. In supervised learning, Theorem 7.3 shows that uniform learnability implies that *every* ERM (AERM) is stable, while in general learning, Theorem 7.5 shows that uniform learnability implies the *existence* of a stable AERM (whose construction is not explicit).

The proof of the above result is given in Theorem 7 in [14]. The hard part of the proof is showing that learnability implies existence of an RO stable AERM. This part of the proof is split into two steps showing that: (1) if there is a uniformly consistent algorithm A, then there exists a uniformly consistent AERM A' (Lemma 20 and Theorem 10); (2) every uniformly consistent AERM is also RO stable (Theorem 9). Note that the results in [14] are given in terms of expectation and with some quantification of how different convergence rates are related. Here we give results in terms of probability to be consistent with the rest of the chapter and state only asymptotic results to simplify the presentation.

7.5 Discussion

In this chapter we reviewed several results concerning the learnability of a hypothesis space. Extensions of these ideas can be found in [5] (and references therein) for multi-category classification, and in [13] for sequential prediction. It would be interesting to devise constructive proofs in general learning suggesting how stable learning algorithms can be designed. Moreover, it would be interesting to study universal consistency and learnability in the case of samples from non-stationary processes.

References

1. Alon, N., Ben-David, S., Cesa-Bianchi, N., Haussler, D.: Scale-sensitive dimensions, uniform convergence, and learnability. J. ACM **44**(4), 615–631 (1997)
2. Anthony, M., Bartlett, P.L.: Neural network learning: theoretical foundations. Cambridge University Press, Cambridge (1999)
3. Bartlett, P., Long, P., Williamson, R.: Fat-shattering and the learnability of real-valued functions. J. Comput. Syst. Sci. **52**, 434–452 (1996)
4. Bousquet, O., Elisseeff, A.: Stability and generalization. J. Mach. Learn. Res. **2**, 499–526 (2002)
5. Daniely, A., Sabato, S., Ben-David, S., Shalev-Shwartz, S.: Multiclass learnability and the ERM principle. J. Mach. Learn. Res. Proc. Track **19**, 207–232 (2011)
6. Devroye, L., Györfi, L., Lugosi, G.: A Probabilistic Theory of Pattern Recognition. Applications of Mathematics 31. Springer, New York (1996)
7. Dudley, R., Giné, E., Zinn, J.: Uniform and universal Glivenko-Cantelli classes. J. Theor. Prob. **4**, 485–510 (1991)
8. Engl, H.W., Hanke, M., Neubauer, A.: Regularization of Inverse Problems. Mathematics and Its Applications, vol. 375. Kluwer, Dordrecht (1996)
9. Kearns, M.J., Schapire, R.E.: Efficient distribution-free learning of probabilistic concepts. In: Computational Learning Theory and Natural Learning Systems. Bradford Books, vol. I, pp. 289–329. MIT, Cambridge (1994)
10. Kutin, S., Niyogi, P.: Almost-everywhere algorithmic stability and generalization error. Technical report TR-2002-03, Department of Computer Science, The University of Chicago (2002)
11. Mukherjee, S., Niyogi, P., Poggio, T., Rifkin, R.: Learning theory: stability is sufficient for generalization and necessary and sufficient for consistency of empirical risk minimization. Adv. Comput. Math. **25**(1–3), 161–193 (2006)
12. Poggio, T., Rifkin, R., Mukherjee, S., Niyogi, P.: General conditions for predictivity in learning theory. Nature **428**, 419–422 (2004)
13. Rakhlin, A., Sridharan, K., Tewari, A.: Online learning: beyond regret. J. Mach. Learn. Res. Proc. Track **19**, 559–594 (2011)
14. Shalev-Shwartz, S., Shamir, O., Srebro, N., Sridharan, K.: Learnability, stability and uniform convergence. J. Mach. Learn. Res. **11**, 2635–2670 (2010)
15. Steinwart, I., Christmann, A.: Support Vector Machines. Information Science and Statistics. Springer, New York (2008)
16. Vapnik, V.: The Nature of Statistical Learning Theory. Springer, New York (1995)
17. Vapnik, V.N., Chervonenkis, A.Y.: Theory of uniform convergence of frequencies of events to their probabilities and problems of search for an optimal solution from empirical data. Avtomatika i Telemekhanika **2**, 42–53 (1971)

Chapter 8
Loss Functions

Robert C. Williamson

Abstract Vapnik described the "three main learning problems" of pattern recognition, regression estimation and density estimation. These are defined in terms of the loss functions used to evaluate performance (0-1 loss, squared loss, and log loss, respectively). But there are many other loss functions one could use. In this chapter I will summarise some recent work by me and colleagues studying the theoretical aspects of loss functions. The results elucidate the richness of the set of loss functions and explain some of the implications of their choice.

8.1 Introduction

If one wishes to give a clear definition of a problem, a good starting point is to define what one means by a solution to the problem. Vapnik's "three main learning problems" [24] are so defined via the loss functions used to measure the quality of their solutions. If y is an observed value and $\hat{y}$ is one's estimate, then pattern recognition is defined via the 0-1 loss

$$\ell_{0-1}(y, \hat{y}) = [\![y = \hat{y}]\!]$$

where $[\![p]\!] = 1$ if p is true and 0 if p is false. Regression is defined in terms of the *squared loss*

$$\ell_{sq}(y, \hat{y}) = (y - \hat{y})^2$$

and probability estimation via the *log loss*

$$\ell_{\log}(\hat{p}) = -\ln(\hat{p}).$$

R.C. Williamson (✉)
Australian National University and NICTA, Canberra, ACT 0200, Australia
e-mail: Bob.Williamson@anu.edu.au

In this chapter I primarily focus on problems where the data (x_i, y_i) is drawn from $\mathcal{X} \times [n]$ where $[n] := \{1, \ldots, n\}$. In this case it is convenient to write the loss as a vector-valued map $\ell \colon [n] \to \mathbb{R}_+^n$ (for pattern recognition or "classification") or $\ell \colon \Delta^n \to \mathbb{R}_+^n$ (for probability estimation), where Δ^n is the n-dimensional simplex. Thus the log loss is $\ell_{\log}(\hat{p}) = (-\ln(\hat{p}_1), \ldots, -\ln(\hat{p}_n))'$, where the prime denotes transpose and $\hat{p} \in \Delta^n$ is one's estimate of the probability (that is, $\hat{p}_i$ is an estimate of the probability that the random variable Y takes on value $i \in [n]$).

These three canonical loss functions that are central to Vapnik's work raise the obvious question of what other loss functions one might choose, and what the implications of those choices are. That is the subject of this chapter, which provides an informal overview of some recent results I have (jointly) obtained. While I focus on my own recent work, taking loss functions seriously has been a research topic for some time [3,4,8,9]. Even within the Bayesian framework where it is often claimed one is only interested in "gaining information" from an experiment, ultimately loss functions arise because in the end one will *do something* with the results of the "information" so obtained [6, 15]. Furthermore, when one acknowledges the impracticality of exact computation of Bayesian inference, loss functions do matter even for Bayesians [12].

Formal and precise statements of the various results can be found in the papers to which I refer, along with detailed historical references. The emphasis of this chapter is the broader picture and intuition and some technical conditions are omitted in the statement of some results.

The rest of this chapter is organised as follows. Section 8.2 introduces the notion of a proper loss for probability estimation and shows two key representations in terms of Bregman divergences induced by the Bayes risk associated with the loss and a Choquet representation of all possible proper losses in terms of weighted combinations of elementary proper losses (0-1 loss is one of these elementary losses). Section 8.3 introduces the notion of an f-divergence between two distributions and shows the 1:1 correspondence between f-divergences and binary proper losses and explains some of the implications. Section 8.4 shows how the representations of proper losses make it much simpler to understand surrogate regret bounds (which is a measure of how much one loses from not following Vapnik's advice of solving the problem directly). Section 8.5 extends the results of Sects. 8.2 and 8.3 (which are for $n = 2$) to a general n and explains how one can thus define a natural f-divergence between several distributions jointly. Section 8.6 studies the parameterisation of losses, and in particular looks at the role of "link functions" and explains when a loss can be written as the composition of a proper loss and a link function, and also explains the convexity of proper losses in terms of the corresponding Bayes risk. Section 8.7 summarises the implications of the choice of loss on the rate of convergence obtainable in two distinct learning settings: worst case online sequence prediction and the traditional statistical batch setting as studied by Vapnik. The chapter concludes (Sect. 8.8) with some remarks on Vladimir Vapnik's impact on the machine learning community in general, and on my scientific work over 20 years in particular and offers a suggestion for a good way forward for the community to effectively build upon his legacy.

8.2 Proper Losses and Their Representations

Consider the problem of class probability estimation where one receives an iid sample $\{(x_i, y_i)\}_{i=1}^m$ of points from $\mathcal{X} \times [n]$. The goal is to estimate, for a given x, the probability $p_i = \Pr(Y = 1|X = x)$, where (X, Y) are random variables drawn from the same distribution as the sample. Given a loss function $\ell: \Delta^n \to \mathbb{R}_+^n$ the *conditional risk* is defined via

$$L: \Delta^n \times \Delta^n \ni (p, q) \mapsto L(p, q) = \mathbb{E}_{Y \sim p} \ell_Y(p) = p' \cdot \ell(q) = \sum_{i=1}^n p_i \ell_i(q) \in \mathbb{R}_+.$$

A natural requirement to impose upon ℓ for this problem is that it be *proper*, which means that $L(p, p) \leq L(p, q)$ for all $p, q \in \Delta^n$. (It is *strictly proper* if the inequality is strict when $p \neq q$.) The *conditional Bayes risk* $\underline{L}: \Delta^n \ni p \mapsto \inf_{q \in \Delta^n} L(p, q)$ and is always concave. If ℓ is proper, $\underline{L}(p) = L(p, p) = p' \cdot \ell(p)$. The full risk $\mathbb{L}(q) = \mathbb{E}_X \mathbb{E}_{Y|X} \ell_Y(q(X))$. Examples of proper losses include 0–1 loss, squared loss, and log loss.

There are many proper losses. A very convenient way to parameterise them arises from an integral representation. The *cost-sensitive* misclassification loss ℓ_c is a generalisation of 0-1 loss and is defined for $c \in (0, 1)$ via

$$\ell_c(q) = (c[\![q > c]\!], (1-c)[\![q \leq c]\!])'.$$

A loss $\ell: \Delta^2 \to \mathbb{R}^+$ is proper if and only if for all $q \in [0, 1]$ and $y \in \{1, 2\}$

$$\ell_y(q) = \int_0^1 \ell_{c,y}(q) w(c) dc, \tag{8.1}$$

where the *weight function* $w: [0, 1] \to \mathbb{R}_+$ is given by $w(c) = -\underline{L}''(c)$. The weight function allows a much easier interpretation of the effect of the different choices of proper losses [4, 18]. Examples of weight functions are $w_{0/1}(c) = 2\delta(c - 1/2)$, $w_{\text{square}}(c) = 1$, and $w_{\log}(c) = \frac{1}{c(1-c)}$.

8.3 Divergences and the Bridge to Proper Losses

An f-divergence [5] is a measure of the closeness of two distributions on some space $\mathcal{X}$ and is defined for a convex function $f: \mathbb{R}_+ \to \mathbb{R}$ (with $f(1) = 0$) via

$$\mathbb{I}_f(P, Q) = \int_{\mathcal{X}} f\left(\frac{dP}{dQ}\right) dQ.$$

Examples of f-divergences include Variational, Kullback–Leibler, and Hellinger.

The *statistical information* [6] $\Delta\mathbb{L}(\pi, P, Q)$ for a binary decision problem ($\mathcal{Y} = \{1, 2\}$) with class conditional probability distributions $P(x) = \Pr(X = x|Y = 2)$ and $Q(x) = \Pr(X = x|Y = 1)$ and prior probability $\pi = \Pr(Y = 2)$ is given by the difference between the prior and posterior Bayes risk (the difference between the best attainable risk when using only π and the distribution on $\mathcal{X}$ and the risk obtainable when using the conditional distributions P and Q).

A key result of [18] is that for any $\pi \in (0, 1)$ and any convex f (satisfying $f(1) = 0$), there exists a proper loss ℓ with associated statistical information $\Delta\mathbb{L}$ such that for all distributions P, Q, $\mathbb{I}_f(P, Q) = \Delta\mathbb{L}(\pi, P, Q)$; and conversely, given a proper loss ℓ, there exists an f such that the same equivalence holds. Thus, in a precise sense, the problem of measuring the divergence between two distributions P and Q is the same as solving the prediction problem relative to some proper loss when P and Q are interpreted as the respective class-conditional distributions. There is also an integral representation for f-divergences [14], and the "weight function" there can be directly related to the weight function for the corresponding proper losses [18].

A *Bregman divergence* is defined in terms of a convex function ϕ as

$$B_\phi(p, q) = \phi(p) - \phi(q) - (p - q)' \cdot D\phi(q)$$

where $D\phi(q)$ is the derivative of ϕ at q. It is also well known that the conditional Bayes risk $\underline{L}$ for a proper loss satisfies $L(p, q) = \underline{L}(q) - (q - p)\underline{L}'(q)$. Consequently the *regret* $L(p, q) - \underline{L}(p) = B_{-\underline{L}}(p, q)$, the Bregman divergence between p and q induced by the convex function $-\underline{L}$. Thus there is an intimate relationship between Bayes risks, f-divergences, and Bregman divergences.

8.4 Surrogate Losses

Working with ℓ_c is computationally difficult, so one often uses a convex surrogate. The question then naturally arises: what additional loss is incurred in doing so? One way of answering this question theoretically is via a "surrogate regret bound" [1]. It turns out that by starting with proper losses, and exploiting the representations presented above, one can derive surrogate regret bounds very simply.

Suppose that for some fixed $c \in (0, 1)$ we know the regret B_c of predicting with q when the true distribution is p is $B_c(p, q) = \alpha$. Then the regret $B(p, q)$ for any proper loss ℓ satisfies

$$B(p, q) \geq \max(\psi(c, \alpha), \psi(c, -\alpha)) \tag{8.2}$$

where $\psi(c, \alpha) = \underline{L}(c) - \underline{L}(c - \alpha) + \alpha \underline{L}'(c)$. Furthermore (8.2) is tight. The above bound can be inverted to obtain the usual results in the literature; see [16, 18] for details.

8.5 Extension to Multiclass Losses and Divergences

Most of the above results extend to the multiclass setting ($n > 2$), although there are some differences [25, 26]. The advantage of the representation of ℓ in terms of $\underline{L}$ becomes greater when $n > 2$, as can be seen by observing in this case $\ell \colon \Delta^n \to \mathbb{R}_+^n$ whereas $\underline{L} \colon \Delta^n \to \mathbb{R}^+$, a considerable simplification. Technically, in the binary case we worked with a projection of Δ^2 onto $[0, 1]$. When $n > 2$ we similarly project $p \in \Delta^n$ into $\tilde{p} \in \tilde{\Delta}^n$ via $\tilde{p}_i = p_i, i = 1, \ldots, n-1$. The induced set $\tilde{\Delta}^n$ then has an open interior and one can thus differentiate.

Multiclass proper losses still have a Choquet integral representation analogous to (8.1) since the set of proper losses is convex. However, in stark contrast to the case where $n = 2$, the set of primitive losses becomes much larger when $n > 2$ (they are dense in the set of all proper losses!); see [26] for details.

Strictly proper losses are quasi-convex, in the sense that if ℓ is strictly proper, then for all $p \in \Delta^n$, $q \mapsto L(p, q) = p' \cdot \ell(q)$ is quasi-convex. Two proper losses $\ell^1, \ell^2 \colon \Delta^n \to \mathbb{R}_+^n$ have the same conditional Bayes risk $\underline{L}$ if and only if $\ell^1 = \ell^2$ almost everywhere. If $\underline{L}$ is differentiable then $\ell^1 = \ell^2$. The proper loss ℓ is continuous at p in the interior of Δ^n if and only if $\underline{L}$ is differentiable at p.

The bridge between proper losses and f-divergences also holds when $n > 2$, but in this case the appropriate definition of $\mathbb{I}_f(P_{[n]}) = \mathbb{I}_f(P_1, \ldots, P_n)$ is not clear in the literature. It turns out [7] that an exact analogue of the binary result holds; for every proper loss $\ell \colon \Delta^n \to \mathbb{R}_+^n$ there exists a convex f such that the induced statistical information for a multiclass experiment with class conditional distributions $P_1, \ldots, P_n$ equals the f divergence $\mathbb{I}_f(P_{[n]})$, and conversely. This multidistribution divergence (which is better understood intuitively as a joint similarity, rather than a joint "distance") satisfies the same properties as traditional (binary) f-divergences. Furthermore, all of these properties are easy to prove by utilising the bridge to loss functions and Bayes risks, and then appealing to the Blackwell-Sherman-Stein theorem.

8.6 Parameterisations, Links and Convexity

There are two key factors one needs to take into account in choosing a loss function: the statistical and computational effects. The statistical effect is controlled by the Bayes risk $\underline{L}$ (this assertion is justified more fully in the next section). The computational effect is (to a first approximation) controlled by the convexity of the loss. In this section I will outline how these two factors can be controlled quite independently using the proper composite representation. Details are in [17] (binary case) and [26] (multiclass case).

We have already seen that proper losses $\ell \colon \Delta^n \to \mathbb{R}_+^n$ are characterised by their Bayes risk. Proper losses are a suitable choice when the predictors are probabilities. Oftentimes (e.g., use of linear predictors) the predictor v may live in some other

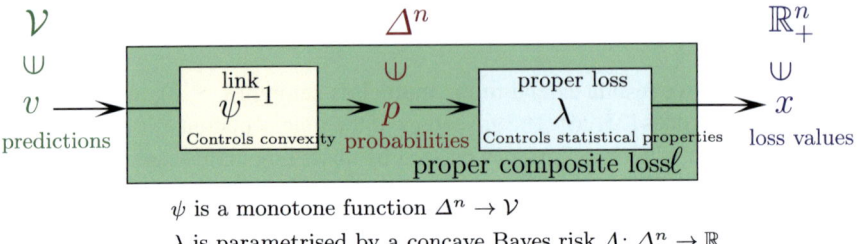

ψ is a monotone function $\Delta^n \to \mathcal{V}$

λ is parametrised by a concave Bayes risk $\underline{\Lambda} \colon \Delta^n \to \mathbb{R}$

Fig. 8.1 The idea of a proper composite loss

set $\mathcal{V}$, such as $\mathbb{R}^n$. In this case it can make sense to use a *proper composite loss* $\ell \colon \mathcal{V} \to \mathbb{R}^n_+$, defined via $\ell(v) = \lambda(\psi^{-1}(v))$, where $\psi \colon \Delta^n \to \mathcal{V}$ is an invertible *link function* and $\lambda \colon \Delta^n \to \mathbb{R}^n_+$ is a proper loss (Fig. 8.1). Since the link preserves quasi-convexity, if ℓ has a strictly proper composite representation then ℓ is quasi-convex.

Suppose ℓ is continuous and has a proper composite representation $\ell = \lambda \circ \psi^{-1} = \mu \circ \phi^{-1}$. Then the proper loss is unique ($\lambda = \mu$ almost everywhere). If ℓ is additionally invertible then the link functions are also unique.

Given a loss ℓ, the existence of a proper composite representation for ℓ is governed by the geometry of the image $\ell(\mathcal{V})$. The precise statement is a little subtle, but roughly speaking the existence is controlled by the "Δ^n-convexity" of $\ell(\mathcal{V})$, which means that $\ell(\mathcal{V})$ should "look convex" from the perspective of supporting hyperplanes with the normal vector in Δ^n; see [26, Sect. 5.2] for details.

A prediction $v \in \mathcal{V}$ is *admissible* if there is no prediction v_1 better than v in the sense that $\ell(v_1) \leq \ell(v)$ and for some $i \in [n]$, $\ell_i(v_1) < \ell_i(v)$. If $\ell \colon \mathcal{V} \to \mathbb{R}^n_+$ is continuous, invertible and has a strictly proper composite representation, then for all $v \in \mathcal{V}$, v is admissible.

All continuous strictly proper losses (and strictly proper composite losses) are quasi-convex, but they are not necessarily convex. The quasi-convexity means that if ℓ is continuous and has a proper composite representation, then it is *minimax*, meaning that

$$\max_{p \in \Delta^n} \min_{v \in \mathcal{V}} L(p, v) = \min_{v \in \mathcal{V}} \max_{p \in \Delta^n} L(p, v).$$

Note that ℓ need not be convex for the above to hold. The convexity of a proper (or proper composite) loss is readily checked, however, in terms of the Hessian of $\underline{L}$ and the gradient of ψ (see [26, Sect. 6.4]). Furthermore, if $\lambda \colon \Delta^n \to \mathbb{R}^n_+$ is a proper loss which is *not convex*, then one can canonically construct a convex composite proper loss with the same Bayes risk $\underline{\Lambda}$ by composing λ with its *canonical link* $\tilde{\psi}_\lambda(\tilde{p}) := -D\underline{\tilde{L}}(\tilde{p})'$, where $\tilde{p} = (p_1, \ldots, p_{n-1})'$ (it is necessary to use this reparameterisation of the simplex to allow the derivative D to be well defined).

In aggregate, these results on proper composite losses justify the separation of concerns mentioned at the beginning of this section: the statistical properties are controlled by the proper loss λ (since it controls the Bayes risk); the geometric properties (convexity) of the composite loss are controlled by the link ψ.

8.7 Effect of Losses on Statistical Convergence

Much of the theoretical literature which analyses the convergence of learning algorithms only looks at loss functions in a crude way. For example, a common trick is to bound the complexity of the class of functions induced by a loss function in terms of the complexity of the hypothesis class by assuming a Lipschitz property of the loss function. Unsurprisingly, such results offer little insight into the consequences of choosing particular loss functions. However, in one setting (the online worst case mixture of experts setting [27]) there is a result that shows the effect of the choice of loss (see [23] and references therein for the detailed background). Roughly speaking, in this setting the learner's task is to aggregate expert predictions such that the aggregated predictor has a cumulative loss not much worse than the cumulative loss of the best expert (which is not known to the learner). There is a very precise result which bounds the additional loss that the learner can incur in terms of the number of experts and a parameter β_ℓ called the mixability constant of the loss. This parameter depends only on the loss, but until recently its value was only known in certain special cases.

By exploiting the structure of proper losses (and in particular their characterisation in terms of their corresponding Bayes risk), it is possible to determine an exact general formula for β_ℓ when ℓ is continuous, smooth and strictly proper:

$$\beta_\ell = \min_{\tilde{p} \in \tilde{\Delta}^n} \lambda_{\max}\left((H\underline{\tilde{L}}(\tilde{p}))^{-1} \cdot H\underline{\tilde{L}}_{\log}(\tilde{p})\right),$$

where H denotes the Hessian, $\lambda_{\max}$ the maximum eigenvalue and $\underline{\tilde{L}}_{\log}$ is the Bayes risk for log loss [23]. If ℓ is suitably smooth [26, Sect. 6.1] then mixability of ℓ implies ℓ has a (strictly) proper composite representation.

Furthermore, it turns out that there is a generalisation of the notion of mixability (called stochastic mixability) [22] that in a special case reduces to mixability of ℓ, but applies in general to the standard statistical learning theory setting; stochastic mixability depends upon ℓ, $\mathcal{F}$ and P^* where $\mathcal{F}$ is the hypothesis space and P^* is the underlying distribution of the data. Analogously to the ordinary mixability result which characterises when fast learning can occur in the online mixture of experts setting, under some conditions, stochastic mixability of $(\ell, \mathcal{F}, P^*)$ also seems to control when learning with fast rates is possible. (I say "seems to" because the theory of stochastic mixability is still incomplete, although there are certainly cases where it can be applied and does indeed guarantee fast learning, including in the special case where $\mathcal{F}$ is convex and ℓ is squared loss [13].)

Thus we see that by taking a detour to simply understand loss functions better, one can obtain new understanding about one of the key problems in learning to which Vapnik made a major contribution—the bounding of the generalisation performance of a learning algorithm when presented with a finite amount of data.

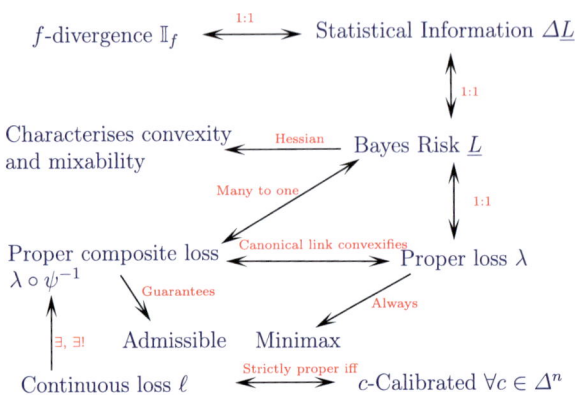

Fig. 8.2 Connections between key objects discussed in this chapter

8.8 Conclusion

Loss functions are central to the definition of machine learning problems. As I hope the above summary shows, there is a lot more to loss functions than one normally sees in the theoretical analysis of learning problems. The results are schematised in Fig. 8.2.

Vladimir Vapnik is renowned for two central contributions: (1) his "fundamental theorem of pattern recognition" [24] (which characterises the difficulty of pattern recognition problems in terms of a central complexity parameter of the class of hypotheses used), and (2) the support vector machine, which vastly expanded the scope of practical machine learning by combining a well-posed and mathematically sound optimisation problem as the basis of learning algorithms, and the use of kernels, which allowed the application of these machines to very diverse data types. More generally he has influenced an entire generation of machine learning researchers in terms of how they go about doing their own research. In my case I can see the clear influence through

Characterisations. The characterisation of the learnability of real-valued functions corrupted by additive noise in terms of the finiteness of the fat-shattering dimension [2] is analogous to the above-mentioned fundamental theorem.

Inductive principles. Vapnik formulated the notion of an inductive principle (how to translate the end goal of learning, such as minimizing expected risk) into an empirically implementable scheme. Thinking of such principles more abstractly was one of the main motivations for the notion of "luckiness" [10, 21].

Practical algorithms and their analysis. The SVM has had widespread impact. Making it easier to tune its regularisation parameter [19], generalising the core idea to novelty detection [20] and the online setting [11], and analysing the influence of the choice of kernel on the generalisation performance [29] were all clearly motivated by Vapnik's original contributions.

What of the future? I believe that the problem-centric approach one sees in Vapnik's work will prevail in the long term. Many techniques come and go. But the core problems will remain. Thus I am convinced that a fruitful way forward is to develop *relations* between different problems [28], akin to some of those I sketched in this chapter. This is hardly a new idea in mathematics (consider the viewpoint of functional analysis). But there is a long way yet to go in doing this for machine learning. Success in doing so will help turn machine learning into a mature engineering discipline.

Acknowledgements This work was supported by the Australian Government through the Australian Research Council and through NICTA, which is co-funded by the Department of Broadband, Communications and the Digital Economy.

References

1. Bartlett, P., Jordan, M., McAuliffe, J.: Convexity, classification, and risk bounds. J. Am. Stat. Assoc. **101**(473), 138–156 (2006)
2. Bartlett, P.L., Long, P.M., Williamson, R.C.: Fat-shattering and the learnability of real-valued functions. J. Comput. Syst. Sci. **52**(3), 434–452 (1996)
3. Berger, J.O.: Statistical Decision Theory and Bayesian Analysis. Springer, New York (1985)
4. Buja, A., Stuetzle, W., Shen, Y.: Loss functions for binary class probability estimation and classification: structure and applications. Technical report, University of Pennsylvania (2005)
5. Csiszár, I.: Information-type measures of difference of probability distributions and indirect observations. Studia Scientiarum Mathematicarum Hungarica **2**, 299–318 (1967)
6. DeGroot, M.H.: Uncertainty, information, and sequential experiments. Ann. Math. Stat. **33**(2), 404–419 (1962)
7. García-García, D., Williamson, R.C.: Divergences and risks for multiclass experiments. In: Conference on Learning Theory (JMLR: W&CP), Edinburgh, vol. 23, pp. 28.1–28.20 (2012)
8. Hand, D.: Deconstructing statistical questions. J. R. Stat. Soc. A (Stat. Soc.) **157**(3), 317–356 (1994)
9. Hand, D., Vinciotti, V.: Local versus global models for classification problems: fitting models where it matters. Am. Stat. **57**(2), 124–131 (2003)
10. Herbrich, R., Williamson, R.: Algorithmic luckiness. J. Mach. Learn. Res. **3**(2), 175–212 (2002)
11. Kivinen, J., Smola, A., Williamson, R.: Online learning with kernels. IEEE Trans. Signal Proc. **52**(8), 2165–2176 (2004)
12. Lacoste-Julien, S., Huszár, F., Gharamani, Z.: Approximate inference for the loss-calibrated Bayesian. In: Proceedings of the 14th International Conference on Artificial Intelligence and Statistics, Fort Lauderdale (2011)
13. Lee, W., Bartlett, P., Williamson, R.: The importance of convexity in learning with squared loss. IEEE Trans. Inf. Theory **44**(5), 1974–1980 (1998)
14. Liese, F., Vajda, I.: On divergences and informations in statistics and information theory. IEEE Trans. Inf. Theory **52**(10), 4394–4412 (2006)
15. Lindley, D.: On a measure of the information provided by an experiment. Ann. Math. Stat. **27**(4), 986–1005 (1956)
16. Reid, M.D., Williamson, R.C.: Surrogate regret bounds for proper losses. In: Proceedings of the International Conference on Machine Learning, Montreal, pp. 897–904 (2009)
17. Reid, M.D., Williamson, R.C.: Composite binary losses. J. Mach. Learn. Res. **11**, 2387–2422 (2010)
18. Reid, M.D., Williamson, R.C.: Information, divergence and risk for binary experiments. J. Mach. Learn. Res. **12**, 731–817 (2011)

19. Schölkopf, B., Smola, A., Williamson, R.C., Bartlett, P.L.: New support vector algorithms. Neural Comput. **12**, 1207–1245 (2000)
20. Schölkopf, B., Platt, J.C., Shawe-Taylor, J., Smola, A.J., Williamson, R.C.: Estimating the support of a high-dimensional distribution. Neural Comput. **13**(7), 1443–1471 (2001)
21. Shawe-Taylor, J., Bartlett, P., Williamson, R., Anthony, M.: Structural risk minimization over data-dependent hierarchies. IEEE Trans. Inf. Theory **44**(5), 1926–1940 (1998)
22. van Erven, T., Grünwald, P., Reid, M.D., Williamson, R.C.: Mixability in statistical learning. In: Neural Information Processing Systems, Lake Tahoe (2012)
23. van Erven, T., Reid, M.D., Williamson, R.C.: Mixability is Bayes risk curvature relative to log loss. J. Mach. Learn. Res. **13**, 1639–1663 (2012)
24. Vapnik, V.N.: Statistical Learning Theory. Wiley, New York (1998)
25. Vernet, E., Williamson, R.C., Reid, M.D.: Composite multiclass losses. In: Neural Information Processing Systems, Granada (2011)
26. Vernet, E., Williamson, R.C., Reid, M.D.: Composite multiclass losses. J. Mach. Learn. Res. **42** (2012) (Submitted)
27. Vovk, V.: A game of prediction with expert advice. In: Proceedings of the 8th Annual Conference on Computational Learning Theory, Santa Cruz, pp. 51–60. ACM (1995)
28. Williamson, R.C.: Introduction. In: Introductory Talk at Workshop on Relations Between Machine Learning Probems. NIPS2011 workshops, Sierra Nevada (2011)
29. Williamson, R.C., Smola, A., Schölkopf, B.: Generalization performance of regularization networks and support-vector machines via entropy numbers of compact operators. IEEE Trans. Inf. Theory **47**(6), 2516–2532 (2001)

Chapter 9
Statistical Learning Theory in Practice

Jason Weston

Abstract In this chapter we discuss the practical application of statistical learning theory: the design of learning algorithms and their use on real datasets. We review some of the most well-known methods and discuss their advantages and disadvantages. Particular emphasis is put on methods that scale well at training and testing time so that they can be used in real-life systems; we discuss their application on large-scale image and text classification tasks. Our goal is to understand what we have done right so far with the models that we have currently built, and what we are missing – that we thus have yet to invent.

9.1 Introduction

In the statistical learning framework, learning means estimating a function

$$y = f(x)$$

where the estimate is constructed given only ℓ examples of the mapping performed by the unknown function (called the training set)

$$(x_1, y_1), \ldots, (x_\ell, y_\ell). \tag{9.1}$$

A learning machine must choose from a given set of functions $\{f(x, \alpha), \alpha \in \Lambda\}$ the one which best approximates the unknown dependency, where Λ are the parameters that define the class of functions. Given this fixed class, the best choice of function is the one that provides the minimum value of the *risk*

J. Weston (✉)
Google Inc., 76 Ninth Avenue, New York, NY 10011, USA
e-mail: jweston@google.com

$$R(\alpha) = \int L(y, f(x,\alpha))dF(x,y) \tag{9.2}$$

where $L(y, f(x,\alpha))$ is the value of the loss function, a measure of difference between the estimate $f(x,\alpha)$ and the actual value y given by the unknown function at a point x. In reality, however, the joint distribution function $F(x, y)$ is unknown and we do not have the value of y for each point x, but only the training set pairs (9.1). We can instead approximate function (9.3) by considering the so-called *empirical risk function*:

$$R_{emp}(\alpha) = \frac{1}{\ell}\sum_{i=1}^{\ell} L(y_i, f(x_i,\alpha)). \tag{9.3}$$

Empirical Risk Minimization involves finding the function that minimizes (9.3), i.e., minimizing the training error. However, the minimum of (9.3) does not necessarily approximate well the minimum of (9.2) when ℓ is small. In the case of pattern recognition, Vapnik showed the following bound holds with probability $1 - \eta$ [25]:

$$R(\alpha) \leq R_{emp}(\alpha) + \sqrt{\frac{h(\log(\frac{2\ell}{h}+1) - \log(\frac{\eta}{4}))}{\ell}}, \tag{9.4}$$

where h is the VC dimension of the set of functions parameterized by α – a measure of the capacity of the set of functions for modelling data. The VC dimension of a set of functions is the maximum number of points that can be separated in all possible ways by that set (see [25]). If the capacity of the set of functions is large, the function that minimizes R_{emp} will yield a very small training error, the first term in the bound. However, the second term of the bound will be large – this matches the intuition that overfitting may be occurring because the set of functions is so powerful that it could fit any small dataset.

To minimize over both terms, the capacity of the set of functions would have to be a parameter in the optimization problem rather than just chosen a priori. Vapnik thus suggested the Structural Risk Minimization (SRM) principle that does just that. He proposed first defining a structure

$$S_1 \subset S_2 \subset \cdots \subset S \tag{9.5}$$

on the set of functions $S = \{f(x,\alpha), \alpha \in \Lambda\}$ whose VC-dimensions satisfy $h_1 \leq h_2 \leq \ldots h_n \leq \ldots$, i.e., they increase in expressive power. One can then choose the S_k that minimizes the bound (9.4), i.e., one can control the trade-off between the degree of fit to the data and the complexity of the set of functions used.

In typical real-life machine learning problems today, practitioners do follow a similar procedure to SRM, except they usually use a validation set instead of a bound to control this trade-off. The structure might be controlled more or less directly (e.g., the size of a neural network) or slightly less directly (e.g., via a regularization hyperparameter).

9.2 Practical Issues with the Classical Approach

The description given in the previous section is at the heart of Vapnik's work [25] and seems like a clear solution to machine learning problems. So, practically speaking, why are all our problems not yet solved?

Well, firstly, a word of encouragement. Machine learning approaches are in use today all over the world, and are making users happy, and making profits for companies. For example, their use is ubiquitous on the Internet, from searching to recommending to serving ads. Vapnik's practical inventions such as large margin methods and kernel machines, or methods directly influenced by them, are in evidence in many applications. However, machine learning is not considered solved. For example there are more papers per year on this topic than ever before.[1] We believe two key reasons researchers continue to search for better methods lie in the details of the described procedure:

1. Which classes of functions in the structure (9.5) to choose is not clear, and there are many possibilities to explore.
2. After choosing S it is still not easy, computationally, to minimize (9.3). Moreover, different solutions can be tailored depending on the choice of S in issue 1.

Choosing a family of functions to optimize is a key ingredient which can make large differences to performance. For example, decision tree models (where the capacity can be controlled by their depth) or neural network models (where capacity can be controlled by the number of neurons) are quite different models. Perhaps more importantly, prior knowledge about the problem can also be encoded into the family, e.g., using convolutional nets for images [19] or by preprocessing the data in a certain way (where we view this as part of $f(x, \alpha)$). In general, if you have more knowledge that you can encode accurately, then you will need less data, but if you have less knowledge you will need more data. However there is a trade-off between the cost of feature/model engineering vs. labelling and learning from more data both in money and time. This trade-off differs based on the task.

The other problem, issue 2, lies in the inconspicuous looking min in Eq. 9.3 – finding the function from the set of functions that gives the best empirical error is in fact very tricky indeed, depending on the family of functions. This leads to choosing families of functions that might not be the best for generalization, but are the best for generalization given the constraints of optimization. It also leads to looking for surrogate functions to approximate Eq. 9.3. For example, one can replace the loss function of interest with the hinge loss in order to provide gradients, which may make the problem convex, depending on the model. This procedure was pioneered in support vector machines [9]. Note that this makes the objective a surrogate of a surrogate as we were already replacing (9.2) with (9.3). Perhaps more importantly,

---

[1] For example, the number of submissions to ICML jumped from last year by about 50 % to around 900 submissions, see http://hunch.net/?p=2289.

this issue also leads to researchers developing a vast set of optimization algorithms, e.g., all the manifestations of gradient-based learning, greedy construction, and so on. If the loss is made approximate it sometimes becomes possible to make the optimization almost exact (e.g., convex optimization for SVMs on small datasets). Vice-versa, to make the loss more exact, the optimization problems get harder and the algorithm to optimize might be more approximate; for example, trying to optimize precision@k for ranking is trickier than optimizing the average rank with pairwise linear constraints (see, e.g., [27, 30]).

Much of the work in the field of machine learning can be divided into dealing with these two key issues. Issue (1) accounts for all the varying general machine learning models like SVMs, decision trees, neural networks, and nearest neighbours, as well as modifications in particular application fields, e.g., different kernels for different types of data in SVMs, different feature representations, and so on. Additionally, all the methods that propose particular regularizers might fall into this area as well, because they restrict the function family in some way, e.g., if the norms of the weights are constrained by some constant. Issue (2) accounts for all the work that proposes modified loss functions (hinge loss, logistic loss,...) and optimization algorithms (serial or parallel, coordinate, stochastic or batch descent, greedy,...). Many works specialize in improving some class of data, e.g., sparse or dense, low- or high-dimensional, binary or continuous, and so on.

Of course, there is much work that does not fit into these two issues, for example, taking advantage of additional data (e.g., semi-supervised learning), and tasks that do not fit directly into the above framework, e.g., clustering. However, even these approaches are often used as tools to solve issues 1 or 2, e.g., clustering is used to make optimization more tractable or define a different family of functions. Therefore, we believe a large body of work comes under these two topics.

In the remainder of this chapter we will discuss some of the most well-known existing machine learning approaches, and discuss their pros and cons with respect to modern large-scale learning tasks. Finally, we will conclude by discussing their practical use on some real-world tasks. We will focus on image- and text-based classification. Our goal is to understand what is really hard about these problems and what is missing in our current approaches, and to try to understand where we should go next.

9.3 A Short Review of Existing Learning Models

9.3.1 Linear Models

Let us start with linear models. A linear model usually requires a layer of preprocessing (for example, various normalization schemes, or a bag-of-words or bag-of-visual-terms transformation) followed by the model:

$$f_c(x) = w_c \cdot \Phi(x) \qquad (9.6)$$

where $\Phi(\cdot)$ performs the preprocessing.

The good points about such a model are that training time is linear in the number of examples and features, as shown in [5], where the authors train linear (and nonlinear) SVMs in the primal. Testing time is also linear. Some work has aimed at making testing time sublinear in the number of classes, as for a large number, e.g., a million classes, this can be far too slow; see, e.g., [22].

Linear Models

Advantages:

- Training and testing time is linear.
- Simple to understand and code.
- Engineers can encode knowledge via features in a straightforward fashion.

Disadvantages:

- The capacity is fixed so performance will saturate as training sizes increase.
- The models can take too much memory for large-scale tasks.
- Performance highly depends on user-specified feature design.

Even such simple models also can get too big in memory. For example, with 100,000 classes and 10,000 features this is around 8 GB; see, e.g., [27]. As we would like to model much more than linear relationships, and as we might want to handle much more than 100,000 classes, this is a serious problem. Further, perhaps more crucially, performance highly depends on the user-specified feature space $\Phi(\cdot)$ and the learner cannot do much to fix this modelling choice if the human engineer has got it drastically wrong. One can also see this as a blessing, if the human is adept at engineering features for linear models. Finally, an additional issue is that the capacity of such models is fixed – as the training set size increases the performance boost will eventually saturate.

9.3.2 Embedding Models

Embedding models aim to fix some of the problems of linear models by employing a smaller factorized model:

$$f_c(x) = \Phi(x)^\top U^\top V_c. \qquad (9.7)$$

where $\Phi(x) \in \mathbb{R}^D$, U is an $N \times D$ matrix and V is an $N \times C$ matrix, where C is the number of classes and N is the embedding dimension, a hyperparameter that is set by the user. These types of models have been used to good effect in a wide range of applications [1, 23, 27, 28].

Embedding Models

Advantages:

- Similar properties to linear models, but take much less memory.
- Hard to overfit. Share weights between classes (multi-tasks).
- The above two points mean they can use larger feature spaces than linear models.

Disadvantages:

- These models can underfit – the capacity is small and saturates like linear models.
- The training is not parallelizable unlike in one-vs-rest linear models.
- The performance still depends strongly on the original feature design.

The main advantage of these models over linear models is that they take much less memory, e.g., 77 MB vs. 8 GB on the same 10,000 feature, 100,000 class problem, by utilizing an embedding dimension of $N = 100$. The factorized model projects the problem into a lower-dimensional subspace which all the classes share. This simultaneously means that there is more memory available to increase the number of features in $\Phi(\cdots)$ and that overfitting from this increase is less likely to occur as the capacity is lower, and is shared between all classes; e.g., consider the case where there are very few examples for some classes. The disadvantage is that underfitting may occur from the low dimensionality of the embedding space and hence there is an inability to memorize certain feature-class pair relationships due to the low rank of the matrix $U^\top V$. We have explored some solutions to this problem in [2] and [29].

9.3.3 Nearest Neighbour

Nearest neighbour, the other extreme, does not suffer from memorization problems as it memorizes all the examples, and then makes a local decision based on the labels of the k nearest neighbours (k-NN).

Nearest Neighbour

Advantages:

- The capacity grows as the training set grows, so it can capture more dependencies.
- There is no training time cost.

Disadvantages:

- Testing time is too slow (linear in the number of examples).
- Storage costs are prohibitive (storing the entire dataset).
- There is a potential inability to generalize due to the local nature of the decision.

As the decision is local in the given feature space, training examples are never used to generalize across the space (indeed, there is really no "learning" step at all). Nevertheless, it still often works remarkably well compared to other methods, and is usually a strong baseline if there are enough computing resources to run it. Because of this, much research has been done in trying to fix its shortcomings. In order to improve its speed and lower its memory consumption, many approximate nearest neighbour methods have been proposed, e.g., via hashing [15]. Unfortunately, generalization can suffer from these approximations. In order to improve generalization several things can be done. Firstly, one can consider nearest neighbours on subsets of features rather than the entire example vector $\Phi(x)$, making the classifier much less local [3]. Secondly, some aspects of the model can be parameterized, and those parameters can be learnt. For example, one can perform a metric learning step to learn a feature representation and then compute the nearest neighbours in the learnt space [26].

9.3.4 SVMs and Kernel Methods

Support vector machines have been a very successful class of models, well supported by theory, that incorporate a number of beneficial practical qualities: the learning of a large margin classifier gives good generalization in the chosen space, and the use of kernels allows the user to efficiently encode high-dimensional spaces (that possibly also encode prior knowledge about the task at hand). These properties were borrowed by many other methods, for example, the linear and embedding methods described earlier can both be trained in a large margin fashion, and many methods have been "kernelized" since the introduction of SVMs. SVMs (and many other kernel methods) take the form:

$$f_c(x) = \sum \alpha_{c,i} \ K(x_i, x)$$

where $K(x, x') = \Phi(x) \cdot \Phi(x')$ is the kernel function. $\Phi(\cdot)$ does not have to be explicitly represented, unlike in Eqs. (9.6) and (9.7), meaning that higher-dimensional spaces, such as polynomials of high degree, can be represented without taking too much memory. However, similarly to nearest neighbour, the model grows as the number of training examples grows (although α can be sparse). If one uses RBF kernels, one can think of SVMs as mixing some of the advantages of linear models and nearest neighbour, as the training points local to the input x will dictate the class prediction, but a weighting for each is learnt with the vector α.

> **SVMs and Kernel Methods**
>
> Advantages:
>
> - Capacity grows with training set size, and does not saturate.
> - Nonlinearities via kernels give superior performance to linear models.
> - Weights are learnt for each example, giving superior performance to k-NN.
>
> Disadvantages:
>
> - Memory grows with training set size and is infeasible for large datasets.
> - Testing time is also slow, due to the summing of kernel functions.

There has been lots of work on speeding up SVMs for large datasets; see, e.g., [4]. Learning the kernel has also become an active topic in order to minimize the dependence on the human engineer.

9.3.5 Neural Networks

Neural Networks are a very wide family of models that technically can encompass all the models described so far, and more. The classical setup, however, is to use several layers of matrix transformations (embedding models can be thought of as one such layer) followed by nonlinearity via a sigmoid function.

> **Neural Networks**
>
> Advantages:
>
> - Capacity fixed, so you know how fast it will be and how much memory it takes.
> - Layers give efficient nonlinearities in terms of memory and speed.
>
> Disadvantages:
>
> - The training time can be slow.
> - The difficulty of optimization (nonconvexity) means that capacity can be wasted.

As in other methods, much work has gone into speeding up training, most recently via parallelism [11, 24]. Note that despite their seemingly large capacity we believe that, as with the deficiencies of embedding models, neural networks can still fail to "memorize" important dependencies, even when the networks are very large. For example, when combining an n-gram model (that memorizes all n-grams) with a very large neural network language model, the performance of the combination is better than that of either one alone.[2] Researchers also point out that due to the

---

[2] Results described by Jeff Dean in his talk "Scaling Deep Learning" at the workshop on Representation Learning, ICML 2012.

difficulty of optimizing the multi-layer network, inherent capacity in the family of functions is actually wasted (the optimizer does not find those weights) [10].

9.3.6 Further Methods

Of course, there are more machine learning methods than can possibly be mentioned here. As hinted in the descriptions above, many of the standard approaches have been modified and extended in various ways. And then, there are many other quite different methods besides. A few notable ones are decision trees, random forests and boosting models (see [13] for a more thorough list). Finally, it has been observed across many datasets that one can obtain improved performance by simply combining *more than one classifier*, i.e., using an ensemble. However, this does not always negate the disadvantages of those methods; for example, the ensemble is only likely to take more memory and be slower still.

9.4 Practical Challenges on Real Datasets

To truly understand what is hard about solving machine learning tasks it is probably best to discuss the attributes of some real datasets. Here, we will focus on image and document annotation. These are almost noise-free tasks in that they can be posed as problems which non-expert humans are extremely good at, so we can collect a huge amount of noise-free data; for example, ImageNet [12] contains millions of images labeled by humans using Mechanical Turk. Even without explicitly employing and paying human labellers, there are many techniques to implicitly collect labels, e.g., via clicks on web applications and search engines [18]; however, in that case the supervised signal may be significantly more noisy. Finally, there is an even larger amount of semi-labeled or completely unlabeled data. (We note that Vapnik's work on transduction and transductive SVMs [17, 25] had a strong influence on the creation of the field of semi-supervised learning [6].)

It is only relatively recently that the scale of these datasets reached the current dataset sizes; for example, in the field of image annotation it was common to use datasets with only thousands of examples or less, e.g., [14]. However, now it is commonplace to have very large datasets, especially in industrial applications. This is why we focused so much on the scalability of algorithms in Sect. 9.3. This recent change in scale should mean that the families of functions, and the optimization algorithms for choosing the best function, should be significantly different today from what they were 10 years ago. This is because on the one hand we should be using higher-capacity models as they are less likely to overfit with the increased data, but on the other we need to utilize optimization procedures that actually work on such large-scale datasets, with millions or even billions of examples. There is also every indication that datasets will get bigger still in the coming years. However, even with the training set sizes we have today, if we have an appropriate high-capacity model that is able to scale at training time, should we not be able to do very well (e.g., human-level performance)? It seems currently that is not the case, so the question is, why?

Text and image tasks do not actually appear to have the same properties at first glance. Assuming *layers* of processing, which every known method does to some degree (e.g., preprocessing followed by learning or multiple layers of learning), the first layer of image processing – transforming a matrix of pixels into features that are more correlated to semantic classes – seems harder than the first layer of processing for text tasks, where words are already discretized and are already strongly discriminative features. For example, a common processing step for images is, via colour and edge histograms followed by clustering, to transform image patches into a dictionary of "visual words" (see, e.g., [20]). This makes image tasks look like text tasks to a machine learning algorithm, and similar approaches can then be used on both. However, the features are much weaker for the image task; for example, compare identifying whether the concept of a sheep is in an image or in a sentence – the image patches are noisy and rarely as clearly discriminative as word features such as the word "sheep" or its synonyms. This is to be expected as language was designed by humans to be processed easily (by their own biological brains), whereas how nature looks is mostly not designed by humans (although we do design visual objects to be specifically easy for machines to read, such as barcodes). Note that when humans design image annotation tasks for computers, humans can be incredibly bad at the task itself (barcodes). This highlights the difference between the skills of computers and humans.

Where text and image tasks seem more similar is at the next level of processing. The way words combine and interact within a sentence is complex; for example, linguists often represent a sentence as a parse tree. Beyond syntactic structure, understanding the higher-order semantics of a sentence requires a great deal of world knowledge, as for example understanding even such seemingly simple sentences as "John flew the kite. He cursed when the string broke." Similarly, the combinations of "visual words" in an image are also complex, e.g., the parts that make up an object and then how all those objects interact within a scene. Here again, knowledge of the world can also be leveraged, for example, knowing that a kite is often attached to a string when annotating an image with a kite flying scene. Note that although text tasks are somehow "easier" due to the strong features mentioned, we have much higher expectations on their performance and how much semantics the computer can extract from the input.

Overall, we believe that to perform very well on these tasks it is required to have a model that has a large "memory" (number of parameters) in order to store all the variations (image patch variability, phrase variability, object or word pair relationships, world knowledge, etc.) and the ability to generalize from memories (to capture the "gestalt"), which may require the ability to chain inference rules in order to come to a final conclusion (cause and effect within a sentence or an image).

In general we believe the models we are using today are too small and unable to perform such inferences.

The problem is that these modelling objectives introduce scalability issues – either from the cost of optimization or the size in memory, and we believe most models fall short of trying to model such complex phenomena. Likely, the functions needed to actually capture all the semantics are exceedingly complex; and while today somewhat complex models that incorporate structured outputs, e.g., for parsing [7], and layers of processing for tagging [8] are employed, they still are often relatively unsophisticated compared to what we could envisage. Just to mention one research direction, still in its infancy, the authors of [21] proposed a method for question answering that learns to predict a program given an input sentence (question) that can then be run to predict the answer. This gives a hint at the kind of complex family of functions that we could be more earnestly investigating to solve these problems, although researchers differ over whether these systems should contain symbolic elements or be entirely subsymbolic.

Perhaps unfortunately, most of the academic field is currently not trying to go in either of those directions. Much of current research aims at small improvements and tweaks on particular subtasks, which can still make a difference in practice to a particular product. For example, in web search retrieval many tweaks can be made, e.g., controlling the diversity of the results, engineering the various bag-of-words and n-gram features, optimizing precision at k, etc. But none of those things gets us closer to a learning machine that can really understand a sentence or document; this would require a much more ambitious model. On many text tasks, however, such as retrieval, the performance is already acceptable, which we believe can inhibit progress, as people are happy with the status quo.

For image tasks, the performance of our models is far worse, we believe mainly because the first layers of processing need to be improved, as described previously. Perhaps this relatively poor performance can at least stimulate further research. However, in that task too it is disappointing how far bag-of-visual-words type models (perhaps combined with a pyramid matching algorithm [16]) can take you, despite not modeling many aspects of the problem – e.g., relations between patches or objects are poorly captured (typically, segmentation is not used); there is no shape detection, no understanding of 3D, no constraints governing the world; and we typically choose simple models with fast learning times (we are very impatient compared to human baby learning or evolution). Presumably, learning systems capable of modelling all those things are what you need to get to human-level performance.

In the short-term, however, we believe some improvements over current solutions could be gained from designing methods that keep the advantages of the methods described in Sect. 9.3 while avoiding their deficiencies. In particular, finding more optimal ways of blending "memorization" and generalization could be an important direction, the challenge being to maintain scalability as well.

References

1. Bai, B., Weston, J., Grangier, D., Collobert, R., Sadamasa, K., Qi, Y., Cortes, C., Mohri, M.: Polynomial semantic indexing. In: Advances in Neural Information Processing Systems (NIPS 2009), Vancouver (2009)
2. Bai, B., Weston, J., Grangier, D., Collobert, R., Sadamasa, K., Qi, Y., Chapelle, O., Weinberger, K.: Learning to rank with (a lot of) word features. Inf. Retr. **13**(3), 291–314 (2010)
3. Boiman, O., Shechtman, E., Irani, M.: In defense of nearest-neighbor based image classification. In: IEEE Conference on Computer Vision and Pattern Recognition, CVPR 2008, pp. 1–8. IEEE (2008)
4. Bottou, L., Chapelle, O., DeCoste, D., Weston, J.: Large-Scale Kernel Machines. MIT, Cambridge (2007)
5. Chapelle, O.: Training a support vector machine in the primal. Neural Comput. **19**(5), 1155–1178 (2007)
6. Chapelle, O., Schölkopf, B., Zien, A., et al.: Semi-Supervised Learning, vol. 2. MIT, Cambridge (2006)
7. Collins, M.: Discriminative training methods for hidden Markov models: theory and experiments with perceptron algorithms. In: Proceedings of the ACL-02 Conference on Empirical Methods in Natural Language Processing, vol. 10, Philadelphia, pp. 1–8. Association for Computational Linguistics (2002)
8. Collobert, R., Weston, J., Bottou, L., Karlen, M., Kavukcuoglu, K., Kuksa, P.: Natural language processing (almost) from scratch. J. Mach. Learn. Res. **12**, 2493–2537 (2011)
9. Cortes, C., Vapnik, V.: Support-vector networks. Mach. Learn. **20**(3), 273–297 (1995)
10. Dauphin, Y., Bengio, Y.: Big neural networks waste capacity. CoRR **abs/1301.3583** (2013)
11. Dean, J., Corrado, G., Monga, R., Chen, K., Devin, M., Le, Q., Mao, M., Senior, A., Tucker, P., Yang, K., et al.: Large scale distributed deep networks. In: Advances in Neural Information Processing Systems 25, pp. 1232–1240 (2012)
12. Deng, J., Dong, W., Socher, R., Li, L.J., Li, K., Fei-Fei, L.: ImageNet: A Large-Scale Hierarchical Image Database. In: IEEE Conference on Computer Vision Pattern Recognition (CVPR), Miami (2009)
13. Duda, R.O., Hart, P.E., Stork, D.G.: Pattern Classification and Scene Analysis, 2nd edn. Wiley, New York (1995)
14. Fei-Fei, L., Fergus, R., Perona, P.: Learning generative visual models from few training examples: an incremental Bayesian approach tested on 101 object categories. Comput. Vis. Image Underst. **106**(1), 59–70 (2007)
15. Gionis, A., Indyk, P., Motwani, R., et al.: Similarity search in high dimensions via hashing. In: Proceedings of the International Conference on Very Large Data Bases, Edinburgh, pp. 518–529 (1999)
16. Grauman, K., Darrell, T.: The pyramid match kernel: discriminative classification with sets of image features. In: Tenth IEEE International Conference on Computer Vision, ICCV 2005, Beijing, vol. 2, pp. 1458–1465. IEEE (2005)
17. Joachims, T.: Transductive inference for text classification using support vector machines. In: Proceedings of the 1999 International Conference on Machine Learning, Bled, pp. 200–209. Morgan Kaufmann (1999)
18. Joachims, T.: Optimizing search engines using clickthrough data. In: Proceedings of the 8th ACM SIGKDD International Conference on Knowledge Discovery and Data Mining, Edmonton, pp. 133–142. ACM (2002)
19. LeCun, Y., Bengio, Y.: Convolutional networks for images, speech, and time series. In: The Handbook of Brain Theory and Neural Networks, vol. 3361. MIT (1995)
20. Leung, T., Malik, J.: Representing and recognizing the visual appearance of materials using three-dimensional textons. Int. J. Comput. Vis. **43**(1), 29–44 (2001)
21. Liang, P., Jordan, M.I., Klein, D.: Learning dependency-based compositional semantics. Comput. Linguist. **39**, 389–446 (2013)

22. Makadia, A., Weston, J., Yee, H.: Label partitioning for sublinear ranking. In: International Conference on Machine Learning, ICML, Atlanta (2013)
23. Melvin, I., Weston, J., Noble, W.S., Leslie, C.: Detecting remote evolutionary relationships among proteins by large-scale semantic embedding. PLoS Comput. Biol. **7**(1), e1001047 (2011)
24. Raina, R., Madhavan, A., Ng, A.Y.: Large-scale deep unsupervised learning using graphics processors. In: Proceedings of the 26th Annual International Conference on Machine Learning, Montreal, vol. 382, pp. 873–880. ACM (2009)
25. Vapnik, V.N.: Statistical Learning Theory. Wiley, New York (1998)
26. Weinberger, K.Q., Blitzer, J., Saul, L.K.: Distance metric learning for large margin nearest neighbor classification. In: NIPS, Citeseer, Vancouver (2006)
27. Weston, J., Bengio, S., Usunier, N.: WSABIE: Scaling up to large vocabulary image annotation. In: International Joint Conference on Artificial Intelligence (IJCAI), Barcelona, pp. 2764–2770 (2011)
28. Weston, J., Bengio, S., Hamel, P.: Multi-tasking with joint semantic spaces for large-scale music annotation and retrieval. J. New Music Res. **40.4**, 337–348 (2011)
29. Weston, J., Weiss, R., Yee, H.: Affinity weighted embedding. arXiv:1301.4171 (2013, preprint)
30. Yue, Y., Finley, T., Radlinski, F., Joachims, T.: A support vector method for optimizing average precision. In: Proceedings of the 30th Annual International ACM SIGIR Conference on Research and Development in Information Retrieval, Amsterdam, pp. 271–278. ACM (2007)

Chapter 10
PAC-Bayesian Theory

David McAllester and Takintayo Akinbiyi

Abstract The PAC-Bayesian framework is a frequentist approach to machine learning which encodes learner bias as a "prior probability" over hypotheses. This chapter reviews basic PAC-Bayesian theory, including Catoni's basic inequality and Catoni's localization theorem.

10.1 Introduction

Vladimir Vapnik pioneered the mathematics of uniform convergence of statistical estimators [15]. Based on the theory of uniform convergence, Vapnik introduced the support vector machine (SVM), which revolutionized the practice of machine learning [1]. The uniform convergence view of learning represents a departure from the Bayesian view. The Bayesian approach models learning bias as a prior probability and models learning as Bayesian conditioning. The uniform convergence approach models learning bias as an a priori commitment to a hypotheses class with uniform convergence properties—a class with finite Vapnik–Chervonenkis (VC) dimension.

PAC-Bayesian theory blends Bayesian and uniform convergence approaches. The acronym PAC comes from Leslie Valiant's introduction of the phrase "probably approximately correct" (PAC) to describe the kind of statement made in a uniform convergence theorem [14]. One can interpret the acronym PAC as describing a broad class of inequalities that hold with high probability (with high confidence) under weak assumptions. The PAC-Bayesian theorem replaces the finite VC dimension class with a "prior probability" on a single hypotheses class which may be of infinite VC dimension.

10.2 Basic PAC-Bayesian Theorems

Let $\mathcal{H}$ be a set of hypotheses, let $\mathcal{O}$ be a set of observations and let L be a loss function such that for $h \in \mathcal{H}$ and $o \in \mathcal{O}$ we have $L(h,o) \in [0,1]$. Let D be a probability distribution (measure) on $\mathcal{O}$ and let P be a distribution (measure) on $\mathcal{H}$. Here we think of D as a data distribution occurring in nature and P as a learning bias expressing the learner's preference for some hypotheses over others. There is no assumed relationship between D and P.

Now let Q be a variable ranging over distributions on the hypothesis space $\mathcal{H}$. For $o \in \mathcal{O}$ we define the loss $L(Q,o)$ to be $E_{h \sim Q}[L(h,o)]$. We have that $L(Q,o)$ is the loss of a stochastic process that selects the hypothesis h according to distribution Q. Now let S be a sequence of N observations drawn IID from the observation distribution D. We define $L(Q,S)$ to be $\frac{1}{N} \sum_{o \in S} L(Q,o)$ and define $L(Q,D)$ to be $E_{o \sim D}[L(Q,o)]$.

The first version of the PAC-Bayesian theorem appeared in [9] and states that for $\delta \in (0,1)$ we have that, with probability at least $1 - \delta$ over the draw of the sample S, the following holds simultaneously for all distributions Q where $\mathcal{D}(Q,P) = E_{h \sim Q}\left[\frac{\ln Q(h)}{\ln P(h)}\right]$ is the Kullback–Leibler divergence from Q to P:

$$L(Q,D) \leq L(Q,S) + \sqrt{\frac{\mathcal{D}(Q,P) + \ln \frac{N}{\delta} + 2}{2N - 1}}. \tag{10.1}$$

The bound (10.1) has been improved by various authors [2,3,5,8,12]. Each improvement gives some bound $B(L(Q,S), \mathcal{D}(Q,P), N, \delta)$ such that with probability $1-\delta$ over the draw of S we have, simultaneously for all Q,

$$L(Q,D) \leq B(L(Q,S), \mathcal{D}(Q,P), N, \delta).$$

Any version of the PAC-Bayesian theorem defines a learning algorithm where one draws a sample S and then constructs $Q^*(S)$—the distribution minimizing the bound $B(L(Q,S), \mathcal{D}(Q,P), N, \delta)$. The theorem provides a guarantee on generalization loss—the loss of the process that draws (test time) observations from D and draws hypotheses from the learned distribution $Q^*(S)$. All versions of the theorem yield learning algorithms of the same form. In all cases we can consider minimizing $\mathcal{D}(Q,P)$ subject to the constraint that $L(Q,S)$ is held constant. An application of the KKT conditions (or Lagrangian duality) then shows that the optimal posterior has the form

$$Q^*(S)(h) = Q_\beta(S)(h) = \frac{1}{Z_\beta(S)} P(h) e^{-N\beta L(h,S)} \qquad (10.2)$$

$$Z_\beta(S) = \mathrm{E}_{h \sim P}\left[e^{-N\beta L(h,S)}\right].$$

Here β is an unknown parameter of the optimal posterior $Q_\beta(S)$. We argue at the end of Sect. 10.4 that β can be taken to be $\Omega(1)$ independent of N.

The PAC-Bayesian theorem can also be used to justify algorithms similar to support vector machines [10, 11, 13]. For this, one takes P to be a unit variance isotropic Gaussian distribution centred at the origin and takes Q_w to be a unit variance isotropic Gaussian centred at weight vector w. In this case we have $\mathcal{D}(Q_w, P) = (1/2)\|w\|^2$. For the case of classification error we have that $L(Q_w, S)$ equals the average profit loss on the training data of the margin of the linear classifier defined by w.

The most commonly referenced version of the PAC-Bayesian theorem is due to Seeger [12] as refined by Maurer [8]. It states that with probability at least $1 - \delta$ over the draw of the sample the following holds simultaneously for all Q, where for real numbers $p, q \in [0, 1]$ we have that $\mathcal{D}(q, p) = q \ln \frac{q}{p} + (1 - q) \ln \frac{1-q}{1-p}$ is the Kullback–Leibler divergence from a binomial distribution with bias q to a binomial distribution with bias p:

$$\mathcal{D}(L(Q, S), L(Q, D)) \leq \frac{1}{N}\left(\mathcal{D}(Q, P) + \ln \frac{2\sqrt{N}}{\delta}\right). \qquad (10.3)$$

A version of (10.1) can be derived from (10.3) using $\mathcal{D}(q, p) \geq 2(q - p)^2$. It is also possible to solve (10.3) exactly for the implicit upper bound on $L(Q, D)$. This can be done with the following equality, which appears implicitly in Hoeffding's paper introducing the exponential moment method [4] and which can be verified by a straightforward optimization over γ:

$$\inf_\gamma (1 - p)e^{-\gamma q} + pe^{\gamma(1-q)} = \inf_\gamma e^{-\gamma q}(1 - p + pe^\gamma) = e^{-\mathcal{D}(q,p)}$$

$$\mathcal{D}(q, p) = \sup_\gamma \gamma q - \ln(1 - p + pe^\gamma). \qquad (10.4)$$

After some calculation we get that for $q \leq p$ we have that $\mathcal{D}(q, p) \leq c$ holds if and only if we have

$$p \leq \inf_{\gamma < 0} \frac{1 - e^{\gamma q - c}}{1 - e^\gamma} = \inf_{\eta > 0} \frac{1 - e^{-(\eta q + c)}}{1 - e^{-\eta}}. \qquad (10.5)$$

For $\eta \in (0, 2)$ the bound (10.5) can be simplified by taking the first-order Taylor expansion of the exponential in the numerator and the second-order Taylor expansion of the exponential in the denominator. We then get that $\mathcal{D}(q, p) \leq c$ implies

$$p \leq \inf_{\eta \in (0,2)} \frac{\eta q + c}{\eta - \frac{1}{2}\eta^2} = \inf_{\lambda > 1/2} \left(\frac{1}{1 - \frac{1}{2\lambda}}\right)(q + \lambda c). \qquad (10.6)$$

One can alternatively derive (10.6) from $\mathcal{D}(q, p) \leq c$ by using $\mathcal{D}(q, p) \geq (p - q)^2/2p$ for $q < p$, which yields $p \leq q + \sqrt{2pc}$, and then using $\sqrt{2pc} = \inf_{\lambda > 0} (p/\lambda + 2\lambda c)/2$. Inequalities (10.3) and (10.6) together yield the following:

$$L(Q, D) \leq \inf_{\lambda > 1/2} \left(\frac{1}{1 - \frac{1}{2\lambda}}\right)\left(L(Q, S) + \frac{\lambda}{N}\left(\mathcal{D}(Q, P) + \ln \frac{2\sqrt{N}}{\delta}\right)\right). \qquad (10.7)$$

Formula (10.7) is perhaps the most insightful form of the PAC-Bayesian theorem. Here λ can be interpreted as defining the trade-off between the complexity term $\mathcal{D}(Q, P)$ and the empirical loss term $L(Q, S)$. For (10.7) there is no point in taking λ to be very large—restricting λ to be less than 50 results in at most a 1 % weakening of the bound. Hence we can take λ to be $O(1)$ independent of N. The difference between (10.3) and (10.7) corresponds to the difference between (10.5) and (10.6). For $c \leq 1/4$ we have that (10.5) is never more than 2.3 times tighter than (10.6).

A fundamental departure point for the proof of (10.3) is the following, which holds for a single fixed $h \in \mathcal{H}$:

$$\mathbb{E}_S \left[e^{N\mathcal{D}(L(h,S),L(h,D))}\right] \leq 2\sqrt{N}. \qquad (10.8)$$

A proof of (10.8) can be found in Maurer [8]. A special case of (10.8) is the statement that for a biased coin with bias p, and for $\hat{p}$ the fraction of heads in N flips, we have $\mathbb{E}\left[e^{N\mathcal{D}(\hat{p},p)}\right] \leq 2\sqrt{N}$. The inequality (10.8) is intuitively a consequence of the well-known concentration inequality in Hoeffding [4] that for $q \leq p$ we have $P[\hat{p} \leq q] \leq e^{-N\mathcal{D}(q,p)}$ and for $q \geq p$ we have $P[\hat{p} \geq q] \leq e^{-N\mathcal{D}(q,p)}$.

To derive (10.3) from (10.8) we first note that (10.8) implies

$$\mathbb{E}_{h \sim P}\left[\mathbb{E}_S\left[e^{N\mathcal{D}(L(h,S),L(h,D))}\right]\right] \leq 2\sqrt{N}$$

$$\mathbb{E}_S\left[\mathbb{E}_{h \sim P}\left[e^{N\mathcal{D}(L(h,S),L(h,D))}\right]\right] \leq 2\sqrt{N}. \qquad (10.9)$$

Applying Markov's inequality to (10.9) we get that with probability at least $1 - \delta$ over the draw of S we have

$$\mathbb{E}_{h \sim P}\left[e^{N\mathcal{D}(L(h,S),L(h,P))}\right] \leq \frac{2\sqrt{N}}{\delta}.$$

Next we observe the following shift of measure lemma:

$$
\begin{aligned}
E_{h\sim Q}\left[f(h)\right] &= E_{h\sim Q}\left[\ln e^{f(h)}\right] \\
&= E_{h\sim Q}\left[\ln \frac{P(h)}{Q(h)} e^{f(h)} + \ln \frac{Q(h)}{P(h)}\right] \\
&\leq \ln E_{h\sim Q}\left[\frac{P(h)}{Q(h)} e^{f(h)}\right] + \mathcal{D}(Q, P) \\
&= \mathcal{D}(Q, P) + \ln E_{h\sim P}\left[e^{f(h)}\right].
\end{aligned}
$$

Setting $f(h) = N\mathcal{D}(L(h, S), L(h, D))$ now gives

$$E_{h\sim Q}\left[\mathcal{D}(L(h, S), L(h, D))\right] \leq \frac{1}{N}\left(\mathcal{D}(Q, P) + \ln \frac{2\sqrt{N}}{\delta}\right).$$

Finally, (10.3) follows from convexity properties of the divergence function.

10.3 Catoni's Basic Inequality

Inequality (10.10) corresponds to Lemma 1.1.1 in [2]. Catoni calls this the basic inequality. We first define $\mathcal{D}_\gamma(q, p)$ by

$$\mathcal{D}_\gamma(q, p) = \gamma q - \ln(1 - p + pe^\gamma).$$

From (10.4) we then have $\mathcal{D}(q, p) = \sup_\gamma \mathcal{D}_\gamma(q, p)$. Now consider a random variable x with $x \in [0, 1]$ and with mean μ. Let $\hat{\mu}$ be the mean of N independent draws of x. Catoni's basic observation is that for any γ we have

$$E\left[e^{N\mathcal{D}_\gamma(\hat{\mu},\mu)}\right] \leq 1. \quad (10.10)$$

To see this, note that $E\left[e^{N\gamma\hat{\mu}}\right] = (E\left[e^{\gamma x}\right])^N$. For $x \in [0, 1]$ we note that the convexity of the exponential function implies $e^{\gamma x} \leq 1 - x + xe^\gamma$. This gives $E\left[e^{N\gamma\hat{\mu}}\right] \leq (1 - \mu + \mu e^\gamma)^N$. Dividing by the right-hand side gives $E\left[e^{N(\gamma\hat{\mu} - \ln(1-\mu+\mu e^\gamma))}\right] \leq 1$. The following is a special case of (10.10).

$$E_S\left[e^{N\mathcal{D}_\gamma(L(h,S),L(h,D))}\right] \leq 1. \quad (10.11)$$

It is useful to compare (10.11) and (10.8). The discrepancy between (10.11) and (10.8) is consistent with $\sup_\gamma E\left[e^{N\mathcal{D}_\gamma(\hat{\mu},\mu)}\right] \leq E\left[\sup_\gamma e^{N\mathcal{D}_\gamma(\hat{\mu},\mu)}\right]$.

Repeating the derivation of the PAC-Bayesian theorem starting from (10.11) rather than (10.8) we get that for any γ (selected before drawing the sample) we have that, with probability at least $1 - \delta$ over the draw of S, the following holds simultaneously for all Q:

$$\mathcal{D}_\gamma(L(Q,S), L(Q,D)) \leq \frac{1}{N}\left(\mathcal{D}(Q,P) + \ln\frac{1}{\delta}\right). \qquad (10.12)$$

For $\lambda = -1/\gamma > 1/2$ (with λ selected before the draw of the sample) we get

$$L(Q,D) \leq \left(\frac{1}{1-\frac{1}{2\lambda}}\right)\left(L(Q,S) + \frac{\lambda}{N}\left(\mathcal{D}(Q,P) + \ln\frac{1}{\delta}\right)\right). \qquad (10.13)$$

One should compare (10.12) and (10.13) with (10.3) and (10.7).

10.4 Catoni's Localization Theorem

Here we describe Catoni's simplest and perhaps most important localization theorem, Corollary 1.3.2 in [2], which corresponds to (10.17) below.

Various authors have observed that it is possible to optimize the prior as a function of the observation distribution D and a fixed learning algorithm $\mathcal{A}$ [2, 6, 7]. We assume the learning algorithm $\mathcal{A}$ takes as input a sample S of observations and returns a "posterior distribution" $Q_\mathcal{A}(S)$. The PAC-Bayesian theorem then provides a generalization guarantee as a function of $L(Q_\mathcal{A}(S), S)$ and $\mathcal{D}(Q_\mathcal{A}(S), P)$. This theorem holds for any P that does not depend on the random variable S. In particular we can now optimize P as a function of D and $\mathcal{A}$. The following calculation, which first appears in [6], shows that $\bar{Q}_\mathcal{A}(h) = \mathrm{E}_S\left[Q_\mathcal{A}(S)(h)\right]$ is the prior minimizing $\mathrm{E}_S\left[\mathcal{D}(Q_\mathcal{A}(S), P)\right]$:

$$\begin{aligned}
P^* &= \mathrm{argmin}_P \; \mathrm{E}_S\left[\mathcal{D}(Q_\mathcal{A}(S), P)\right] \\
&= \mathrm{argmin}_P \; \mathrm{E}_{h \sim \bar{Q}_\mathcal{A}}\left[\ln \frac{1}{P(h)}\right] + \mathrm{E}_{S,\, h \sim Q_\mathcal{A}(S)}\left[\ln Q_\mathcal{A}(S)(h)\right] \\
&= \mathrm{argmin}_P \; \mathcal{D}(\bar{Q}_\mathcal{A}, P) \;=\; \bar{Q}_\mathcal{A}. \qquad (10.14)
\end{aligned}$$

Catoni observes that $\mathrm{E}_S\left[\mathcal{D}(Q_\mathcal{A}(S), \bar{Q}_\mathcal{A})\right]$ equals the mutual information between S and h under the joint distribution defined by $S \sim D^N$ and $h \sim Q_\mathcal{A}(S)$.

In light of (10.2) we consider the learning algorithm with parameter β that maps a sample S to the distribution $Q_\beta(S)$ defined by (10.2). For this algorithm (with β fixed) we can then apply (10.14) and replace the prior P with $\bar{Q}_\beta = \mathrm{E}_S\left[Q_\beta(S)\right]$. Catoni approximates $\bar{Q}_\beta$ with $\ddot{Q}_\beta(h) = (1/\ddot{Z}_\beta) P(h) e^{-N\beta L(h,D)}$ with

$\ddot{Z}_\beta = \mathrm{E}_{h\sim P}\left[e^{-N\beta L(h,D)}\right]$. To analyze this situation we first observe that for any fixed sample S the shift of measure lemma yields

$$\mathcal{D}_\gamma(L(Q_\beta(S),S), L(Q_\beta(S),D))$$
$$\leq \frac{1}{N}\left(\mathcal{D}(Q_\beta(S),\ddot{Q}_\beta) + \ln \mathrm{E}_{h\sim \ddot{Q}_\beta}\left[e^{N\mathcal{D}_\gamma(L(h,S),L(h,D))}\right]\right).$$

Rather than apply a Markov inequality to (10.11) we can simply take the expectation with respect to the draw of S and note that, after an application of Jensen's inequality and a reversal of expectations, (10.11) implies that the last term vanishes. This gives

$$\mathcal{D}_\gamma(\mathrm{E}_S\left[L(Q_\beta(S),S)\right], \mathrm{E}_S\left[L(Q_\beta(S),D)\right]) \leq \frac{1}{N}\mathrm{E}_S\left[\mathcal{D}(Q_\beta(S),\ddot{Q}_\beta)\right].$$

Taking $\lambda = -1/\gamma > 1/2$ we get

$$\mathrm{E}_S\left[L(Q_\beta(S),D)\right] \leq \frac{1}{1-\frac{1}{2\lambda}}\left(\mathrm{E}_S\left[L(Q_\beta(S),S)\right] + \frac{\lambda}{N}\mathrm{E}_S\left[\mathcal{D}(Q_\beta(S),\ddot{Q}_\beta)\right]\right).$$
(10.15)

Catoni then analyzes $\mathrm{E}_S\left[\mathcal{D}(Q_\beta(S),\ddot{Q}_\beta)\right]$ as follows:

$$\mathrm{E}_S\left[\mathcal{D}(Q_\beta(S),\ddot{Q}_\beta)\right] = \mathrm{E}_{S,\,h\sim Q_\beta(S)}\left[\ln \frac{Q_\beta(S)(h)}{\ddot{Q}_\beta(h)}\right]$$
$$= \mathrm{E}_{S,\,h\sim Q_\beta(S)}\left[N\beta(L(h,D)-L(h,S)) - \ln Z_\beta(S) + \ln \ddot{Z}_\beta\right]$$

From convexity properties of the log partition function we have

$$\mathrm{E}_S\left[\ln Z_\beta(S)\right] = \mathrm{E}_S\left[\ln \mathrm{E}_{h\sim P}\left[e^{-N\beta L(h,S)}\right]\right]$$
$$\geq \ln \mathrm{E}_{h\sim P}\left[e^{-N\beta\, \mathrm{E}_S[L(h,S)]}\right]$$
$$= \ln \mathrm{E}_{h\sim P}\left[e^{-N\beta\, L(h,D)}\right]$$
$$= \ln \ddot{Z}_\beta.$$

We now have

$$\mathrm{E}_S\left[\mathcal{D}(Q_\beta(S),\ddot{Q}_\beta)\right] \leq N\beta\, \mathrm{E}_{S,\,h\sim Q_\beta(S)}\left[L(h,D)-L(h,S)\right]. \quad (10.16)$$

Let L_S abbreviate $\mathrm{E}_S\left[L(Q_\beta(S),S)\right]$ and L_D abbreviate $\mathrm{E}_S\left[L(Q_\beta(S),D)\right]$. From (10.15) and (10.16) we get

$$L_D \leq \frac{1}{1 - \frac{1}{2\lambda}} (L_S + \lambda \beta (L_D - L_S))$$

$$\left(1 - \frac{1}{2\lambda} - \lambda \beta\right) L_D \leq (1 - \lambda \beta) L_S$$

$$\left(1 - \frac{1}{2\lambda(1 - \lambda \beta)}\right) L_D \leq L_S.$$

Setting $\lambda = 1/2\beta$ we get the following for $\beta \in (0, 1/2)$:

$$E_S\left[L(Q_\beta(S), D)\right] \leq \frac{1}{1 - 2\beta} E_S\left[L(Q_\beta(S), S)\right]. \qquad (10.17)$$

As β get larger (as the temperature gets colder) the distribution $Q_\beta(S)$ becomes more closely fit to the training sample S and $E_S\left[L(Q_\beta(S), S)\right]$ gets smaller. But the leading factor $1/(1 - 2\beta)$ prevents β from getting too large. There is no point in making β very small—restricting β to be larger than $1/200$ results in at most a 1 % weakening of the bound. Hence we can take β to $\Omega(1)$ independent of N.

10.5 Discussion

PAC-Bayesian theory is a frequentist approach to machine learning which bases learner bias on a prior probability. Here we have discussed the fundamental equations with an emphasis on Catoni's localization methods. Although localization was introduced almost immediately into PAC-Bayesian theory, and is intuitively important, it has been difficult to unambiguously demonstrate its value either theoretically or empirically. It seems that more theoretical work in this direction is warranted.

References

1. Boser, B.E., Guyon, I.M., Vapnik, V.N.: A training algorithm for optimal margin classifiers. In: Proceedings of the 5th Annual Workshop on Computational Learning Theory, Pittsburgh, pp. 144–152. ACM (1992)
2. Catoni, O.: PAC-Bayesian supervised classification: the thermodynamics of statistical learning. arXiv:0712.0248 (2007, preprint)
3. Germain, P., Lacasse, A., Laviolette, F., Marchand, M.: PAC-Bayesian learning of linear classifiers. In: Proceedings of the 26th Annual International Conference on Machine Learning, Montreal, pp. 353–360. ACM (2009)
4. Hoeffding, W.: Probability inequalities for sums of bounded random variables. J. Am. Stat. Assoc. **58**(301), 13–30 (1963)

5. Langford, J.: Tutorial on practical prediction theory for classification. J. Mach. Learn. Res. **6**(1), 273 (2006)
6. Langford, J., Blum, A.: Microchoice bounds and self bounding learning algorithms. In: Proceedings of the 12th Annual Conference on Computational Learning Theory, Santa Cruz, pp. 209–214. ACM (1999)
7. Lever, G., Laviolette, F., Shawe-Taylor, J.: Distribution-dependent PAC-Bayes priors. In: Algorithmic Learning Theory, pp. 119–133. Springer, Berlin/Heidelberg (2010)
8. Maurer, A.: A note on the PAC Bayesian theorem. arXiv:cs/0411099 (2004, preprint)
9. McAllester, D.A.: PAC-Bayesian model averaging. In: COLT, Santa Cruz, pp. 164–170 (1999)
10. McAllester, D.: Simplified PAC-Bayesian margin bounds. In: Learning Theory and Kernel Machines, Washington, DC, pp. 203–215 (2003)
11. McAllester, D., Keshet, J.: Generalization bounds and consistency for latent structural probit and ramp loss. In: Proceedings of the 25th Annual Conference on Neural Information Processing Systems (NIPS), Granada (2011)
12. Seeger, M.: PAC-Bayesian generalisation error bounds for Gaussian process classification (English). J. Mach. Learn. Res. **3**(2), 233–269 (2003)
13. Shawe-Taylor, J., Langford, J.: PAC-Bayes & margins. In: Advances in Neural Information Processing Systems 15: Proceedings of the 2002 Conference, Vancouver, vol. 15, p. 439. MIT (2003)
14. Valiant, L.G.: A theory of the learnable. Commun. ACM **27**(11), 1134–1142 (1984)
15. Vapnik, V.N., Chervonenkis, A.Y.: On the uniform convergence of relative frequencies of events to their probabilities. Theory Probab. Appl. **16**(2), 264–280 (1971)

Chapter 11
Kernel Ridge Regression

Vladimir Vovk

Abstract This chapter discusses the method of Kernel Ridge Regression, which is a very simple special case of Support Vector Regression. The main formula of the method is identical to a formula in Bayesian statistics, but Kernel Ridge Regression has performance guarantees that have nothing to do with Bayesian assumptions. I will discuss two kinds of such performance guarantees: those not requiring any assumptions whatsoever, and those depending on the assumption of randomness.

11.1 Introduction

This chapter is based on my talk at the Empirical Inference Symposium (see p. x). It describes some developments influenced by Vladimir Vapnik, which are related to, but much less well known than, the Support Vector Machine. The Support Vector Machine is a powerful combination of the idea of generalized portrait (1962; see Chap. 3) and the kernel methods, and from the very beginning the performance guarantees for it were non-Bayesian, depending only on the *assumption of randomness*: the data are generated independently from the same probability distribution. An example of such a performance guarantee is (3.2); numerous other examples are given in Vapnik [15]. Kernel Ridge Regression (KRR) is a special case of Support Vector Regression, which has been known in Bayesian statistics for a long time. However, the advent of the Support Vector Machine encouraged non-Bayesian analyses of KRR, and this chapter presents two examples of such analyses. The first example is in the tradition of prediction with expert advice

V. Vovk (✉)
Dept. of Computer Science, Royal Holloway, University of London, Egham, Surrey,
United Kingdom
e-mail: v.vovk@rhul.ac.uk

[3] and involves no statistical assumptions whatsoever. The second example belongs to the area of conformal prediction [17] and only depends on the assumption of randomness.

11.2 Kernel Ridge Regression

It appears that the term "Kernel Ridge Regression" was coined in 2000 by Cristianini and Shawe-Taylor [5] to refer to a simplified version of Support Vector Regression; this was an adaptation of the earlier "ridge regression in dual variables" [12]. Take the usual Support Vector Regression in primal variables

$$\text{minimize} \quad \|w\|^2 + C \sum_{t=1}^{T} \left((\xi_t)^k + (\xi_t')^k \right)$$

$$\text{subject to} \quad (w \cdot x_t + b) - y_t \leq \epsilon + \xi_t, \quad t = 1, \ldots, T,$$
$$y_t - (w \cdot x_t + b) \leq \epsilon + \xi_t', \quad t = 1, \ldots, T,$$
$$\xi_t, \xi_t' \geq 0, \quad t = 1, \ldots, T,$$

where $(x_t, y_t) \in \mathbb{R}^n \times \mathbb{R}$ are the training examples, w is the weight vector, b is the bias term, ξ_t, ξ_t' are the slack variables, and T is the size of the training set; $\epsilon, C > 0$ and $k \in \{1, 2\}$ are the parameters. Simplify the problem by ignoring the bias term b (it can be partially recovered by adding a dummy attribute 1 to all x_t), setting $\epsilon := 0$, and setting $k := 2$. The optimization problem becomes

$$\text{minimize} \quad a \|w\|^2 + \sum_{t=1}^{T} (y_t - w \cdot x_t)^2$$

(where $a := 1/C$), the usual Ridge Regression problem. And Vapnik's usual method ([15], Sect. 11.3.2) then gives the prediction

$$\hat{y} = \hat{w} \cdot x = Y'(K + aI)^{-1} k \qquad (11.1)$$

for the label of a new object x, where Y is the vector of labels (with components $Y^t := y_t$), K is the Gram matrix $K_{s,t} := x_s \cdot x_t$, and k is the vector with components $k^t := x_t \cdot x$. The kernel trick replaces x_t by $F(x_t)$, and so K by the kernel matrix $K_{s,t} := \mathcal{K}(x_s, x_t)$ and k by the vector $k^t := \mathcal{K}(x_t, x)$, where $\mathcal{K}$ is the kernel $\mathcal{K}(x, x') := F(x) \cdot F(x')$.

This simple observation was made in [12], where this simplified SVR method was called "ridge regression in dual variables". There is no doubt that this calculation has been done earlier as well, but the result does not appear useful. First, compared to the "full" SVM, there is no sparsity of examples (and there is no sparsity in attributes, as in the case of the Lasso). Having an explicit formula is an advantage, but the formula is not new: mathematically, the formula for KRR coincides with one of the formulas in kriging [4], an old method in geostatistics for predicting values of a Gaussian random field; this formula had been widely used in Bayesian statistics.

However, there is a philosophical and practical difference:

- In kriging, the kernel is estimated from the results of observations and in Bayesian statistics it is supposed to reflect the statistician's beliefs;
- In KRR, as in Support Vector Regression in general, the kernel is not supposed to reflect any knowledge or beliefs about reality, and the usual approach is pragmatic: one consults standard libraries of kernels and uses whatever works.

In the remaining sections of this chapter we will explore KRR in the SVM style, without making Bayesian assumptions. The practical side of this non-Bayesian aspect of KRR is that it often gives good results on real-world data, despite the Bayesian assumptions being manifestly wrong. We will, however, concentrate on its theoretical side: non-Bayesian performance guarantees for KRR.

An important special case of KRR is (ordinary) Ridge Regression (RR): it is a special case (as far as the output is concerned) of KRR for $\mathcal{K}$ as the dot product. However, in the case of RR the usual representation of the prediction is

$$\hat{y} = \hat{w} \cdot x = x'(X'X + aI)^{-1}X'Y \qquad (11.2)$$

rather than (11.1), where X is the matrix whose rows are $x'_1, \ldots, x'_T$; there are many ways to show that (11.2) and (11.1) indeed coincide when $\mathcal{K}$ is the dot product.

Under a standard Bayesian assumption (which we do not state explicitly in general; see, e.g., [17], Sect. 10.3), the conditional distribution of the label y of a new example (x, y) given $x_1, \ldots, x_T, x$ and $y_1, \ldots, y_T$ is

$$N\left(Y'(K + aI)^{-1}k, \sigma^2 + \frac{\sigma^2}{a}\mathcal{K}(x,x) - \frac{\sigma^2}{a}k'(K + aI)^{-1}k\right), \qquad (11.3)$$

where K and k are as before (the postulated probability distribution generating the examples depends on $\mathcal{K}$ and a, and we parameterize a normal probability distribution $N(\mu, \sigma^2)$ by its mean μ and standard deviation σ^2). The mean of the distribution (11.3) is the KRR prediction, but now we have not only a point prediction but also an estimate of its accuracy.

When $\mathcal{K}$ is the dot product, (11.3) can be rewritten as

$$N\left(x'(X'X + aI)^{-1}X'Y, \sigma^2 x'(X'X + aI)^{-1}x + \sigma^2\right). \qquad (11.4)$$

In this case the Bayesian assumption can be stated as follows: $x_1, x_2, \ldots$ are fixed vectors in $\mathbb{R}^n$ (alternatively, we can make our analysis conditional on their values) and

$$y_t = \theta \cdot x_t + \xi_t, \tag{11.5}$$

where $\theta \sim N(0, (\sigma^2/a)I)$ and $\xi_t \sim N(0, \sigma^2)$ are all independent.

Equations (11.3) and (11.4) give exhaustive information about the next observation; the Bayesian assumption, however, is rarely satisfied.

11.3 Kernel Ridge Regression Without Probability

It turns out that KRR has interesting performance guarantees even if we do not make any stochastic assumptions whatsoever. Due to lack of space no proofs will be given; they can be found in the technical report [21].

In this section we consider the following perfect-information protocol of on-line regression:

Protocol 1 On-line regression protocol

   for $t := 1, 2, \ldots$ **do**
      Reality announces $x_t \in \mathbf{X}$
      Learner predicts $\hat{y}_t \in \mathbb{R}$
      Reality announces $y_t \in \mathbb{R}$
   end for

First we consider the case where the space $\mathbf{X}$ from which the objects x_t are drawn is a Euclidean space, $\mathbf{X} := \mathbb{R}^n$, and our goal is to compete with linear functions; in this case ordinary Ridge Regression is a suitable strategy for Learner. Then we move on to the case of an arbitrary $\mathbf{X}$ and replace RR by KRR.

11.3.1 Ordinary Ridge Regression

In this section, $\mathbf{X} = \mathbb{R}^n$. The RR strategy for Learner is given by the formula $\hat{y}_t := b'_{t-1} A_{t-1}^{-1} x_t$, where $b_0, b_1, \ldots$ is the sequence of vectors and $A_0, A_1, \ldots$ is the sequence of matrices defined by

$$b_T := \sum_{t=1}^{T} y_t x_t, \qquad A_T := aI + \sum_{t=1}^{T} x_t x'_t$$

(cf. (11.2)), where $a > 0$ is a parameter. The incremental update of the matrix A_t^{-1} can be done effectively by the Sherman–Morrison formula. The following performance guarantee is proved in [21], Sect. 2.

11 Kernel Ridge Regression

Theorem 11.1. *The Ridge Regression strategy for Learner with parameter $a > 0$ satisfies, at any step T,*

$$\sum_{t=1}^{T} \frac{(y_t - \hat{y}_t)^2}{1 + x_t' A_{t-1}^{-1} x_t} = \min_{\theta \in \mathbb{R}^n} \left(\sum_{t=1}^{T} (y_t - \theta' x_t)^2 + a\|\theta\|^2 \right). \quad (11.6)$$

The part $x_t' A_{t-1}^{-1} x_t$ in the denominator of (11.6) is usually close to 0 for large t.

Theorem 11.1 has been adapted to the Bayesian setting by Zhdanov and Kalnishkan [20], who also notice that it can be extracted from [1] (by summing their (4.21) in an exact rather than an estimated form).

Theorem 11.1 and its kernel version (Theorem 11.2 below) imply surprisingly many well-known inequalities.

Corollary 11.1. *Assume $|y_t| \leq \mathbf{y}$ for all t, clip the predictions of the Ridge Regression strategy to $[-\mathbf{y}, \mathbf{y}]$, and denote them by $\hat{y}_t^{\mathbf{y}}$. Then*

$$\sum_{t=1}^{T} (y_t - \hat{y}_t^{\mathbf{y}})^2 \leq \min_{\theta} \left(\sum_{t=1}^{T} (y_t - \theta' x_t)^2 + a\|\theta\|^2 \right)$$

$$+ 4\mathbf{y}^2 \ln \det \left(I + \frac{1}{a} \sum_{t=1}^{T} x_t x_t' \right). \quad (11.7)$$

The bound (11.7) is exactly the bound obtained in [16] (Theorem 4) for the algorithm merging linear experts with predictions clipped to $[-\mathbf{y}, \mathbf{y}]$, which does not have a closed-form description and so is less interesting than clipped RR. The bound for the strategy called the AAR in [16] has $\mathbf{y}^2$ in place of $4\mathbf{y}^2$ ([16], Theorem 1). (The AAR is very similar to RR: its predictions are $b_{t-1}' A_t^{-1} x_t$ rather than $b_{t-1}' A_{t-1}^{-1} x_t$; it is called the Vovk–Azoury–Warmuth algorithm in [3].) The regret term in (11.7) has the logarithmic order in T if $\|x_t\|_\infty \leq X$ for all t, because

$$\ln \det \left(I + \frac{1}{a} \sum_{t=1}^{T} x_t x_t' \right) \leq n \ln \left(1 + \frac{TX^2}{a} \right) \quad (11.8)$$

(the determinant of a positive definite matrix is bounded by the product of its diagonal elements; see [2], Chap. 2, Theorem 7). From Theorem 11.1 we can also deduce Theorem 11.7 in [3], which is somewhat similar to Corollary 11.1. That theorem implies (11.7) when RR's predictions happen to be in $[-\mathbf{y}, \mathbf{y}]$ without clipping (but this is not what Corollary 11.1 asserts).

RR is not as good as the AAR in the setting where $\sup_t |y_t| \leq \mathbf{y}$ and the goal is to obtain bounds of the form (11.7) (since the AAR is to some degree optimized for this setting), but is still very good; and we can achieve an interesting equality (rather than inequality) for it.

The upper bound (11.7) does not hold for the RR strategy if the coefficient 4 is replaced by any number less than $\frac{3}{2\ln 2} \approx 2.164$, as can be seen from an example given in Theorem 3 [16], where the left-hand side of (11.7) is $4T + o(T)$, the minimum on the right-hand side is at most T, $\mathbf{y} = 1$, and the logarithm is $2T \ln 2 + O(1)$. It is also known that there is no strategy achieving (11.7) with the coefficient less than 1 instead of 4, even in the case where $\|x_t\|_\infty \le X$ for all t: see Theorem 2 in [16].

There is also an upper bound on the cumulative square loss of the RR strategy without a logarithmic part and without assuming that the labels are bounded.

Corollary 11.2. *If $\|x_t\|_2 \le Z$ for all t then the Ridge Regression strategy for Learner with parameter $a > 0$ satisfies, at any step T,*

$$\sum_{t=1}^T (y_t - \hat{y}_t)^2 \le \left(1 + \frac{Z^2}{a}\right) \min_{\theta \in \mathbb{R}^n} \left(\sum_{t=1}^T (y_t - \theta' x_t)^2 + a\|\theta\|^2\right).$$

This bound is better than the bound in Corollary 3.1 of [8], which has an additional regret term of the logarithmic order in time.

Asymptotic properties of the RR strategy can be further studied using Corollary A.1 of Kumon et al. [9]. Kumon et al.'s result states that when $\|x_t\|_2 \le 1$ for all t, then $x_t' A_{t-1}^{-1} x_t \to 0$ as $t \to \infty$. It is clear that we can replace $\|x_t\|_2 \le 1$ for all t by $\sup_t \|x_t\|_2 < \infty$. This gives the following corollary, which can be summarized as follows. If there exists a very good expert (asymptotically), then RR also predicts very well. If there is no such very good expert, RR performs asymptotically as well as the best regularized expert.

Corollary 11.3. *Let $a > 0$ and $\hat{y}_t$ be the predictions output by the Ridge Regression strategy with parameter a. Suppose $\sup_t \|x_t\|_2 < \infty$. Then*

$$\left(\exists \theta \in \mathbb{R}^n : \sum_{t=1}^\infty (y_t - \theta' x_t)^2 < \infty\right) \implies \sum_{t=1}^\infty (y_t - \hat{y}_t)^2 < \infty$$

and

$$\left(\forall \theta \in \mathbb{R}^n : \sum_{t=1}^\infty (y_t - \theta' x_t)^2 = \infty\right)$$

$$\implies \lim_{T \to \infty} \frac{\sum_{t=1}^T (y_t - \hat{y}_t)^2}{\min_{\theta \in \mathbb{R}^n} \left(\sum_{t=1}^T (y_t - \theta' x_t)^2 + a\|\theta\|^2\right)} = 1.$$

11.3.2 Kernel Ridge Regression

In this section, **X** is an arbitrary set. Let $\mathcal{F}$ be the RKHS with kernel $\mathcal{K}$ of functions on **X**. The KRR strategy for Learner with parameter $a > 0$ is defined by the formula (11.1) applied to the past examples.

The following version of Theorem 11.1 for KRR can be derived from Theorem 11.1 itself (see [21], Sect. 6, for details).

Theorem 11.2. *The KRR strategy with parameter $a > 0$ for Learner satisfies, at any step T,*

$$\sum_{t=1}^{T} \frac{(y_t - \hat{y}_t)^2}{1 + \frac{1}{a}\mathcal{K}(x_t, x_t) - \frac{1}{a}k_t'(K_{t-1} + aI)^{-1}k_t}$$
$$= \min_{f \in \mathcal{F}} \left(\sum_{t=1}^{T} (y_t - f(x_t))^2 + a\|f\|_\mathcal{F}^2 \right).$$

The denominator on the left-hand side tends to 1 under some regularity conditions:

Lemma 11.1 ([20], Lemma 2). *Let $\mathcal{K}$ be a continuous kernel on a compact metric space. Then*

$$\mathcal{K}(x_t, x_t) - k_t'(K_{t-1} + aI)^{-1}k_t \to 0 \text{ as } t \to \infty.$$

Again, we can derive several interesting corollaries from Theorem 11.2.

Corollary 11.4. *Assume $|y_t| \leq \mathbf{y}$ for all t and let $\hat{y}_t^\mathbf{y}$ be the predictions of the KRR strategy clipped to $[-\mathbf{y}, \mathbf{y}]$. Then*

$$\sum_{t=1}^{T} (y_t - \hat{y}_t^\mathbf{y})^2 \leq \min_{f \in \mathcal{F}} \left(\sum_{t=1}^{T} (y_t - f(x_t))^2 + a\|f\|_\mathcal{F}^2 \right)$$
$$+ 4\mathbf{y}^2 \ln \det \left(I + \frac{1}{a} K_T \right). \quad (11.9)$$

But now we have a problem: in general, the ln det term is not small compared to T. However, we still have the analogue of Corollary 11.3 (for a detailed derivation, see [20]).

Corollary 11.5 ([20], Corollary 4). *Let **X** be a compact metric space and $\mathcal{K}$ be a continuous kernel on **X**. Then*

$$\left(\exists f \in \mathcal{F} : \sum_{t=1}^{\infty} (y_t - f(x_t))^2 < \infty \right) \Longrightarrow \sum_{t=1}^{\infty} (y_t - \hat{y}_t)^2 < \infty$$

and

$$\left(\forall f \in \mathcal{F} : \sum_{t=1}^{\infty}(y_t - f(x_t))^2 = \infty\right)$$

$$\implies \lim_{T \to \infty} \frac{\sum_{t=1}^{T}(y_t - \hat{y}_t)^2}{\min_{f \in \mathcal{F}}\left(\sum_{t=1}^{T}(y_t - f(x_t))^2 + a\|f\|_{\mathcal{F}}^2\right)} = 1.$$

To obtain a non-asymptotic result of this kind under the assumption $\sup_t |y_t| \leq \mathbf{y}$, let us first assume that the number of steps T is known in advance. We will need the notation $c_{\mathcal{F}} := \sqrt{\sup_{x \in \mathbf{X}} \mathcal{K}(x, x)}$. Bounding the logarithm of the determinant in (11.9) we have

$$\ln \det\left(I + \frac{1}{a}\mathbf{K}_T\right) \leq T \ln\left(1 + \frac{c_{\mathcal{F}}^2}{a}\right)$$

(cf. (11.8)). Since we know the number T of steps in advance, we can choose a specific value for a; let $a := c_{\mathcal{F}}\sqrt{T}$. This gives us an upper bound with the regret term of the order $O(\sqrt{T})$ for any $f \in \mathcal{F}$:

$$\sum_{t=1}^{T}(y_t - \hat{y}_t^{\mathbf{y}})^2 \leq \sum_{t=1}^{T}(y_t - f(x_t))^2 + c_{\mathcal{F}}(\|f\|_{\mathcal{F}}^2 + 4\mathbf{y}^2)\sqrt{T}.$$

If we do not know the number of steps in advance, it is possible to achieve a similar bound using a mixture of KRR over the parameter a with a suitable prior over a:

$$\sum_{t=1}^{T}(y_t - \hat{y}_t^{\mathbf{y}})^2 \leq \sum_{t=1}^{T}(y_t - f(x_t))^2 + 8\mathbf{y} \max\left(c_{\mathcal{F}}\|f\|_{\mathcal{F}}, \mathbf{y}\delta T^{-1/2+\delta}\right)\sqrt{T+2}$$

$$+ 6\mathbf{y}^2 \ln T + c_{\mathcal{F}}^2 \|f\|_{\mathcal{F}}^2 + O(\mathbf{y}^2) \quad (11.10)$$

for any arbitrarily small $\delta > 0$, where the constant implicit in $O(\mathbf{y}^2)$ depends only on δ. (No proof of this result has been published.) The inequality (11.10) still looks asymptotic in that it contains an O term; however, it is easy to obtain an explicit (but slightly messier) non-asymptotic inequality.

In particular, (11.10) shows that if $\mathcal{K}$ is a universal kernel [14] on a topological space $\mathbf{X}$, KRR is competitive with all continuous functions on $\mathbf{X}$: for any continuous $f : \mathbf{X} \to \mathbb{R}$,

$$\limsup_{T \to \infty} \frac{1}{T}\left(\sum_{t=1}^{T}(y_t - \hat{y}_t^{\mathbf{y}})^2 - \sum_{t=1}^{T}(y_t - f(x_t))^2\right) \leq 0 \quad (11.11)$$

(assuming $|y_t| \leq \mathbf{y}$ for all t). For example, (11.11) holds for $\mathbf{X}$ a compact set in $\mathbb{R}^n$, $\mathcal{K}$ an RBF kernel, and $f : \mathbf{X} \to \mathbb{R}$ any continuous function (see Example 1 in [14]). For continuous universal kernels on compact spaces, (11.11) also follows from Corollary 11.5.

11.4 Kernel Ridge Regression in Conformal Prediction

Suppose we would like to have prediction intervals rather than point predictions, and we would like them to have guaranteed coverage probabilities. It is clear that to achieve this we need a stochastic assumption; it turns out that the randomness assumption is often sufficient to obtain informative prediction intervals. In general, our algorithms will output prediction sets (usually intervals, but not always); to obtain prediction intervals we will apply convex closure (which can only improve coverage probability).

The special case of conformal prediction discussed in this section works as follows. Suppose we have an "underlying algorithm" (such as KRR) producing point predictions in $\mathbb{R}$. Let $(x_1, y_1), \ldots, (x_T, y_T)$ be a training set and x_{T+1} be a new object. To find the prediction set for y_{T+1} at a significance level $\epsilon \in (0, 1)$:

- For each possible label $z \in \mathbb{R}$:

    - Set $y_{T+1} := z$;
    - For each $t \in \{1, \ldots, T+1\}$ compute the *nonconformity score* $\alpha_t^z := |y_t - \hat{y}_t^z|$, where $\hat{y}_t^z$ is the point prediction for the label of x_t computed by the underlying algorithm from the extended training set $(x_1, y_1), \ldots, (x_{T+1}, y_{T+1})$;
    - Compute the p-value

    $$p(z) := \frac{1}{T+1} \sum_{t=1}^{T+1} \mathbf{1}_{\{\alpha_t^z \geq \alpha_{T+1}^z\}},$$

    where $\mathbf{1}_{\{\cdot\}}$ is the indicator function;

- Output the prediction set $\{z \in \mathbb{R} \mid p(z) > \epsilon\}$, where ϵ is the given significance level.

This set predictor is the *conformal predictor* based on the given underlying algorithm. Conformal predictors have a guaranteed coverage probability:

Theorem 11.3. *The probability that the prediction set output by a conformal predictor is an error (i.e., fails to include y_{T+1}) does not exceed the significance level ϵ.*

Moreover, in the on-line prediction protocol (Protocol 1, in which Reality outputs (x_t, y_t) independently from the same probability distribution), the long-run frequency of errors also does not exceed ϵ almost surely. For a proof, see [17] (Theorem 8.1).

The property of conformal predictors asserted by Theorem 11.3 is their *validity*. Validity being achieved automatically, the remaining desiderata for conformal predictors are their "efficiency" (we want the prediction sets to be small, in a suitable sense) and "conditional validity" (we might want to have prespecified coverage probabilities conditional on the training set or some property of the new example).

The idea of conformal prediction is inspired by the Support Vector Machine (and the notation α for nonconformity scores is adapted from Vapnik's Lagrange multipliers). The immediate precursor of conformal predictors was described in the paper [7] co-authored by Vapnik, which is based on the idea that a small number of support vectors warrants a high level of confidence in the SVM's prediction. This idea was present in Vapnik and Chervonenkis's thinking in the 1960s: see, e.g., (3.2) and [15], Theorem 10.5. The method was further developed in [17]; see [13] for a tutorial.

In the case where the conformal predictor is built on top of RR or KRR, there is no need to go over all potential labels $z \in \mathbb{R}$. The set prediction for the example (x_{T+1}, y_{T+1}) can be computed in time $O(T \log T)$ (in the case of RR) or $O(T^2)$ (in the case of KRR). This involves only solving linear equations and sorting; the simple resulting algorithm is called the Ridge Regression Confidence Machine (RRCM) in [11] and [17]. There is an R package implementing the RRCM (in the case of RR), PredictiveRegression, available from CRAN.

The Bayes predictions (11.3) and (11.4) can be easily converted into prediction intervals. But they are valid only under the postulated probability model, whereas the prediction intervals output by the RRCM are valid under the randomness assumption (as is common in machine learning). This is illustrated by Fig. 11.1, which is a version of Wasserman's Fig. 1 in [19]. We consider the simplest case, where $x_t = 1$ for all t; therefore, the examples (x_t, y_t) can be identified with their labels $y_t \in \mathbb{R}$, which we will call *observations*. The chosen significance level is 20 % and the kernel $\mathcal{K}$ is the dot product. In the top plot, the four observations are generated from $N(1, 1)$; in the middle plot, from $N(10, 1)$; and in the bottom plot, from $N(100, 1)$. The blue lines are the prediction intervals computed by the RRCM with $a = 1$ and the red lines are the Bayes prediction intervals computed as the shortest intervals containing 80 % of the mass (11.4) with $a = 1$ and $\sigma = 1$.

All observations are generated from $N(\theta, 1)$ for various constants θ. When $\theta = 1$ (and so the Bayesian assumption (11.5) can be regarded as satisfied), the Bayes prediction intervals are on average only slightly shorter than the RRCM's (the Bayes prediction interval happens to be wider in Fig. 11.1; for a random seed of the random number generator, the Bayes prediction intervals are shorter in about 54 % of cases). But as θ grows, the RRCM's prediction intervals also grow (in order to cover the observations), whereas the width of the Bayes prediction intervals remains constant. When $\theta = 100$ (and so (11.5) is clearly violated), the Bayes prediction intervals give very misleading results.

11 Kernel Ridge Regression

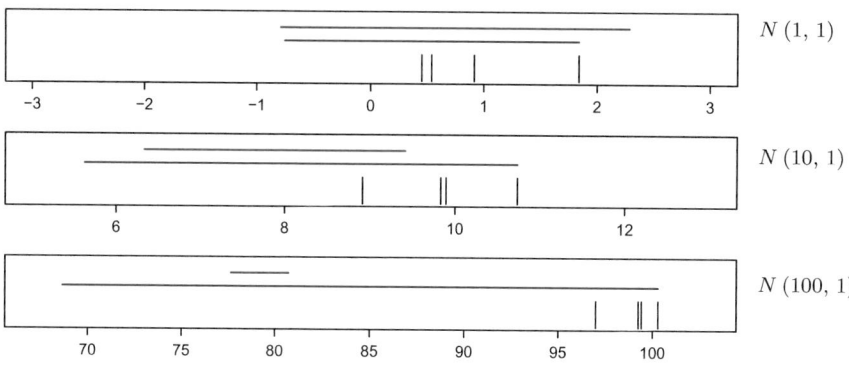

Fig. 11.1 In the *top plot*, the four observations (shown as *short vertical lines*) are generated from $N(1, 1)$; in the *middle plot*, from $N(10, 1)$; and in the *bottom plot*, from $N(100, 1)$. The *blue lines* are prediction intervals computed by a conformal predictor, and the *red lines* are Bayes prediction intervals

In parametric statistics, it is widely believed that the choice of the prior does not matter much: the data will eventually swamp the prior. However, even in parametric statistics the model (such as $N(\theta, 1)$) itself may be wrong.

In nonparametric statistics, the situation is much worse:

the prior can swamp the data, no matter how much data you have

(Diaconis and Freedman [6], Sect. 4). In this case, using Bayes prediction intervals becomes particularly problematic. The RRCM can be interpreted as an example of *renormalized Bayes*, as discussed in [18] and later papers.

As mentioned earlier, the RRCM is valid under the assumption of randomness; no further assumptions are required. However, conditional validity and, especially, efficiency do require extra assumptions. For example, [10] uses standard statistical assumptions used in density estimation to demonstrate the conditional validity and efficiency of a purpose-built conformal predictor. It remains an open problem to establish whether similar results hold for the RRCM.

Acknowledgements I am deeply grateful to Vladimir Vapnik for numerous discussions and support over the years, starting from our first meetings in the summer of 1996. Many thanks to Alexey Chervonenkis, Alex Gammerman, Valya Fedorova, and Ilia Nouretdinov for their advice and help. This work has been supported in part by the Cyprus Research Promotion Foundation (TPE/ORIZO/0609(BIE)/24) and EPSRC (EP/K033344/1).

References

1. Azoury, K.S., Warmuth, M.K.: Relative loss bounds for on-line density estimation with the exponential family of distributions. Mach. Learn. **43**, 211–246 (2001)
2. Beckenbach, E.F., Bellman, R.: Inequalities. Springer, Berlin (1965)

3. Cesa-Bianchi, N., Lugosi, G.: Prediction, Learning, and Games. Cambridge University Press, Cambridge (2006)
4. Cressie, N.: The origins of kriging. Math. Geol. **22**, 239–252 (1990)
5. Cristianini, N., Shawe-Taylor, J.: An Introduction to Support Vector Machines and Other Kernel-Based Methods. Cambridge University Press, Cambridge (2000)
6. Diaconis, P., Freedman, D.: On the consistency of Bayes estimates (with discussion). Ann. Stat. **14**, 1–67 (1986)
7. Gammerman, A., Vovk, V., Vapnik, V.: Learning by transduction. In: Proceedings of the Fourteenth Conference on Uncertainty in Artificial Intelligence, Madison, pp. 148–155. Morgan Kaufmann, San Francisco (1998)
8. Kakade, S.M., Ng, A.Y.: Online bounds for Bayesian algorithms. In: Proceedings of the Eighteenth Annual Conference on Neural Information Processing Systems, Vancouver (2004)
9. Kumon, M., Takemura, A., Takeuchi, K.: Sequential optimizing strategy in multi-dimensional bounded forecasting games. Stoch. Process. Appl. **121**, 155–183 (2011)
10. Lei, J., Wasserman, L.: Distribution free prediction bands. Tech. Rep. `arXiv:1203.5422` [`stat.ME`], `arXiv.org` e-Print archive (2012). To appear in the Journal of the Royal Statistical Society B
11. Nouretdinov, I., Melluish, T., Vovk, V.: Ridge regression confidence machine. In: Proceedings of the Eighteenth International Conference on Machine Learning, Williamstown, pp. 385–392. Morgan Kaufmann, San Francisco (2001)
12. Saunders, C., Gammerman, A., Vovk, V.: Ridge regression learning algorithm in dual variables. In: Shavlik, J.W. (ed.) Proceedings of the Fifteenth International Conference on Machine Learning, Madison, pp. 515–521. Morgan Kaufmann, San Francisco (1998)
13. Shafer, G., Vovk, V.: A tutorial on conformal prediction. J. Mach. Learn. Res. **9**, 371–421 (2008)
14. Steinwart, I.: On the influence of the kernel on the consistency of support vector machines. J. Mach. Learn. Res. **2**, 67–93 (2001)
15. Vapnik, V.N.: Statistical Learning Theory. Wiley, New York (1998)
16. Vovk, V.: Competitive on-line statistics. Int. Stat. Rev. **69**, 213–248 (2001)
17. Vovk, V., Gammerman, A., Shafer, G.: Algorithmic Learning in a Random World. Springer, New York (2005)
18. Wasserman, L.: Frequentist Bayes is objective (comment on articles by Berger and by Goldstein). Bayesian Anal. **1**, 451–456 (2006)
19. Wasserman, L.: Frasian inference. Stat. Sci. **26**, 322–325 (2011)
20. Zhdanov, F., Kalnishkan, Y.: An identity for kernel ridge regression. Theor. Comput. Sci. **473**, 157–178 (2013)
21. Zhdanov, F., Vovk, V.: Competing with Gaussian linear experts. Tech. Rep. `arXiv:0910.4683` [`cs.LG`], `arXiv.org` e-Print archive (2009). Revised in May 2010

Chapter 12
Multi-task Learning for Computational Biology: Overview and Outlook

Christian Widmer, Marius Kloft, and Gunnar Rätsch

Abstract We present an overview of the field of regularization-based multi-task learning, which is a relatively recent offshoot of statistical machine learning. We discuss the foundations as well as some of the recent advances of the field, including strategies for learning or refining the measure of task relatedness. We present an example from the application domain of Computational Biology, where multi-task learning has been successfully applied, and give some practical guidelines for assessing a priori, for a given dataset, whether or not multi-task learning is likely to pay off.

C. Widmer (✉)
Computational Biology Center, Memorial Sloan-Kettering Cancer Center, 415 E 68th Street, New York, NY 10065, USA

Machine Learning Group, Technische Universität Berlin, Franklinstr. 28/29, 10587 Berlin, Germany
e-mail: cwidmer@cbio.mskcc.org

M. Kloft
Computational Biology Center, Memorial Sloan-Kettering Cancer Center, 415 E 68th Street, New York, NY 10065, USA

Courant Institute of Mathematical Sciences, New York University, 251 Mercer Street, New York, NY 10012, USA
e-mail: kloftm@mskcc.org; mkloft@cs.nyu.edu

G. Rätsch
Computational Biology Center, Memorial Sloan-Kettering Cancer Center, 415 E 68th Street, New York, NY 10065, USA
e-mail: raetsch@cbio.mskcc.org

12.1 Introduction

A series of seminal papers has greatly changed the way we view the field of machine learning. In their 1971 paper *On the Uniform Convergence of Relative Frequencies of Events to Their Probabilities* [23], Vapnik and Chervonenkis laid out the foundations of statistical learning theory, which led to the development of support vector machines (SVMs) in the 1990s [6, 9]. Subsequently, with the coming of the information age, machine learning – and computer science in general – has diverged into multiple fascinating and manifold branches, many of which can be traced back to these classical papers, for example, preference learning, multiple kernel learning, structured output learning, and transfer learning, to name just a few subfields. Meanwhile, established machine learning methods such as SVMs have matured to a degree that, nowadays, they are frequently employed out-of-the-box in science and technology, for their favorable generalization performance while maintaining computational feasibility. In this chapter, we present an overview of one of the recent branches of machine learning, that is, *multi-task learning* (MTL) [7, 8], and discuss interesting applications in the domain of Computational Biology.

In science – and in biology in particular – supervised learning methods are often used to model complex mechanisms in order to describe and ultimately understand them. These models have to be rich enough to capture the considerable complexity of these mechanisms, which requires an enormous amount of training data. We frequently observe that the prediction accuracy does not saturate even when employing millions of training data points, which indicates that using even much more data could still help accuracy (cf., e.g., [21]). But how can we obtain this massive amount of training data?

Especially in the biomedical domain, obtaining additional labelled training examples can be either very costly or – e.g., due to technological limitations – even impossible. Multi-task learning overcomes this requirement by incorporating information from several related tasks in order to increase the accuracy of the target task at hand. For example, in genetics, we have a good understanding of how close two organisms are in terms of their evolutionary relationship; this information is summarized in the *tree of life*. Because basic genetic mechanisms tend to be relatively well conserved throughout evolution, we can benefit from combining data from several species for the detection of, for example, splice sites or promoter regions.

The relevance of MTL to Computational Biology goes beyond the setting where we view organisms as tasks; we may also view different tasks corresponding to different related protein variants [13], cell lines, pathways [19], tissues, genes [17], technology platforms such as microarrays, experimental conditions, individuals, and tumour subtypes, to name just a few. In this chapter, we provide an overview of selected MTL approaches that have been successfully applied in Computational Biology. In this respect, our presentation is based on [24], but goes beyond it by covering also some very recent developments that are not yet systematically investigated in biology.

12.2 Multi-task Learning

In this section we describe the problem setting of multi-task learning. We also present particular instances of multi-task learning machines, focusing on formulations that are appealing for Computational Biology. For a detailed overview, see the survey of [18].

12.2.1 Relation to Transfer Learning

Transfer learning very generally refers to learning methods that transfer information from one or multiple source tasks to a target task with the aim of improving the prediction accuracy of the target task. Multi-task learning is a specific branch of transfer learning that is characterized by *simultaneously* learning the prediction models of all given tasks. Typically this is achieved by optimizing a joint learning criterion with respect to the various prediction functions. There exist several general strategies for multi-task learning [18]:

1. *Instance-based transfer*, where data points from different domains are incorporated into the learning problem, typically in combination with some form of re-weighting
2. *Feature representation transfer*, where the instances from the various domains are mapped to a joint feature representation
3. *Parameter transfer*, a form of parametric learning paradigm assuming that closely related tasks should also yield similar parameters in the learning model.

The main focus of this chapter is on parameter transfer, where the parameters of similar tasks are often coupled by a regularizer. This approach is often called *regularization-based multi-task learning*.

12.2.2 Regularization-Based Multi-task Learning

From a historical perspective, regularization-based MTL has its foundation in regularized risk minimization [23] and supervised learning methods such as the Support Vector Machine (SVM) [6, 9] or Logistic Regression. In regularized risk minimization, we aim at computing a model Θ minimizing an objective $J(\Theta)$ consisting of a loss term that captures the error with respect to the training data (X, Y) and a regularizer that penalizes model complexity:

$$J(\Theta) = L(\Theta|X, Y) + R(\Theta).$$

This formulation can easily be generalized to the MTL setting, where we are interested in obtaining several models parameterized by $\Theta_1, \ldots, \Theta_T$, where T is the number of tasks. The above formulation can be extended by introducing an additional regularization term R_{MTL} that penalizes the discrepancy between the individual models:

$$J(\Theta_1, \ldots, \Theta_T) = \sum_{t=1}^{T} L(\Theta_t | X, Y) + \sum_{t=1}^{T} R(\Theta_t) + R_{\text{MTL}}(\Theta_1, \ldots, \Theta_T).$$

A highly relevant and active line of research in this context is finding a good regularizer R_{MTL}. A proven approach to this end is to introduce a parameter matrix Q in the regularizer, giving rise to

$$J(\Theta_1, \ldots, \Theta_T, Q) = \sum_{t=1}^{T} L(\Theta_t | X, Y) + \sum_{t=1}^{T} R(\Theta_t) + R_{\text{MTL}}(\Theta_1, \ldots, \Theta_T | Q).$$

Learning Q from the data is often referred to as *learning the task similarities*.

12.2.2.1 Common Approaches

In the following, we denote the training examples by (x_i, y_i), $i = 1, \ldots, n$, each of which is associated with a task $\tau(i) \in \{1, \ldots, T\}$. We denote the set of indices of training points of the tth task by $I_t := \{i \in \{1, \ldots, n\} : \tau(i) = t\}$ and their number by $n_t := \#I_t$. One of the first works on regularization-based MTL is by Evgeniou and Pontil [11], where at optimization time all parameter vectors are "pulled" towards their average $\bar{w} = \frac{1}{T} \sum_{t=1}^{T} w_t$,

$$\min_{b, w_1, \ldots, w_T} \frac{1}{2} \sum_{t=1}^{T} ||w_t||^2 + \sum_{t=1}^{T} ||w_t - \bar{w}||^2 + C \sum_{i=1}^{n} \ell\left(\langle x_i, w_{\tau(i)} \rangle + b, y_i\right).$$

Here ℓ denotes the hinge loss $\ell(z, y) = \max\{1 - yz, 0\}$. Note that all tasks are treated equally in the above formulation; however, often we are given the a priori information that some tasks are more closely related than others. To penalize the differences between the parameter vectors accordingly, the above setting was extended by Evgeniou et al. [12],

$$\min_{b, w_1, \ldots, w_T} \frac{1}{2} \sum_{t=1}^{T} ||w_t||^2 + \frac{1}{2} \sum_{s=1}^{T} \sum_{t=1}^{T} A_{st} ||w_s - w_t||^2 + C \sum_{i=1}^{n} \ell\left(\langle x_i, w_{\tau(i)} \rangle + b, y_i\right),$$

(12.1)

where the graph adjacency matrix $A = (A_{st})$, captures the task similarities. We can rewrite the above formulation using the graph Laplacian $L = (L_{st})$,

$$\min_{b,w_1,\ldots,w_T} \frac{1}{2}\sum_{t=1}^{T}||w_t||^2 + \sum_{s=1}^{T}\sum_{t=1}^{T} L_{st} w_s^T w_t + C \sum_{i=1}^{n} \ell\left(\langle x_i, w_{\tau(i)}\rangle + b, y_i\right),$$

where $L = D - A$, where $D_{s,t} = \delta_{s,t} \sum_k A_{s,k}$. Finally, it can be shown that this gives rise to the following *multi-task* kernel to be used in the corresponding dual:

$$K((x,s),(z,t)) = H_{st}^+ \cdot K_B(x,z),$$

where the K_B is a kernel defined on examples and $H^+ = (H_{st}^+)$ denotes the pseudo-inverse of $H := I + L$, where I is the identity matrix. A closely related formulation was successfully used in the context of Computational Biology by Jacob and Vert [14], where a kernel on tasks K_T is used instead of the pseudo-inverse, giving rise to

$$K((x,s),(z,t)) = K_T(s,t) \cdot K_B(x,z). \tag{12.2}$$

Note that the corresponding joint feature space between task t and feature vector x can be written as a tensor product $\phi(t,x) = \phi_T(t) \cdot \phi_B(x)$ [14]. A "frustratingly easy" special case of (12.2) is studied in [10] in the context of Domain Adaptation, where $\phi_T(t) = (1,1,0)$ was used as the source task descriptor and $\phi_T(t) = (1,0,1)$ was used for the target task, corresponding to $K_T(s,t) = (1 + \delta_{s,t})$.

12.2.3 Learning Task Similarities

The above exposition assumes that we are a priori given a task similarity measure; but how can we access the relatedness of tasks? Although we are often provided with external information on task relatedness (e.g., an evolutionary history of organisms), the given task similarity measure is not necessarily informative of how tightly tasks should be coupled in the MTL algorithm in order to achieve better predictive performance; therefore we are in need of strategies to automatically learn or adjust the degree of coupling between tasks. In the following, we discuss several approaches to this problem – including our own method, Multi-task Multiple Kernel Learning (MT-MKL) [26].

12.2.3.1 A Simple Approach

Very recently, Blanchard et al. [5] presented a simple method to very generally compute a task similarity matrix A from the data at hand only. Their approach is based on the concept of Hilbert space embedding of probability distributions [22] and consists of two steps: first, computing an average similarity of the examples of a pair of tasks,

$$\tilde{A}_{st} = \frac{1}{n_s n_t} \sum_{i \in I_s, j \in I_t} k(x_i, x_j),$$

and then applying a non-linear transformation such as

$$A_{st} = \left(1 + \tilde{A}_{st}\right)^d.$$

The authors show that under a hierarchical frequentist i.i.d. setup this method enjoys favorable theoretical guarantees such as consistency when $\forall t = 1, \ldots, T : n_t \to \infty$ and $T \to \infty$.

We would like to remark at this point that the parameter d may be selected by cross-validation. Generally, we may choose non-linear transformations of the form $A_{st} = \lambda_1 \cdot exp(\tilde{A}_{s,t}/\lambda_2)$ and select the parameters λ_1, λ_2 per cross-validation.

12.2.3.2 Multi-task Relationship Learning (MTRL)

The authors in [2, 27] propose a convex method of jointly learning a task similarity measure along with the individual parameter vectors. Their method extends the graph-regularized MTL formulation given in (12.1).

$$\min_{W=(w_1,\ldots,w_t),\Omega} tr(W\Omega^{-1}W^T) + C \sum_{i=1}^{n} \ell\left(\langle x_i, w_{\tau(i)}\rangle + b, y_i\right),$$

$$\text{s.t.} \quad tr(\Omega) = 1, \Omega \succeq 0$$

In [27] this formulation is solved by alternatingly optimizing the objective with respect to W and Ω, where in each optimization step, Ω is updated according to

$$\Omega = (W^T W)^{1/2}/tr(W^T W).$$

The above approach is especially appealing when only little a priori information about the task similarities is present. An advantage over the method described in the previous section is that the task similarities and the weights w_t are learned simultaneously so that interactions can be captured well.

12.2.3.3 Multi-task Multiple Kernel Learning (MT-MKL)

In the following section, we present our own approach, MT-MKL. The formulation given below is an extension of the ideas presented in [26]. An in-depth presentation, including a theoretical and empirical analysis as well as details on our large-scale implementation, will be made in a forthcoming journal publication.

Problem 12.1 (Primal MTL problem). Solve

$$\inf_{\theta:\|\theta\|_p \leq 1, W} \frac{1}{2} \sum_{m=1}^{M} \frac{\|W_m\|^2_{Q_m}}{\theta_m} + C \sum_{i=1}^{n} l\left(y_i \sum_{m=1}^{M} \langle w_{m\tau(i)}, \varphi_m(x_i) \rangle \right)$$

where $W = (W_m)_{1 \leq m \leq M}$, $W_m = (w_{m1}, \ldots, w_{mT})$ and

$$\|W_m\|_{Q_m} := \operatorname{tr}(W_m Q_m W_m^\top) = \sqrt{\sum_{s,t=1}^{T} q_{mst} \langle w_{ms}, w_{mt} \rangle}.$$

□

In the above problem, we assume being given a number M of external task similarity measures Q_m and kernel feature maps φ_m, each being associated with a weight θ_m. As in multiple kernel learning [15, 16], we automatically learn these weights, which allows us to fine-tune the given task similarities. In contrast to MTRL, our method is in need of some external information (which is often a reasonable assumption), but has in turn fewer free parameters to be learnt. Furthermore, the above method lets us associate different feature maps ϕ_m with different task similarity measures Q_m, which gives flexibility in encoding prior information.

12.2.3.4 Hierarchical MT-MKL

There are many ways in which the set of Q_m may be chosen. One valid strategy is to define a set of task groups, where information is shared within each group. In the setting of hierarchical task relations (e.g., the evolutionary history between different organisms), these groups come naturally from the inner node of our tree. Tasks corresponds to leaf nodes, or *taxa*, in this context, whereas each inner nodes defines a task group (see Fig. 12.1). Let $G_m :=\{l \mid l \text{ is descendant of } m\}$ be the set of leaves below the sub-tree rooted at node m. Then, we can give the following definition for the hierarchically constructed task adjacency matrix:

$$A_m(s,t) = \begin{cases} 1 & \text{if } s \in G_m \text{ and } t \in G_m \\ 0 & \text{else}. \end{cases}$$

As an example, consider the kernel defined by a hierarchical decomposition according to Fig. 12.1. We seek a non-sparse weighting of the task sets defined by the hierarchy and will therefore use ℓ_2-norm MKL [15].

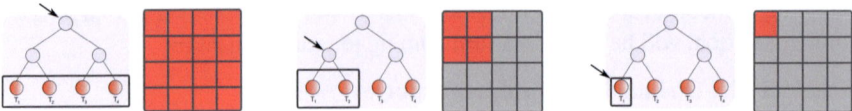

Fig. 12.1 Example of a hierarchical decomposition. According to a simple binary tree, it is shown that each node defines a subset of tasks (a *block* in the corresponding adjacency matrix on the *left*). Here, the decomposition is shown for only one path: the subset defined by the root node contains all tasks, the subset defined by the left inner node contains t_1 and t_2 and the subset defined by the leftmost leaf only contains t_1. Accordingly, each of the remaining nodes defines a subset S_i that contains all descendant tasks

12.2.4 When Does MTL Pay Off?

In this section, we give some practical guidelines about when it is promising to use MTL algorithms. First, the tasks should be neither too similar nor too different [25]. If the tasks are too different, one will not be able to transfer information, and may even end up with *negative transfer* [18]. On the other hand, if tasks are almost identical, it might suffice to pool the training examples and train a single combined classifier. Another integral factor that needs to be considered is whether the problem is *easy* or *hard* with respect to the available training data. In this context, the problem can be considered *easy* if the performance of a classifier saturates as a function of the available training data. In that case using more out-of-domain information will not improve classification performance.

In order to investigate the problem difficulty in the sense defined above, we can compute a learning curve (e.g., auROC as a function of the number of training examples). If the curve saturates when n is large, this indicates that multi-task learning will not considerably help performance, as model performance is most likely limited only by label noise. The same methodology can be employed to empirically measure the similarity between two tasks: we can compute saturation curves for various pairs of tasks, gaining us a useful measure of whether or not transferring information between two tasks may be beneficial.

12.3 Application in Computational Biology

In this section, we give a brief example for an application in Computational Biology, where we have successfully employed Multitask Learning. The recognition of splice sites is an important problem in genome biology. By now it is fairly well understood and there exist experimentally confirmed labels for a broad range of organisms. In previous work, we have investigated how well information can be transferred between source and target organisms in different evolutionary distances (i.e. one-to-many) and training set sizes [20]. We identified TL algorithms that are particularly well suited for this task. In a follow-up project we investigated how our results

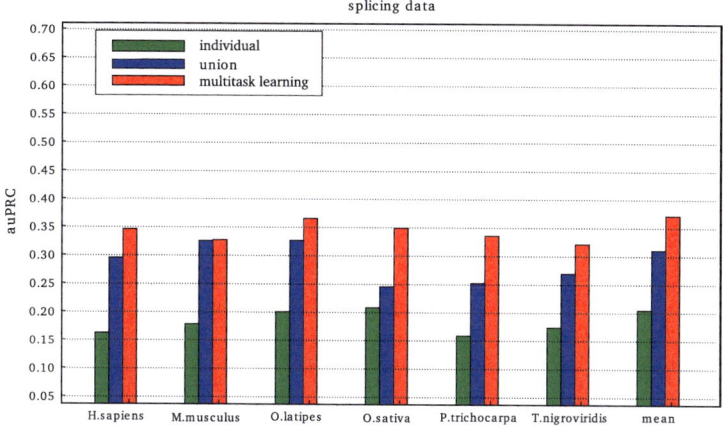

Fig. 12.2 Results of the RNA splicing experiment (Figure taken from [25])

generalize to the MTL scenario (i.e. many-to-many) and showed that exploiting prior information about task similarity provided by taxonomy can be very valuable [25]. An example of how MTL can improve performance compared to baseline methods *individual* (i.e., learn a classifier for each task independently) and *union* (i.e., pool examples from all tasks and obtain a global classifier) is given in Fig. 12.2.

The figure shows results for 6 out of 15 organisms for the baseline methods *individual* and *union* and the multitask learning algorithm described in Sect. 12.2.2. The mean performance is shown in the last column. For each task, we obtained 10,000 training examples and an additional test set of 5,000 examples. We normalized the data sets such that there are 100 negative examples per positive example. We report the area under the precision recall curve (auPRC), which is an appropriate measure for unbalanced classification problems (i.e., detection problems). For an elaborate discussion of our experiments with splice-site prediction, please consider the original publications [20, 25].

12.4 Conclusion

We have presented a brief overview of regularization-based multi-task learning methods and their application in the field of Computational Biology. Especially in the context of biomedical data, where generating training labels can be very expensive, multi-task learning can be viewed as an appealing means to obtain more cost-effective predictors. Accessing – or even learning – the similarity or relatedness of tasks is of central importance when applying multi-task learning methods, especially when we have prior knowledge of the hierarchical task structure, e.g., in the form of a taxonomy. To this end, we have discussed several approaches such as multi-task multiple kernel learning to exploit task relationships in multi-task learning. We have reviewed some basic insights obtained from our experiments on MTL over the past few years and give some practical guidelines for assessing, for a given dataset, whether or not multi-task learning is likely to improve performance over more straightforward baseline approaches. Lastly, we would like to mention that multi-task learning enjoys deep theoretical foundations. This has not been a focus of this chapter, though, but we refer the interested reader to, e.g., [1, 3, 4]. A common approach in MTL theory is to phrase multi-task learning within a hierarchical frequentist i.i.d. setup. This approach is taken, e.g., in Ando and Zhang [1] and Baxter [3], who extend the classical statistical learning theory of Vapnik and Chervonenkis [23] to multiple tasks.

Acknowledgements We thank Klaus-Robert Müller and Mehryar Mohri for inspiring and helpful discussions. This work was supported by the German Research Foundation (DFG) under MU 987/6-1 and RA 1894/1-1 as well as by the European Community's 7th Framework Programme under the PASCAL2 Network of Excellence (ICT-216886). Marius Kloft acknowledges a postdoctoral fellowship by the German Research Foundation (DFG).

References

1. Ando, R., Zhang, T.: A framework for learning predictive structures from multiple tasks and unlabeled data. J. Mach. Learn. Res. **6**, 1817–1853 (2005)
2. Argyriou, A., Evgeniou, T., Pontil, M.: Multi-task feature learning. In: Advances in Neural Information Processing Systems 19, Vancouver. MIT Press, Cambridge (2007)
3. Baxter, J.: A model of inductive bias learning. J. Artif. Intell. Res. **2777**, 149–198 (2000)
4. Ben-David, S., Schuller, R.: Exploiting task relatedness for multiple task learning. Lect. Notes Comput. Sci. **2777**, 567–580 (2003)
5. Blanchard, G., Lee, G., Scott, C.: Generalizing from several related classification tasks to a new unlabeled sample. In: Advances in Neural Information Processing Systems, Granada, vol. 24 (2011)
6. Boser, B., Guyon, I., Vapnik, V.: A training algorithm for optimal margin classifiers. In: Proceedings of the Fifth Annual Workshop on Computational Learning Theory, COLT'92, Pittsburgh, pp. 144–152. ACM, New York (1992)
7. Caruana, R.: Multitask learning: a knowledge-based source of inductive bias. In: ICML, Amherst, pp. 41–48. Morgan Kaufmann (1993)
8. Caruana, R.: Multitask learning. Mach. Learn. **28**(1), 41–75 (1997)
9. Cortes, C., Vapnik, V.: Support vector networks. Mach. Learn. **20**, 273–297 (1995)
10. Daumé, H.: Frustratingly easy domain adaptation. In: Annual Meeting—Association for Computational Linguistics, Prague, vol. 45, p. 256 (2007)
11. Evgeniou, T., Pontil, M.: Regularized multi-task learning. In: International Conference on Knowledge Discovery and Data Mining, Chicago, p. 109 (2004)
12. Evgeniou, T., Micchelli, C., Pontil, M.: Learning multiple tasks with kernel methods. J. Mach. Learn. Res. **6**(1), 615–637 (2005)
13. Heckerman, D., Kadie, C., Listgarten, J.: Leveraging information across HLA alleles/supertypes improves epitope prediction. J. Comput. Biol. **14**(6), 736–746 (2007)
14. Jacob, L., Vert, J.: Efficient peptide-MHC-I binding prediction for alleles with few known binders. Bioinformatics (Oxford, England) **24**(3), 358–366 (2008)
15. Kloft, M., Brefeld, U., Sonnenburg, S., Zien, A.: Lp-norm multiple kernel learning. J. Mach. Learn. Res. **12**, 953–997 (2011)

16. Lanckriet, G., Cristianini, N., Ghaoui, L.E., Bartlett, P., Jordan, M.I.: Learning the kernel matrix with semi-definite programming. JMLR **5**, 27–72 (2004)
17. Mordelet, F., Vert, J.: Prodige: prioritization of disease genes with multitask machine learning from positive and unlabeled examples. BMC Bioinf. **22**, 389 (2011)
18. Pan, S., Yang, Q.: A survey on transfer learning. IEEE Trans. Knowl. Data Eng. **22**, 1345–1359 (2009)
19. Park, C., Hess, D., Huttenhower, C., Troyanskaya, O.: Simultaneous genome-wide inference of physical, genetic, regulatory, and functional pathway components. PLoS Comput. Biol. **6**(11), e1001,009 (2010)
20. Schweikert, G., Widmer, C., Schölkopf, B., Rätsch, G.: An empirical analysis of domain adaptation algorithms for genomic sequence analysis. In: Koller, D., Schuurmans, D., Bengio, Y., Bottou, L. (eds.) Advances in Neural Information Processing Systems (NIPS), Vancouver, vol. 21, pp. 1433–1440 (2009)
21. Sonnenburg, S., Zien, A., Rätsch, G.: ARTS: accurate recognition of transcription starts in human. Bioinformatics **22**(14), e472–e480 (2006)
22. Sriperumbudur, B., Gretton, A., Fukumizu, K., Lanckriet, G., Schölkopf, B.: Injective Hilbert space embeddings of probability measures. In: Servedio, R.A., Zhang, T. (eds.) Proceedings of the 21st Annual Conference on Learning Theory, Helsinki, pp. 111–122. Omnipress (2008)
23. Vapnik, V.N., Chervonenkis, A.Y.: On the uniform convergence of relative frequencies of events to their probabilities. Theory Probab. Appl. **16**(2), 264–280 (1971)
24. Widmer, C., Rätsch, G.: Multitask learning in computational biology. In: JMLR W&CP. ICML 2011 Unsupervised and Transfer Learning Workshop, Bellevue, vol. 27, pp. 207–216 (2012)
25. Widmer, C., Leiva, J., Altun, Y., Rätsch, G.: Leveraging sequence classification by taxonomy-based multitask learning. In: Berger, B. (ed.) Research in Computational Molecular Biology, Lisbon, pp. 522–534. Springer (2010)
26. Widmer, C., Toussaint, N., Altun, Y., Rätsch, G.: Inferring latent task structure for multitask learning by multiple kernel learning. BMC Bioinf. **11**(Suppl 8), S5 (2010)
27. Zhang, Y., Yeung, D.: A convex formulation for learning task relationships in multi-task learning. In: Proceedings of the 26th Annual Conference on Uncertainty in Artificial Intelligence (UAI-10), Catalina Island, pp. 733–742. AUAI Press, Corvallis (2010)

Chapter 13
Semi-supervised Learning in Causal and Anticausal Settings

Bernhard Schölkopf, Dominik Janzing, Jonas Peters, Eleni Sgouritsa, Kun Zhang, and Joris Mooij

Abstract We consider the problem of learning in the case where an underlying causal model can be inferred. Causal knowledge may facilitate some approaches for a given problem, and rule out others. We formulate the hypothesis that semi-supervised learning can help in an anti-causal setting, but not in a causal setting, and corroborate it with empirical results.

13.1 Introduction

Es gibt keinen gefährlicheren Irrtum, als die Folge mit der Ursache zu verwechseln: ich heiße ihn die eigentliche Verderbnis der Vernunft.[1]

Friedrich Nietzsche, Götzen-Dämmerung

It has been argued that statistical dependencies are always due to underlying causal structures [12]. Machine learning has been very successful in exploiting these dependencies [19]. However, could it also benefit from knowledge of the underlying causal structures? We assay this in the simplest possible setting, where the causal structure only consists of cause and effect, with a focus on the case of semi-supervised learning. This follows the work presented at the Festschrift symposium,

---

[1] *There is no more dangerous mistake than confusing cause and effect: I call it the actual corruption of reason.*

B. Schölkopf (✉) · D. Janzing · J. Peters · E. Sgouritsa · K. Zhang
Max Planck Institute for Intelligent Systems, Spemannstrasse, 72076 Tübingen, Germany
e-mail: bs@tuebingen.mpg.de; dominik.janzing@tuebingen.mpg.de; peters@stat.math.ethz.ch; eleni.sgouritsa@tuebingen.mpg.de; kun.zhang@tuebingen.mpg.de

J. Mooij
Institute for Computing & Information Sciences, Radboud University, Nijmegen, Netherlands
e-mail: j.mooij@cs.ru.nl

and it draws heavily from a conference paper published since then [13]. The latter provides less detail on the experiments for the case of semi-supervised learning, but it discusses the cases of covariate shift and transfer learning; see also [18]. Pearl and Bareinboim [11] introduce a variable S that labels different domains or datasets and explains how the way in which S is causally linked to variables of interest is relevant for transferring causal or statistical statements across domains. Its authors' notion of transportability employs conditional independencies to express invariance of mechanisms. The paper [13] discusses a type of invariance where the function in a structural equation remains the same, but the distribution of the noise changes across datasets. Finally, note that the issue is also related to the distinction between generative and discriminative learning; see, for instance, [15].

13.2 Causal Inference

We briefly summarize some aspects of causal graphical models as pioneered by Pearl [10] and Spirtes et al. [17]. These are usually thought of as joint probability distributions over a set of variables $X_1, \ldots, X_n$, along with a directed acyclic graph with vertices X_i and arrows indicating direct causal influences. The *causal Markov condition*, linking causal semantics to statistics, states that each vertex X_i is independent of its non-descendants in the graph, given its parents. Here, independence is usually meant in a statistical sense, although alternative views have been developed, e.g., using algorithmic independence [7]. Given observations from a joint distribution, this allows us to test conditional independence statements and thus infer (subject to a genericity assumption referred to as *faithfulness*) what causal models are consistent with an observed distribution. This will typically not lead us to a unique causal model though, and in the case of graphs with only two variables, there are no conditional independence statements to test and we cannot do anything.

An alternative approach, referred to as a functional causal model (a.k.a. structural causal model or nonlinear structural equation model), starts with a set of jointly independent noise variables, one per vertex, and each vertex computes a deterministic function of its noise variable and its parents. These functions describe not only relations between observations, but also how the system behaves under interventions: by changing the input of some of the functions, one can compute the effect of *setting* some variables to specific values. One can prove that a functional causal model entails a joint distribution which along with the graph satisfies the causal Markov condition [10]. Vice versa, each causal graphical model can be expressed as a functional causal model.

Notation. We consider the causal structure shown in Fig. 13.1, with two observables, modelled by random variables. The variable C stands for the cause and E for the effect. We denote their distributions by $P(C)$ and $P(E)$ (overloading the notation P), and their domains by calligraphic symbols $\mathcal{C}$ and $\mathcal{E}$. The variable X will always be the input and Y the output (or prediction), but input and output can be either cause or effect—more below. For simplicity, we assume that their

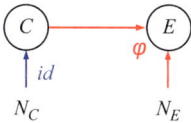

Fig. 13.1 A functional causal model, where C is the cause variable, φ a deterministic mechanism, and E the effect. N_C is a noise variable influencing C (w.l.o.g., we identify this with C), and N_E influences E via $E = \varphi(C, N_E)$. We assume that N_C and N_E are independent

distributions have a joint density with respect to some product measure. In some places, we will use conditional densities, always implicitly assuming that they exist.

The following assumptions are used throughout the chapter.

Causal sufficiency. We assume that there are two independent noise variables N_C and N_E, modelled as random variables with distributions $P(N_C)$ and $P(N_E)$, respectively. The function φ and the noise N_E then jointly determine $P(E|C)$ via $E = \varphi(C, N_E)$. We think of $P(E|C)$ as the *mechanism* transforming cause C into effect E.[2]

Independence of mechanism and input. We finally assume that the mechanism is "independent" of the distribution of the cause (i.e., independent of $P(C) = P(N_C)$; cf. Fig. 13.1), in the sense that $P(E|C)$ contains no information about $P(C)$, and vice versa; in particular, if $P(E|C)$ changes at some point in time, there is no reason to believe that $P(C)$ changes at the same time.[3]

This assumption has been used by Janzing and Schölkopf [7], inspired by Lemeire and Dirkx [8]. It is plausible if we are dealing with a mechanism of nature that does not care what we feed into it. For instance, in the problem of predicting splicing patterns from genomic sequences, the basic splicing mechanism may be assumed to be evolutionarily stable and thus independent of the species [14], even though the genomic sequences and their statistical properties differ. Intuitively, if we learn a *causal* model of splicing, we can hope to be robust with respect to changes to the input statistics.

The independence assumption introduces an asymmetry between cause and effect, since it is violated in the backward direction, i.e., $P(E)$ and $P(C|E)$ are dependent because both inherit properties from $P(E|C)$ and $P(C)$ [3, 7]. Intuitively, the mechanism has left a trace that is visible in the effect's distribution. We expect that this kind of information, which is not used by traditional approaches, may also be useful in inferring larger causal graphs.

Richness of functional causal models. It turns out that the two-variable functional causal model is so rich that the causal direction cannot be inferred. To understand

---

[2] Note that we will use the term "mechanism" both for the function φ and for the conditional $P(E|C)$, but not for $P(C|E)$.

[3] This "independence" condition is closely related to the concept of exogeneity in economics [10]. Given two variables C and E, we say C is exogenous if $P(E|C)$ remains invariant to changes in the process that generates C.

the richness of the class intuitively, consider the simple case where the noise N_E can take only a finite number of values, say $\{1,\ldots,v\}$. This noise could affect φ, for instance, as follows: there is a set of functions $\{\varphi_n : n = 1,\ldots,v\}$, and the noise randomly switches one of them on at any point, i.e., $\varphi(c, n) = \varphi_n(c)$. The functions φ_n could implement arbitrarily different mechanisms, and it would thus be hard to identify φ from empirical data sampled from such a complex model. In view of this, it is surprising that conditional independence alone does allow us to perform causal inference of practical significance, as implemented by the PC and FCI algorithms [10, 17]. However, additional assumptions that prevent the noise switching construction can significantly facilitate the task of inferring causal graphs from data. Intuitively, such assumptions need to control the sensitivity of the mechanism φ to the change in the noise N_E, and thus the complexity of $P(E|C)$.

Additive noise models. One such assumption is referred to as ANM, standing for *additive noise model* [6]. This model assumes $\varphi(C, N_E) = \phi(C) + N_E$ for some function ϕ, so

$$E = \phi(C) + N_E,$$

and then that ϕ and N_E can be inferred in the generic case, provided that N_E has 0 mean (see also [20], including the post-nonlinear case $E = \psi(\phi(C) + N_E)$). Apart from some exceptions, such as the case where ϕ is linear and C and N_E are Gaussian, a given joint distribution of two real-valued random variables X and Y can be fit by an ANM model in at most one direction (which we then consider the causal one). In practice, an ANM model can be fit by regressing the hypothesized effect on the hypothesized cause while enforcing that the residual noise variable is independent of the cause [9]. If this is impossible, the model is incorrect (e.g., cause and effect are interchanged, the noise is not additive, or there are confounders; in the latter two cases the method cannot find the causal direction).

Let us assume we have correctly identified the causal direction. What does this have to do with machine learning? Perhaps somewhat surprisingly, learning problems need not always predict effect from cause; rather often, things are reverse. It turns out that the direction of the prediction has consequences on which tasks are easy and which tasks are hard.

13.3 Semi-supervised Learning

Let us first consider the case where we are trying to estimate a function $f : \mathcal{X} \to \mathcal{Y}$ or a conditional distribution $P(Y|X)$ in the causal direction, i.e., where X is the cause and Y the effect. Intuitively, this situation of *causal prediction* should be the 'easy' case since there exists a functional mechanism φ which f should try to mimic.

In semi-supervised learning (SSL), we are given training points sampled from $P(X, Y)$ and an additional set of inputs sampled from $P(X)$. Our goal is to estimate $P(Y|X)$, or properties thereof (e.g., its expectation).

Fig. 13.2 Causal prediction: Predicting effect Y from cause X

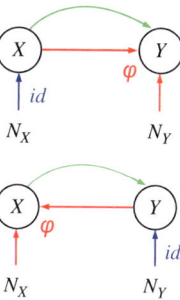

Fig. 13.3 Anticausal prediction: Predicting cause Y from effect X

However, by independence of the mechanism, $P(X)$ contains no information about $P(Y|X)$. A more accurate estimate of $P(X)$, as may be possible by the addition of the test inputs, does thus not influence an estimate of $P(Y|X)$, and SSL is pointless for the scenario in Fig. 13.2. In [13] we argue that while SSL is hard, covariate shift adaptation is easy in this case. Below, it will be the other way round.

We now turn to the opposite direction, where we consider the effect as input and we try to predict the value of the cause variable that led to it. This situation, which we refer to as *anticausal prediction*, may seem unnatural, but it is actually ubiquitous in machine learning. Consider, for instance, the task of predicting the class label of a handwritten digit from its image. The causal structure is as follows: a person intends to write the digit 7, say, and this intention causes a motor pattern producing an image of the digit 7—in that sense the class label Y causes the image X (Fig. 13.3).[4]

As above, we are given training points sampled from $P(X, Y)$ and an additional set of inputs sampled from $P(X)$, and our goal is to estimate $P(Y|X)$, or properties thereof (e.g., its expectation).

Now that we are in the setting of anticausal prediction, $P(X)$ and $P(Y|X)$ are *dependent* [3,7] and thus contain information about each other. The additional inputs hence *may* allow a more accurate estimate of $P(Y|X)$.[5]

Known assumptions for SSL, as discussed by Chapelle et al. [2], can indeed be viewed as linking properties of $P(X)$ to properties of $P(Y|X)$:

- The *cluster assumption* stipulates that points lying in the same cluster of $P(X)$ have the same Y;
- The *low density separation assumption* states that the decision boundary of a classifier (i.e., the point where $P(Y|X)$ crosses 0.5) should lie in a region where $P(X)$ is small;

---

[4]Note that anticausal prediction has also been called *inverse inference*, as opposed to *direct inference* from cause to effects [4]. However, these terms have been used rather broadly, and may also refer to inference relating hypotheses and consequences [4], or inference from population to sample (direct) vs. the other way round (inverse) [16].

[5]Note that a weak form of SSL could roughly work as follows: after learning a generative model for $P(X, Y)$ from the first part of the sample, we can use the additional samples from $P(X)$ to double-check whether our model generates the right distribution for $P(X)$.

- The *semi-supervised smoothness assumption* says that the estimated function (which we may think of as the expectation of $P(Y|X)$) should be smooth in regions where $P(X)$ is large.

Note also that some SSL algorithms assume a model for $P(X|Y)$ (e.g., a mixture of Gaussians) and learn it on both labelled and unlabelled data [21].

In conclusion, we expect that under the assumption of independence of mechanism and input, SSL is impossible in the causal direction, but it may be helpful in the anticausal one. Let's look at empirical results, drawing from a number of earlier benchmark studies.

13.4 Empirical Results

We compare the performance of SSL algorithms with that of base classifiers using only labelled data. For many datasets, X is vector-valued. We assign each dataset to one of three categories:

1. *Anticausal/confounded:* (a) datasets in which at least one predictor X_i is an effect of the target Y to be predicted (Anticausal) (includes also cyclic causal relations between X_i and Y) and (b) datasets in which at least one predictor X_i has an unobserved common cause with the target Y to be predicted (Confounded). In both (a) and (b), the mechanism $P(Y|X_i)$ can be dependent on $P(X_i)$. For these datasets, additional data from $P(X)$ may thus improve prediction.
2. *Causal:* datasets in which some features are causes of the target, and there is no predictor which (a) is an effect of the target or (b) has a common cause with the target. If our assumption on independence of cause and mechanism holds, then SSL should be futile on these datasets.
3. *Unclear:* datasets which were difficult to categorize into one of the aforementioned categories. Some of the reasons for that are incomplete documentation and lack of domain knowledge.

In practice, we count a dataset as causal when we believe that the dependence between X and Y is *mainly* due to X causing Y, although additional confounding effects may be possible.

Semi-supervised classification. We first analyze the results in the benchmark chapter of a book on SSL (Tables 21.11 and 21.13 of [2]) for the case of 100 labelled training points. The chapter compares 11 SSL methods to the base classifiers 1-NN and SVM. In Table 13.1, we give details on our subjective categorization of the eight datasets used in the chapter.

In view of our hypothesis, it is encouraging to see (Fig. 13.4) that SSL does not significantly improve the accuracy in the one causal dataset, but it helps in most of the anticausal/confounded datasets. However, it is difficult to draw conclusions from this small collection of datasets; moreover, three additional issues may confound things: (1) the experiments were carried out in a *transductive* setting. Inductive

13 Semi-supervised Learning in Causal and Anticausal Settings

Table 13.1 Categorization of eight benchmark datasets as anticausal/confounded, Causal or Unclear

Category	Dataset	Reason of categorization
Anticausal/confounded	g241c	The class causes the 241 features
	g241d	The class (binary) and the features are confounded by a variable with four states
	Digit1	The positive or negative angle and the features are confounded by the variable of continuous angle.
	USPS	The class and the features are confounded by the 10-state variable of all digits
	COIL	The 6-state class and the features are confounded by the 24-state variable of all objects
Causal	SecStr	The amino acid is the cause of the secondary structure
Unclear	BCI, Text	Unclear which is the cause and which the effect

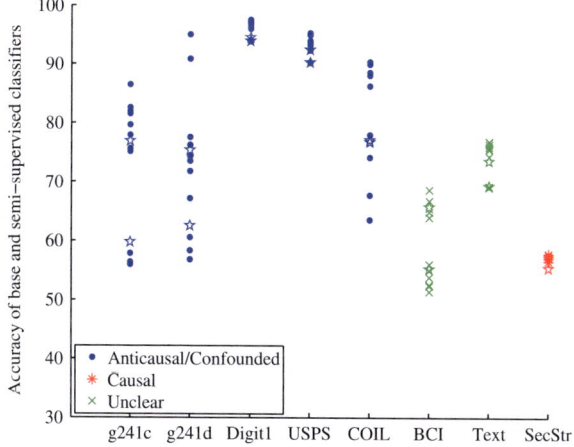

Fig. 13.4 Accuracy of base classifiers (*star shape*) and different SSL methods on eight benchmark datasets

methods use unlabelled data to arrive at a classifier which is subsequently applied to an unknown test set; in contrast, transductive methods use the test inputs to make predictions. This could potentially allow performance improvements independent of whether a dataset is causal or anticausal; (2) the SSL methods used cover a broad range, and are not extensions of the base classifiers; (3) moreover, the results on the SecStr dataset are based on a different set of methods than the rest of the benchmarks.

Table 13.2 Categorization of 26 UCI datasets as anticausal/confounded, Causal or Unclear

Categ.	Dataset	Reason of categorization
Anticausal/confounded	breast-w	The class of the tumour (benign or malignant) causes some of the features of the tumour (e.g., thickness, size, shape, etc.)
	diabetes	Whether or not a person has diabetes affects some of the features (e.g., glucose concentration, blood pressure), but also has an effect of some others (e.g., age, number of times pregnant)
	hepatitis	The class (die or survive) and many of the features (e.g., fatigue, anorexia, liver big) are confounded by the presence or absence of hepatitis. Some of the features, however, may also cause death
	iris	The size of the plant is an effect of the category it belongs to
	labor	Cyclic causal relationships: good or bad labour relations can cause or be caused by many features (e.g., wage increase, number of working hours per week, number of paid vacation days, employer's help during employee's long term disability). Moreover, the features and the class may be confounded by elements of the character of the employer and the employee (e.g., ability to cooperate)
	letter	The class (letter) is a cause of the produced image of the letter
	mushroom	The attributes of the mushroom (shape, size) and the class (edible or poisonous) are confounded by the taxonomy of the mushroom (23 species)
	segment	The class of the image is the cause of the features of the image
	sonar	The class (Mine or Rock) causes the sonar signals
	vehicle	The class of the vehicle causes the features of its silhouette
	vote	This dataset may contain causal, anticausal, confounded and cyclic causal relations. E.g., having handicapped infants or being part of religious groups in school can cause one's vote; being Democrat or Republican can causally influence whether one supports Nicaraguan contras; and immigration may have a cyclic causal relation with the class. Crime and the class may be confounded, e.g., by the environment in which one grew up
	vowel	The class (vowel) causes the features
	waveform-5000	The class of the wave causes its attributes
Causal	balance-scale	The features (weight and distance) cause the class
	kr-vs-kp	The board-description causally influences whether white will win
	splice	The DNA sequence causes the splice sites
Unclear	breast-cancer, colic, colic.ORIG, credit-a, credit-g, heart-c, heart-h, heart-statlog, ionosphere, sick	In some of the datasets, it is unclear whether the class label may have been generated or defined based on the features (e.g., ionoshpere, credit, sick)

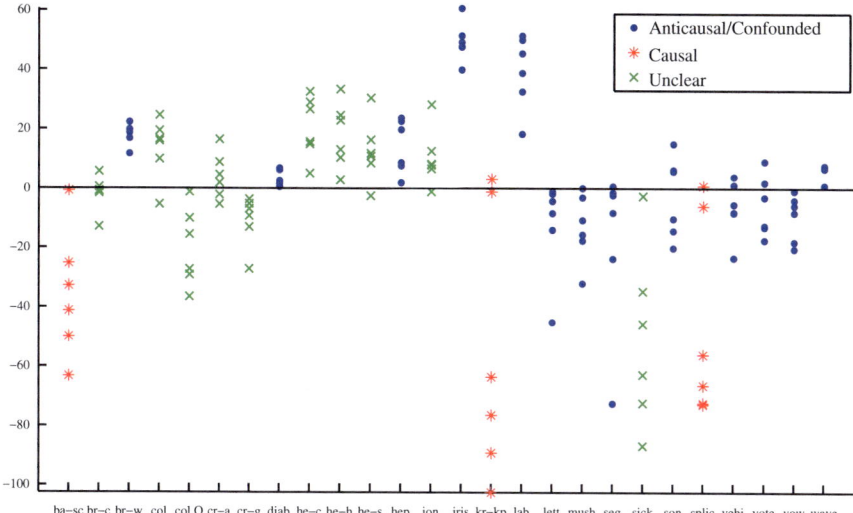

Fig. 13.5 Plot of the relative decrease of error when using self-training, for six base classifiers on 26 UCI datasets. Here, relative decrease is defined as (error(base) − error(self-train))/error(base). Self-training, a method for SSL, overall does not help for the causal datasets, but it does help for several of the anticausal/confounded datasets

We next consider 26 UCI datasets and six different base classifiers. The original results are from Tables III and IV in [5], and are presently re-analyzed in terms of the above dataset categories. The comprehensive results of [5] allow us the luxury of (1) considering only self-training, which is an extension of supervised learning to unlabelled data in the sense that if the set of unlabelled data is empty, we recover the results of the base method (in this case, self-training would stop at the first iteration). This lets us compare an SSL method to its corresponding base algorithm. Moreover, (2) we included only the *inductive* methods considered by Guo et al. [5], and not the *transductive* ones (cf. our discussion above).

Table 13.2 describes our subjective categorization of the 26 UCI datasets into anticausal/confounded, Causal, or Unclear.

In Fig. 13.5, we observe that SSL does not significantly decrease the error rate in the three causal datasets, but it does increase the performance in several of the anticausal/confounded datasets. This is again consistent with our hypothesis that if mechanism and input are independent, SSL will not help for causal datasets.

Semi-supervised regression (SSR). Classification problems are often inherently asymmetric in that the inputs are continuous and the outputs categorical. It is reassuring that we obtain similar results in the case of regression. To this end, we consider the co-regularized least squares regression (co-RLSR) algorithm, compared to regular RLSR on 32 real-world datasets by Brefeld et al. [1] (two of which are identical, so 31 datasets are considered). We categorized them into causal, anticausal/confounded, unclear, as in Table 13.3, prior to the subsequent

Table 13.3 Categorization of 31 UCI datasets as anticausal/confounded, Causal or Unclear

Categ.	Dataset	Target variable	Reason of categorization
Anticausal/confounded	breastTumor	Tumour size	Causing predictors such as inv-nodes and deg-malig
	cholesterol	Cholesterol	Causing predictors such as resting blood pressure and fasting blood sugar
	cleveland	Presence of heart disease in the patient	Causing predictors such as chest pain type, resting blood pressure, and fasting blood sugar
	lowbwt	Birth weight	Causing the predictor indicating low birth weight
	pbc	Histologic stage of disease	Causing predictors such as Serum bilirubin, Prothrombin time, and Albumin
	pollution	Age-adjusted mortality rate per 100,000	Causing the predictor number of 1960 SMSA population aged 65 or older
	wisconsin	Time to recur of breast cancer	Causing predictors such as perimeter, smoothness, and concavity
Causal	autoMpg	City-cycle fuel consumption in miles per gallon	Caused by predictors such as horsepower and weight
	cpu	cpu relative performance	Caused by predictors such as machine cycle time, maximum main memory, and cache memory
	fishcatch	Fish weight	Caused by predictors such as fish length and fish width
	housing	Housing values in suburbs of Boston	Caused by predictors such as pupil-teacher ratio and nitric oxides concentration
	machine_cpu	cpu relative performance	See remark on "cpu"
	meta	Normalized prediction error	Caused by predictors such as number of examples, number of attributes, and entropy of classes
	pwLinear	Value of piecewise linear function	Caused by all ten involved predictors
	sensory	Wine quality	Caused by predictors such as trellis
	servo	Rise time of a servomechanism	Caused by predictors such as gain settings and choices of mechanical linkages
Unclear	auto93 (target: midrange price of cars); bodyfat (target: percentage of body fat); autoHorse (target: price of cars); autoPrice (target: price of cars); baskball (target: points scored per minute); cloud (target: period rainfalls in the east target); echoMonths (target: number of months patient survived); fruitfly (target: longevity of mail fruitflies); pharynx (target: patient survival); pyrim (quantitative structure activity relationships); sleep (target: total sleep in hours per day); stock (target: price of one particular stock); strike (target: strike volume); triazines (target: activity); veteran (survival in days)		

Fig. 13.6 RMSE for anticausal/confounded datasets

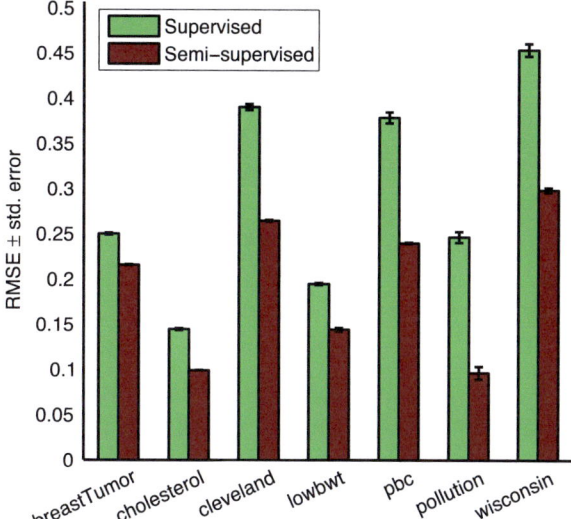

Fig. 13.7 RMSE for Causal datasets

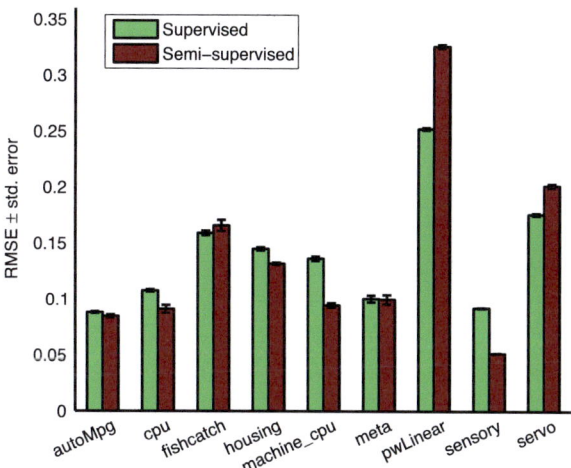

analysis. Note that the categorization of Tables 13.2 and 13.3 is subjective and was made independently. That is the reason why the heart-c dataset (which coincides with the Cleveland dataset) was categorized as Unclear in Table 13.2 and as anticausal/confounded in Table 13.3. Nevertheless, this does not create any conflict with our claims.

We deemed seven of the datasets anticausal, i.e., the target variable can be considered as the cause of (some of) the predictors; Fig. 13.6 shows that SSR reduces the root mean square errors (RMSEs) in all these cases. Nine of the remaining datasets can be considered causal, and Fig. 13.7 shows that there is usually little performance improvement for those. Like [1], we used the Wilcoxon signed rank test to assess whether SSR outperforms supervised regression in the anticausal and causal cases. The null hypothesis is that the distribution of the difference between the RMSE produced by SSR and that by supervised regression is symmetric around 0 (i.e., that SSR does not help). On the anticausal datasets, the p-value is 0.0156, while it is 0.6523 on the causal datasets. Therefore, we reject the null hypothesis in the anticausal case at a 5 % significance level, but not in the causal case.

13.5 Conclusion

If one is interested in predicting one variable from another one, it helps to know the causal structure underlying the variables. In particular, this leads to the hypothesis that under an independence assumption for causal mechanism and input, semi-supervised learning works better in anticausal or confounded problems than in causal problems, which is consistent with our analysis on empirical data.

Acknowledgements We thank Ulf Brefeld and Stefan Wrobel who kindly shared their detailed experimental results with us, allowing for our meta-analysis. We thank Bob Williamson, Vladimir Vapnik, and Jakob Zscheischler for helpful discussions.

References

1. Brefeld, U., Gärtner, T., Scheffer, T., Wrobel, S.: Efficient co-regularised least squares regression. In: ICML, Pittsburgh (2006)
2. Chapelle, O., Schölkopf, B., Zien, A.: Semi-supervised Learning. MIT, Cambridge (2006)
3. Daniušis, P., Janzing, D., Mooij, J., Zscheischler, J., Steudel, B., Zhang, K., Schölkopf, B.: Inferring deterministic causal relations. In: UAI, Catalina Island (2010)
4. Fetzer, J.H., Almeder, R.F.: Glossary of Epistemology/Philosophy of Science. Paragon House, New York (1993)
5. Guo, Y., Niu, X., Zhang, H.: An extensive empirical study on semi-supervised learning. In: ICDM, Sydney (2010)
6. Hoyer, P.O., Janzing, D., Mooij, J.M., Peters, J., Schölkopf, B.: Nonlinear causal discovery with additive noise models. In: NIPS, Vancouver (2008)
7. Janzing, D., Schölkopf, B.: Causal inference using the algorithmic Markov condition. IEEE Trans. Inf. Theory **56**(10), 5168–5194 (2010)
8. Lemeire, J., Dirkx, E.: Causal models as minimal descriptions of multivariate systems. http://parallel.vub.ac.be/~jan/ (2007)
9. Mooij, J., Janzing, D., Peters, J., Schölkopf, B.: Regression by dependence minimization and its application to causal inference in additive noise models. In: ICML, Montreal (2009)
10. Pearl, J.: Causality. Cambridge University Press, New York (2000)
11. Pearl, E., Bareinboim, E.: Transportability of causal and statistical relations: a formal approach. In: Proceedings of the 25th AAAI Conference on Artificial Intelligence, Menlo Park, pp. 247–254 (2011)
12. Reichenbach, H.: The Direction of Time. University of California Press, Berkeley (1956)

13. Schölkopf, B., Janzing, D., Peters, J., Sgouritsa, E., Zhang, K., Mooij, J.: On causal and anticausal learning. In: Langford, J., Pineau, J. (eds.) Proceedings of the 29th International Conference on Machine Learning, Edinburgh, pp. 1255–1262. Omnipress, New York (2012)
14. Schweikert, G., Widmer, C., Schölkopf, B., Rätsch, G.: An empirical analysis of domain adaptation algorithms for genomic sequence analysis. In: NIPS, Vancouver (2009)
15. Seeger, M.: Learning with labeled and unlabeled data. Technical Report (Tech. rep.), University of Edinburgh (2001)
16. Seidenfeld, T.: Direct inference and inverse inference. J. Philos. **75**(12), 709–730 (1978)
17. Spirtes, P., Glymour, C., Scheines, R.: Causation, Prediction, and Search. Springer, New York (1993). (2nd edn.: MIT, Cambridge, 2000)
18. Storkey, A.: When training and test sets are different: characterizing learning transfer. In: Dataset Shift in Machine Learning. MIT, Cambridge (2009)
19. Vapnik, V.: Estimation of Dependences Based on Empirical Data (in Russian). Nauka, Moscow (1979). English translation: Springer, New York, 1982
20. Zhang, K., Hyvärinen, A.: On the identifiability of the post-nonlinear causal model. In: UAI, Montreal (2009)
21. Zhu, X., Goldberg, A.: Introduction to semi-supervised learning. In: Synthesis Lectures on Artificial Intelligence and Machine Learning, vol. 3, pp. 1–130. Morgan & Claypool, San Rafael (2009)

Chapter 14
Strong Universal Consistent Estimate of the Minimum Mean Squared Error

Luc Devroye, Paola G. Ferrario, László Györfi, and Harro Walk

Abstract Consider the regression problem with a response variable Y and a feature vector $\mathbf{X}$. For the regression function $m(\mathbf{x}) = \mathbf{E}\{Y \mid \mathbf{X} = \mathbf{x}\}$, we introduce new and simple estimators of the minimum mean squared error $L^* = \mathbf{E}\{(Y - m(\mathbf{X}))^2\}$, and prove their strong consistencies. We bound the rate of convergence, too.

14.1 Introduction

Let the label Y be a real-valued random variable and let the feature vector $\mathbf{X} = (X_1, \ldots, X_d)$ be a d-dimensional random vector. The regression function m is defined by

$$m(\mathbf{x}) = \mathbf{E}\{Y \mid \mathbf{X} = \mathbf{x}\}.$$

L. Devroye (✉)
School of Computer Science, McGill University, 3480 University Street, Montreal, H3A OE9, Canada
e-mail: lucdevroye@gmail.com

P.G. Ferrario · H. Walk
Department of Mathematics, University of Stuttgart, Pfaffenwaldring 57, 70569 Stuttgart, Germany
e-mail: paola.ferrario@mathematik.uni-stuttgart.de; walk@mathematik.uni-stuttgart.de

L. Györfi
Department of Computer Science and Information Theory, Budapest University of Technology and Economics, Stoczek u. 2, 1521 Budapest, Hungary
e-mail: gyorfi@cs.bme.hu

The minimum mean squared error, called also variance of the residual $Y - m(\mathbf{X})$, is denoted by

$$L^* := \mathbf{E}\{(Y - m(\mathbf{X}))^2\} = \min_f \mathbf{E}\{(Y - f(\mathbf{X}))^2\}.$$

The regression function m and the minimum mean squared error L^* cannot be calculated when the distribution of $(\mathbf{X}, Y)$ is unknown. Assume, however, that we observe data $D_n = \{(\mathbf{X}_1, Y_1), \ldots, (\mathbf{X}_n, Y_n)\}$ consisting of independent and identically distributed copies of $(\mathbf{X}, Y)$. D_n can be used to produce an estimate of L^*.

For nonparametric estimates of the minimum mean squared error $L^* = \mathbf{E}\{(Y - m(\mathbf{X}))^2\}$ see, e.g., Dudoit and van der Laan [4], Liitiäinen et al. [9–11], Müller and Stadtmüller [12], Neumann [14], Stadtmüller and Tsybakov [15], and Müller, Schick and Wefelmeyer [13] and the literature cited there.

Devroye et al. [3] proved that without any tail and smoothness condition, L^* cannot be estimated with a guaranteed rate of convergence. They introduced a modified nearest neighbour cross-validation estimate

$$\hat{L}_n = \frac{1}{2n} \sum_{i=1}^n (Y_i - Y_{j(i)})^2, \quad n \geq 2,$$

where $Y_{j(i)}$ is the label of the modified first nearest neighbour of $\mathbf{X}_i$ from among $\mathbf{X}_1, \ldots, \mathbf{X}_{i-1}, \mathbf{X}_{i+1}, \ldots, \mathbf{X}_n$. If Y and $\mathbf{X}$ are bounded, and m is Lipschitz continuous

$$|m(\mathbf{x}) - m(\mathbf{z})| \leq C \|\mathbf{x} - \mathbf{z}\|, \tag{14.1}$$

then for $d \geq 3$, they proved that

$$\mathbf{E}\{|\hat{L}_n - L^*|\} \leq c_1 n^{-1/2} + c_2 n^{-2/d}. \tag{14.2}$$

Liitiäinen et al. [9,11] introduced another estimate of the minimum mean squared error L^* by the first and second nearest neighbour cross-validation

$$L_n = \frac{1}{n} \sum_{i=1}^n (Y_i - Y_{n,i,1})(Y_i - Y_{n,i,2}), \quad n \geq 3,$$

where $Y_{n,i,1}$ and $Y_{n,i,2}$ are the labels of the first and second nearest neighbours $\mathbf{X}_{n,i,1}$ and $\mathbf{X}_{n,i,2}$ of $\mathbf{X}_i$ from among $\mathbf{X}_1, \ldots, \mathbf{X}_{i-1}$, and $\mathbf{X}_{i+1}, \ldots, \mathbf{X}_n$, resp. (In the sequel, assume that for calculating the first and second nearest neighbours, ties occur with probability 0. When $\mathbf{X}$ has a density, the case of ties among nearest neighbour distances occurs with probability 0.) If Y and $\mathbf{X}$ are bounded and m is Lipschitz continuous, then for $d \geq 2$, they proved the rate of convergence of order in the inequality (14.2).

14 Strong Universal Consistent Estimate of the Minimum Mean Squared Error

In this chapter we introduce a non-recursive and a recursive estimator of the minimum mean squared error L^*, and prove their distribution-free strong consistencies. Under some mild conditions on the regression function m and on the distribution of $(\mathbf{X}, Y)$, we bound the rate of convergence of the non-recursive estimate.

14.2 Strong Universal Consistency

One can derive a new and simple estimator of L^*, considering the definition

$$L^* = \mathbf{E}\{(Y - m(\mathbf{X}))^2\} = \mathbf{E}\{Y^2\} - \mathbf{E}\{m(\mathbf{X})^2\}.$$

Obviously, $\mathbf{E}\{Y^2\}$ can be estimated by $\frac{1}{n}\sum_{i=1}^{n} Y_i^2$, while we estimate the term $\mathbf{E}\{m(\mathbf{X})^2\}$ by $\frac{1}{n}\sum_{i=1}^{n} Y_i Y_{n,i,1}$. Thus we estimate L^* by

$$\tilde{L}_n := \frac{1}{n}\sum_{i=1}^{n} Y_i^2 - \frac{1}{n}\sum_{i=1}^{n} Y_i Y_{n,i,1}.$$

Theorem 14.1. *Assume that ties occur with probability 0. If $|Y|$ is bounded then*

$$\tilde{L}_n \to L^* \quad a.s.$$

If $\mathbf{E}\{Y^2\} < \infty$ then

$$\bar{L}_n := \frac{1}{n}\sum_{k=1}^{n} \tilde{L}_k \to L^* \quad a.s.$$

Proof. This theorem says that, for bounded $|Y|$, the estimate $\tilde{L}_n$ is strongly consistent, while the estimate $\bar{L}_n$ is strongly universally consistent. The theorem is an easy consequence of Ferrario and Walk [6] (Theorems 2.1 and 2.5), who proved that, for bounded Y,

$$L_n \to L^* \tag{14.3}$$

a.s., and moreover, under the only condition $\mathbf{E}\{Y^2\} < \infty$,

$$\frac{1}{n}\sum_{k=1}^{n} L_k \to L^* \tag{14.4}$$

a.s. We simply use the decomposition

$$L_n = \tilde{L}_n - \frac{1}{n}\sum_{i=1}^{n} Y_i Y_{n,i,2} + \frac{1}{n}\sum_{i=1}^{n} Y_{n,i,1} Y_{n,i,2}.$$

Then, as in the proof of Theorem 2.1 in Ferrario and Walk [6], on the basis of (21)–(25) in [9], one can show that, for bounded Y,

$$\frac{1}{n}\sum_{i=1}^{n} Y_i Y_{n,i,2} \to \mathbf{E}\{m(\mathbf{X})^2\} \tag{14.5}$$

a.s. and

$$\frac{1}{n}\sum_{i=1}^{n} Y_{n,i,1} Y_{n,i,2} \to \mathbf{E}\{m(\mathbf{X})^2\} \tag{14.6}$$

a.s. Similarly, as in the proof of Theorem 2.5 in [6], for $\mathbf{E}\{Y^2\} < \infty$, one can show that

$$\frac{1}{n}\sum_{k=1}^{n} \frac{1}{k}\sum_{i=1}^{k} Y_i Y_{k,i,2} \to L^* \tag{14.7}$$

a.s. and

$$\frac{1}{n}\sum_{k=1}^{n} \frac{1}{k}\sum_{i=1}^{k} Y_{k,i,1} Y_{k,i,2} \to L^* \tag{14.8}$$

a.s. Now the statements of the theorem follow from (14.3), (14.5), and (14.6), and from (14.4), (14.7) and (14.8), respectively. □

Next we consider a recursive estimate

$$L'_n := \frac{1}{n}\sum_{i=1}^{n} Y_i^2 - \frac{1}{n}\sum_{i=1}^{n} Y_i Y_{i,i,1}, \quad n \geq 2,$$

where $Y_{1,1,1} := 0$. It is really recursive since

$$L'_n = \left(1 - \frac{1}{n}\right) L'_{n-1} + \frac{1}{n}(Y_n^2 - Y_n Y_{n,n,1}).$$

Theorem 14.2. *Assume that ties occur with probability 0. If* $\mathbf{E}\{Y^2\} < \infty$ *then*

$$L'_n \to L^* \quad a.s.$$

Proof. We have to show that

$$\frac{1}{n}\sum_{i=1}^{n} Y_i Y_{i,i,1} \to \mathbf{E}\{m(\mathbf{X})^2\} \quad \text{a.s.} \tag{14.9}$$

For $a > 0$, introduce the truncation function

$$T_a(z) = \begin{cases} a & \text{if } z > a; \\ z & \text{if } |z| < a; \\ -a & \text{if } z < -a. \end{cases}$$

As in to the proof of Theorem 2.5 in Ferrario and Walk [6], one can check that in order to show (14.9), it suffices to prove that

$$\frac{1}{n}\sum_{i=1}^{n} T_{\sqrt{i}}(Y_i) T_{\sqrt{i}}(Y_{i,i,1}) \to \mathbf{E}\{m(\mathbf{X})^2\}$$

a.s. Let $\mathcal{F}_{i-1}$ be the σ-algebra generated by $(\mathbf{X}_1, Y_1), \ldots, (\mathbf{X}_{i-1}, Y_{i-1})$. Introduce the decomposition

$$\frac{1}{n}\sum_{i=1}^{n} T_{\sqrt{i}}(Y_i) T_{\sqrt{i}}(Y_{i,i,1}) = I_n + J_n,$$

where

$$I_n = \frac{1}{n}\sum_{i=1}^{n}(T_{\sqrt{i}}(Y_i) T_{\sqrt{i}}(Y_{i,i,1}) - \mathbf{E}\{T_{\sqrt{i}}(Y_i) T_{\sqrt{i}}(Y_{i,i,1}) \mid \mathcal{F}_{i-1}\})$$

and

$$J_n = \frac{1}{n}\sum_{i=1}^{n} \mathbf{E}\{T_{\sqrt{i}}(Y_i) T_{\sqrt{i}}(Y_{i,i,1}) \mid \mathcal{F}_{i-1}\}.$$

I_n is an average of martingale differences such that the a.s. convergence

$$I_n \to 0 \tag{14.10}$$

can be derived from the Chow [1] theorem if

$$\sum_{n=1}^{\infty} \frac{\mathbf{Var}\{T_{\sqrt{n}}(Y_n) T_{\sqrt{n}}(Y_{n,n,1})\}}{n^2} < \infty. \tag{14.11}$$

We have that

$$\mathbf{Var}\{T_{\sqrt{n}}(Y_n)T_{\sqrt{n}}(Y_{n,n,1})\} \leq \mathbf{E}\{(T_{\sqrt{n}}(Y_n))^2(T_{\sqrt{n}}(Y_{n,n,1}))^2\}$$

$$\leq \frac{\mathbf{E}\{(T_{\sqrt{n}}(Y_n))^4\} + \mathbf{E}\{(T_{\sqrt{n}}(Y_{n,n,1}))^4\}}{2}.$$

Because of $\mathbf{E}\{Y_1^2\} < \infty$,

$$\sum_{n=1}^{\infty} \frac{\mathbf{E}\{(T_{\sqrt{n}}(Y_n))^4\}}{n^2} = \sum_{n=1}^{\infty} \frac{\mathbf{E}\{T_{n^2}(Y_n^4)\}}{n^2} = \sum_{n=1}^{\infty} \frac{\mathbf{E}\{T_{n^2}(Y_1^4)\}}{n^2} < \infty.$$

Recall now the following useful lemma.

Lemma 14.1. *(Györfi et al. [7], Corollary 6.1) Under the assumption that ties occur with probability 0,*

$$\sum_{i=1}^{n} \mathbf{I}_{\{\mathbf{X} \text{ is the first NN of } \mathbf{X}_i \text{ in } \{\mathbf{X}_1,\ldots,\mathbf{X}_{i-1},\mathbf{X},\mathbf{X}_{i+1},\ldots,\mathbf{X}_n\}\}} \leq \gamma_d$$

a.s., where $\mathbf{I}$ *denotes the indicator and* $\gamma_d < \infty$ *depends only on* d.

Lemma 14.1 implies that

$$\mathbf{E}\{(T_{\sqrt{n}}(Y_{n,n,1}))^4\}$$

$$= \mathbf{E}\left\{\sum_{j=1}^{n-1}(T_{\sqrt{n}}(Y_j))^4 \mathbf{I}_{\{\mathbf{X}_j \text{ is the first NN of } \mathbf{X}_n \text{ in } \{\mathbf{X}_1,\ldots,\mathbf{X}_{n-1}\}\}}\right\}$$

$$= (n-1)\mathbf{E}\left\{(T_{\sqrt{n}}(Y_1))^4 \mathbf{I}_{\{\mathbf{X}_1 \text{ is the first NN of } \mathbf{X}_n \text{ in } \{\mathbf{X}_1,\ldots,\mathbf{X}_{n-1}\}\}}\right\}$$

$$= (n-1)\mathbf{E}\left\{(T_{\sqrt{n}}(Y_n))^4 \mathbf{I}_{\{\mathbf{X}_n \text{ is the first NN of } \mathbf{X}_1 \text{ in } \{\mathbf{X}_2,\ldots,\mathbf{X}_n\}\}}\right\}$$

$$= \mathbf{E}\left\{(T_{\sqrt{n}}(Y_n))^4 \sum_{j=1}^{n-1} \mathbf{I}_{\{\mathbf{X}_n \text{ is the first NN of } \mathbf{X}_j \text{ in } \{\mathbf{X}_1,\ldots,\mathbf{X}_{j-1},\mathbf{X}_{j+1},\ldots,\mathbf{X}_n\}\}}\right\}$$

$$\leq \mathbf{E}\left\{(T_{\sqrt{n}}(Y_n))^4 \gamma_d\right\}.$$

Therefore

$$\sum_{n=1}^{\infty} \frac{\mathbf{E}\{(T_{\sqrt{n}}(Y_{n,n,1}))^4\}}{n^2} \leq \sum_{n=1}^{\infty} \frac{\gamma_d \mathbf{E}\{T_{n^2}(Y_n^4)\}}{n^2} = \gamma_d \sum_{n=1}^{\infty} \frac{\mathbf{E}\{T_{n^2}(Y_1^4)\}}{n^2} < \infty,$$

and so (14.11) is verified, which implies (14.10). Concerning the term J_n, the derivations below are based on the fact that the ordinary 1-NN regression estimate is not universally consistent; however, it is strongly Cesàro convergent in the weak topology, and for noiseless observations ($Y_i = m(\mathbf{X}_i)$) it is strongly convergent in L_2. Introduce the notation

$$m_i(\mathbf{x}) := \mathbf{E}\{T_{\sqrt{i}}(Y) \mid \mathbf{X} = \mathbf{x}\}.$$

Let $\mathbf{X}_{i-1,1}(\mathbf{x})$ denote the 1-NN (first nearest neighbour) of $\mathbf{x}$ from among $\{\mathbf{X}_1, \ldots, \mathbf{X}_{i-1}\}$ and $Y_{i-1,1}(\mathbf{x})$ denote the corresponding label ($x \in \mathbf{R}^d$, $i \geq 2$); then

$$Y_{i-1,1}(\mathbf{X}_i) = Y_{i,i,1} \quad \text{and} \quad \mathbf{X}_{i-1,1}(\mathbf{X}_i) = \mathbf{X}_{i,i,1}.$$

The representation

$$J_n = \frac{1}{n} \sum_{i=1}^{n} \int_{\mathbf{R}^d} m_i(\mathbf{x}) T_{\sqrt{i}}(Y_{i-1,1}(\mathbf{x})) \mu(d\mathbf{x})$$

holds, where $Y_{0,1}(\mathbf{x}) := 0$. It remains to show that

$$J_n \to \mathbf{E}\{m(\mathbf{X})^2\} \quad \text{a.s.} \tag{14.12}$$

Before proving (14.12) we use two lemmas. Let μ denote the distribution of $\mathbf{X}$.

Lemma 14.2. *If* $\mathbf{E}\{Y^2\} < \infty$ *then*

$$\int | m(\mathbf{X}_{n-1,1}(\mathbf{x})) - m(\mathbf{x}) |^2 \mu(d\mathbf{x}) \to 0 \quad \text{a.s.}$$

Proof. The proof is in the spirit of the proof of Theorem 4.1 and Problems 4.5 and 6.3 in Györfi et al. [7]. □

The following lemma is a reformulation of a classic deterministic Tauberian theorem of Landau [8] in summability theory. For a proof and further references, see Lemma 1 in Walk [16].

Lemma 14.3. *If the sequence* a_n, $n = 1, 2, \ldots$ *of real numbers is bounded from below and satisfies*

$$\sum_{n=1}^{\infty} \frac{(\sum_{i=1}^{n} a_i)^2}{n^3} < \infty,$$

then

$$\frac{1}{n} \sum_{i=1}^{n} a_i \to 0.$$

Proof of (14.12). It suffices to show

$$J_n^* := \frac{1}{n}\sum_{i=1}^n \int_{R^d} m(\mathbf{x}) T_{\sqrt{i}}(Y_{i-1,1}(\mathbf{x})) \mu(d\mathbf{x}) \to \mathbf{E}\{m(\mathbf{X})^2\} \quad a.s. \qquad (14.13)$$

In fact, we notice that for each $\alpha > 0$

$$\int_{R^d} |m_i(\mathbf{x}) - m(\mathbf{x})| |T_{\sqrt{i}}(Y_{i-1,1}(\mathbf{x}))| \mu(d\mathbf{x})$$
$$\le \frac{1}{2}\frac{1}{\alpha}\int_{R^d} |m_i(\mathbf{x}) - m(\mathbf{x})|^2 \mu(d\mathbf{x}) + \frac{1}{2}\alpha \int_{R^d} T_i(Y_{i-1,1}(\mathbf{x})^2) \mu(d\mathbf{x}).$$

If we can show

$$\limsup \frac{1}{n}\sum_{i=1}^n \int T_i(Y_{i-1,1}(\mathbf{x})^2) \mu(d\mathbf{x}) \le c\mathbf{E}\{Y^2\} \quad a.s., \qquad (14.14)$$

for some constant c, then this together with $\int_{R^d} |m_i(\mathbf{x}) - m(\mathbf{x})|^2 \mu(d\mathbf{x}) \to 0$ implies

$$\limsup \frac{1}{n}\sum_{i=1}^n \int_{R^d} |m_i(\mathbf{x}) - m(\mathbf{x})| |T_{\sqrt{i}}(Y_{i-1,1}(\mathbf{x}))| \mu(d\mathbf{x}) \le \frac{1}{2}\alpha c\mathbf{E}\{Y^2\} \quad a.s.$$

But $\alpha \to 0$ yields that left-hand side equals 0 a.s. This, together with (14.13), implies (14.12). Therefore, to complete the proof it remains to show (14.14) and (14.13). In the first part we show (14.14). Set $r(\mathbf{x}) := \mathbf{E}\{Y^2|\mathbf{X} = \mathbf{x}\}$, $r_i(\mathbf{x}) := \mathbf{E}\{T_i(Y^2)|\mathbf{X} = \mathbf{x}\}$. In order to get (14.14) it is enough to show

$$\frac{1}{n}\sum_{i=1}^n \int \left(T_i(Y_{i-1,1}(\mathbf{x})^2) - r_i(\mathbf{X}_{i-1,1}(\mathbf{x}))\right) \mu(d\mathbf{x}) \to 0 \quad a.s., \qquad (14.15)$$

where $r_1(\mathbf{X}_{0,1}(\mathbf{x})) := 0$, and

$$\limsup \frac{1}{n}\sum_{i=1}^n \int r_i(\mathbf{X}_{i-1,1}(\mathbf{x})) \mu(d\mathbf{x}) \le c\mathbf{E}\{Y^2\} \quad a.s.$$

The latter follows from

$$\limsup \int r_n(\mathbf{X}_{n-1,1}(\mathbf{x})) \mu(d\mathbf{x}) \le \lim \int r(\mathbf{X}_{n-1,1}(\mathbf{x})) \mu(d\mathbf{x}) = \mathbf{E}\{r(\mathbf{X})\} = \mathbf{E}\{Y^2\}$$

a.s. (where the first equality holds by Lemma 14.2), which further yields that the sequence

$$\int (T_n(Y_{n-1,1}(\mathbf{x})^2) - r_n(\mathbf{X}_{n-1,1}(\mathbf{x})))\mu(d\mathbf{x}),$$

$n = 1, 2, \ldots$, is a.s. bounded from below. In order to get (14.15) and therefore (14.14) by Lemma 14.3, it suffices to show

$$\sum_{n=1}^{\infty} \frac{\mathbf{E}\left\{\left[\sum_{i=1}^{n} \int \left(T_i(Y_{i-1,1}(\mathbf{x})^2) - r_i(\mathbf{X}_{i-1}(\mathbf{x}))\right)\mu(d\mathbf{x})\right]^2\right\}}{n^3} < \infty. \tag{14.16}$$

We now show (14.16). Set for it $A_{i,j} := \{\mathbf{x}; \mathbf{X}_{i-1,1}(\mathbf{x}) = \mathbf{X}_j\}$. We note

$$\sum_{i=1}^{n} \int_{\mathbf{R}^d} \left(T_i(Y_{i-1,1}(\mathbf{x})^2) - r_i(\mathbf{X}_{i-1,1}(\mathbf{x}))\right)\mu(d\mathbf{x})$$

$$= \sum_{i=1}^{n} \int_{\mathbf{R}^d} \sum_{j=1}^{i-1} \mathbf{I}_{\{\mathbf{X}_{i-1,1}(\mathbf{x}) = \mathbf{X}_j\}} \mu(d\mathbf{x}) \left(T_i(Y_j^2) - r_i(\mathbf{X}_j)\right)$$

$$= \sum_{i=1}^{n} \sum_{j=1}^{i-1} \mu(A_{i,j}) \left(T_i(Y_j^2) - r_i(\mathbf{X}_j)\right)$$

$$= \sum_{j=1}^{n-1} \left(\sum_{i=j+1}^{n} \mu(A_{i,j}) \left(T_i(Y_j^2) - r_i(\mathbf{X}_j)\right)\right),$$

where the $(n-1)$ summands in brackets are orthogonal, because $\mathbf{E}\{T_i(Y_j^2) - r_i(\mathbf{X}_j) \mid \mathbf{X}_1, \ldots, \mathbf{X}_{n-1}, Y_j'\} = 0$ for all i and all $j' \neq j$ $(j, j' \in \{1, \ldots, n-1\})$. Thus (14.16) is equivalent to

$$\sum_{n=1}^{\infty} \frac{1}{n^3} \sum_{j=1}^{n-1} \mathbf{E}\left\{\left(\sum_{i=j+1}^{n} \mu(A_{i,j}) \left(T_i(Y_j^2) - r_i(\mathbf{X}_j)\right)\right)^2\right\} < \infty.$$

Let the cones $C_1, \ldots, C_{\gamma_d}$ have top $\mathbf{0}$ and angle $\frac{\pi}{3}$, which cover $\mathbf{R}^d$, and let $B_{i,j,l}$ be the subset of $C_{j,l} := \mathbf{X}_j + C_l$ $(j = 1, \ldots, i-1; l = 1, \ldots, \gamma_d)$ consisting of all $\mathbf{x}$ that are closer to $\mathbf{X}_j$ than the 1-NN of $\mathbf{X}_j$ in $\{\mathbf{X}_1, \ldots, \mathbf{X}_{j-1}, \mathbf{X}_{j+1}, \ldots, \mathbf{X}_{i-1}\} \cap C_{j,l}$. For $j \leq i - 1$, a covering result of Devroye et al. [2], and also of pp. 489 and 490 in Györfi et al. [7], holds as follows:

$$\mu(A_{i,j}) \leq \sum_{l=1}^{\gamma_d} \mu(B_{i,j,l}).$$

It suffices to show, for each $l \in \{1, \ldots, \gamma_d\}$,

$$\sum_{n=1}^{\infty} \frac{1}{n^3} \sum_{j=1}^{n-1} \mathbf{E}\left\{\left(\sum_{i=j+1}^{n} \mu(B_{i,j,l})\left(T_i(Y_j^2) - r_i(\mathbf{X}_j)\right)\right)^2\right\} < \infty. \qquad (14.17)$$

We have that

$$\mathbf{E}\left\{\left(\sum_{i=j+1}^{n} \mu(B_{i,j,l})\left(T_i(Y_j^2) - r_i(\mathbf{X}_j)\right)\right)^2\right\}$$

$$= \mathbf{E}\left\{\sum_{i,i'=j+1}^{n} \mathbf{E}\left\{\mu(B_{i,j,l})\mu(B_{i',j,l})(T_i(Y_j^2) - r_i(\mathbf{X}_j))(T_{i'}(Y_j^2) - r_{i'}(\mathbf{X}_j)) \mid \mathbf{X}_j\right\}\right\}$$

$$= \mathbf{E}\left\{\sum_{i,i'=j+1}^{n} \mathbf{E}\left\{\mu(B_{i,j,l})\mu(B_{i',j,l}) \mid \mathbf{X}_j\right\} \mathbf{E}\left\{(T_i(Y_j^2) - r_i(\mathbf{X}_j))(T_{i'}(Y_j^2) - r_{i'}(\mathbf{X}_j)) \mid \mathbf{X}_j\right\}\right\}$$

$$\leq \int \sum_{i,i'=j+1}^{n} \sqrt{\mathbf{E}\{\mu(B_{i,j,l})^2 \mid \mathbf{X}_j = \mathbf{x}\}} \sqrt{\mathbf{E}\{\mu(B_{i',j,l})^2 \mid \mathbf{X}_j = \mathbf{x}\}}$$

$$\cdot \sqrt{\mathbf{E}\{(T_i(Y^2))^2 \mid \mathbf{X} = \mathbf{x}\}} \sqrt{\mathbf{E}\{(T_{i'}(Y^2))^2 \mid \mathbf{X} = \mathbf{x}\}} \mu(d\mathbf{x})$$

$$= \int \left(\sum_{i=j+1}^{n} \sqrt{\mathbf{E}\{\mu(B_{i,j,l})^2 \mid \mathbf{X}_j = \mathbf{x}\}} \sqrt{\mathbf{E}\{(T_i(Y^2))^2 \mid \mathbf{X} = \mathbf{x}\}}\right)^2 \mu(d\mathbf{x}).$$

According to [2] and pp. 489 and 490 in [7], one has that $\mathbf{P}\{\mu(B_{i,j,l}) > \sqrt{p}\}$ equals the probability that a *Binom*$(i-2, \sqrt{p})$-distributed random variable takes the value 0, i.e., $(1 - \sqrt{p})^{i-2}$ $(0 < p < 1)$. Thus,

$$\mathbf{E}\left\{\left(\sum_{i=j+1}^{n} \mu(B_{i,j,l})\left(T_i(Y_j^2) - r_i(\mathbf{X}_j)\right)\right)^2\right\}$$

$$\leq \int \left(\sum_{i=j+1}^{n} \sqrt{\int_0^1 \mathbf{P}\{\mu(B_{i,j,l}) > \sqrt{p} \mid \mathbf{X}_j = \mathbf{x}\}dp} \sqrt{\mathbf{E}\{(T_i(Y^2))^2 \mid \mathbf{X} = \mathbf{x}\}}\right)^2 \mu(d\mathbf{x})$$

$$= \int \left(\sum_{i=j+1}^{n} \sqrt{\int_0^1 (1-\sqrt{p})^{i-2}dp} \sqrt{\mathbf{E}\{(T_i(Y^2))^2 \mid \mathbf{X} = \mathbf{x}\}}\right)^2 \mu(d\mathbf{x})$$

$$= \int \left(\sum_{i=j+1}^{n} \sqrt{\frac{2}{(i-1)i}} \sqrt{\mathbf{E}\{(T_i(Y^2))^2 \mid \mathbf{X} = \mathbf{x}\}}\right)^2 \mu(d\mathbf{x})$$

$$\leq 8 \int \left(\sum_{i=j+1}^{n} \frac{1}{i} \sqrt{\mathbf{E}\{(T_n(Y^2))^2 | \mathbf{X} = \mathbf{x}\}} \right)^2 \mu(d\mathbf{x})$$

$$\leq 8 \left(\ln \frac{n}{j} \right)^2 \int \mathbf{E}\{(T_n(Y^2))^2 | \mathbf{X} = \mathbf{x}\} \mu(d\mathbf{x})$$

$$= 8 \left(\ln \frac{n}{j} \right)^2 \mathbf{E}\{(T_n(Y^2))^2\}.$$

Thus the left-hand side of (14.17) is bounded by

$$8 \sum_{n=1}^{\infty} \frac{1}{n^3} \sum_{j=1}^{n-1} \left(\ln \frac{n}{j} \right)^2 \mathbf{E}\{(T_n(Y^2))^2\} \leq 8 \, const \sum_{n=1}^{\infty} \frac{\mathbf{E}\{(T_n(Y^2))^2\}}{n^2} < \infty$$

(because of $\mathbf{E}\{Y^2\} < \infty$), our having used the fact that $\frac{1}{n} \sum_{j=1}^{n-1} \left(\ln \frac{n}{j} \right)^2 \to \int_0^1 \left(\ln \frac{1}{t} \right)^2 dt = \int_0^1 (\ln t)^2 dt < \infty$. Thus (14.17), and therefore (14.14) is proved. In the second part, it remains to show (14.13). In order to get it, according to the proof of Lemma 23.3 in Györfi et al. [7] it suffices to show

$$\limsup \frac{1}{n} \sum_{i=1}^{n} \int | m(\mathbf{x}) T_{\sqrt{i}}(Y_{i-1,1}(\mathbf{x})) | \mu(d\mathbf{x}) \leq c^* \mathbf{E}\{Y^2\} \quad \text{a.s.} \quad (14.18)$$

for some constant c^* and to show (14.13) for bounded Y. We prove first (14.18). Notice that

$$\int | m(\mathbf{x}) T_{\sqrt{i}}(Y_{i-1,1}(\mathbf{x})) | \mu(d\mathbf{x}) \leq \frac{1}{2} \int m(\mathbf{x})^2 \mu(d\mathbf{x}) + \frac{1}{2} \int T_i (Y_{i-1,1}(\mathbf{x}))^2 \mu(d\mathbf{x}).$$

From $\int m(\mathbf{x})^2 \mu(d\mathbf{x}) \leq \mathbf{E}\{Y^2\}$ and from (14.14) we obtain (14.18), with $c^* = \frac{1}{2} + \frac{1}{2}c$. By boundedness of Y, from some index on we have that $T_{\sqrt{i}}(Y) = Y$. Therefore, and because of Lemma 14.2, it suffices to show

$$\frac{1}{n} \sum_{i=1}^{n} \int_{R^d} m(\mathbf{x}) \left(Y_{i-1,1}(\mathbf{x}) - m(\mathbf{X}_{i-1,1}(\mathbf{x})) \right) \mu(d\mathbf{x}) \to 0 \quad a.s.,$$

where $m(\mathbf{X}_{0,1}(\mathbf{x})) := 0$. By boundedness, because of Lemma 14.3 it is enough to show

$$\sum_{n=1}^{\infty} \frac{\mathbf{E}\{[\sum_{i=1}^{n} \int_{R^d} m(\mathbf{x}) (Y_{i-1,1}(\mathbf{x}) - m(\mathbf{X}_{i-1,1}(\mathbf{x}))) \mu(d\mathbf{x})]^2\}}{n^3} < \infty. \quad (14.19)$$

Noticing

$$\int_{R^d} m(\mathbf{x}) \left(Y_{i-1,1}(\mathbf{x}) - m(\mathbf{X}_{i-1,1}(\mathbf{x}))\right) \mu(d\mathbf{x})$$

$$= \int_{R^d} m(\mathbf{x}) \sum_{j=1}^{i-1} \mathbf{I}_{\{\mathbf{X}_{i-1,1}(\mathbf{x}) = \mathbf{X}_j\}} \mu(d\mathbf{x}) \left(Y_j - m(\mathbf{X}_j)\right)$$

$$= \sum_{j=1}^{i-1} \int_{A_{i,j}} m(\mathbf{x}) \mu(d\mathbf{x}) \left(Y_j - m(\mathbf{X}_j)\right),$$

we obtain, with suitable constants c' and c'', that the left-hand side of (14.19) equals

$$\sum_{n=1}^{\infty} \frac{1}{n^3} \mathbf{E} \left\{ \left[\sum_{i=1}^{n} \sum_{j=1}^{i-1} \int_{A_{i,j}} m(\mathbf{x}) \mu(d\mathbf{x}) \left(Y_j - m(\mathbf{X}_j)\right) \right]^2 \right\}$$

$$= \sum_{n=1}^{\infty} \frac{1}{n^3} \mathbf{E} \left\{ \left[\sum_{j=1}^{n-1} \left(\sum_{i=j+1}^{n} \int_{A_{i,j}} m(\mathbf{x}) \mu(d\mathbf{x}) \right) \left(Y_j - m(\mathbf{X}_j)\right) \right]^2 \right\}$$

$$= \sum_{n=1}^{\infty} \frac{1}{n^3} \sum_{j=1}^{n-1} \mathbf{E} \left\{ \left(\sum_{i=j+1}^{n} \int_{A_{i,j}} m(\mathbf{x}) \mu(d\mathbf{x}) \left(Y_j - m(\mathbf{X}_j)\right) \right)^2 \right\}$$

(by orthogonality of the $(n-1)$ summands in the brackets)

$$\leq c' \sum_{n=1}^{\infty} \frac{1}{n^3} \sum_{j=1}^{n-1} \mathbf{E} \left\{ \left(\sum_{i=j+1}^{n} \mu(A_{i,j}) \right)^2 \right\}$$

$$\leq c'' \sum_{n=1}^{\infty} \frac{1}{n^3} \sum_{j=1}^{n-1} \left(\ln \frac{n}{j} \right)^2 < \infty;$$

the latter is as in the proof of (14.14). Thus (14.13) is proved for bounded Y. Therefore (14.12) and thus the assertion have been verified. □

14.3 Rate of Convergence

Next we bound the rate of convergence:

Theorem 14.3. *Assume that Y and $\mathbf{X}$ are bounded ($|Y| < L$, $\|\mathbf{X}\| < K$) and m is Lipschitz continuous and ties occur with probability 0. In addition, suppose that*

(i) μ has a Lipschitz continuous density f,
(ii) For any $\mathbf{x}$ from the support of μ and $0 < r < 2K$,

$$\mu(S_{\mathbf{x},r}) \geq \gamma r^d,$$

with $\gamma > 0$.

Then for $d \geq 2$, we have that

$$\mathbf{E}\{|\tilde{L}_n - L^*|\} \leq c_1 n^{-1/2} + c_2 n^{-2/d}.$$

Proof. Apply the decomposition

$$\mathbf{E}\{|\tilde{L}_n - L^*|\} \leq \mathbf{E}\{|\tilde{L}_n - \mathbf{E}\{\tilde{L}_n\}|\} + |\mathbf{E}\{\tilde{L}_n\} - L^*| \leq \sqrt{\mathbf{Var}(\tilde{L}_n)} + |\mathbf{E}\{\tilde{L}_n\} - L^*|.$$

For the variance term $\mathbf{Var}(\tilde{L}_n)$, introduce the notation

$$R_n := -\frac{1}{n} \sum_{i=1}^{n} Y_i Y_{n,i,1}.$$

For bounded Y ($|Y| \leq L$), we show that

$$\mathbf{Var}(R_n) \leq \frac{2(1 + 2\gamma_d)^2 L^4}{n}, \qquad (14.20)$$

from which we get that

$$\mathbf{Var}(\tilde{L}_n) = \mathbf{Var}\left(\frac{1}{n} \sum_{i=1}^{n} Y_i^2 + R_n\right)$$

$$\leq 2\mathbf{Var}\left(\frac{1}{n} \sum_{i=1}^{n} Y_i^2\right) + 2\mathbf{Var}(R_n) \leq \frac{2L^4}{n} + \frac{4(1 + 2\gamma_d)^2 L^4}{n},$$

and thus,

$$\sqrt{\mathbf{Var}(\tilde{L}_n)} \leq \frac{c_1}{\sqrt{n}}.$$

In the same way as in Liitiäinen et al. [9], we show (14.20) using the Efron–Stein inequality [5]. Replacement of $(\mathbf{X}_j, Y_j)$ by $(\mathbf{X}'_j, Y'_j)$ for fixed $j \in \{1, \ldots, n\}$ (where $(\mathbf{X}_1, Y_1), \ldots, (\mathbf{X}_n, Y_n), (\mathbf{X}'_1, Y'_1), \ldots, (\mathbf{X}'_n, Y'_n)$ are independent and identically distributed) leads to the estimator

$$R_{n,j} = \frac{1}{n}\left(Y_j' Y_{n,j,1}' + \sum_{i \neq j} Y_i Y_{n,i,1}'\right).$$

According to the Efron–Stein inequality we have that

$$\mathbf{Var}(R_n) \leq \frac{1}{2} \sum_{i=1}^{n} \mathbf{E}\{(R_n - R_{n,i})^2\} = \frac{n}{2} \mathbf{E}\{(R_n - R_{n,1})^2\}.$$

Evaluate the difference $R_n - R_{n,1}$:

$$R_n - R_{n,1} = \frac{1}{n}\left(Y_1 Y_{n,1,1} + \sum_{i \neq 1} Y_i Y_{n,i,1}\right) - \frac{1}{n}\left(Y_1' Y_{n,1,1}' + \sum_{i \neq 1} Y_i Y_{n,i,1}'\right)$$

$$= \frac{1}{n}\left(Y_1 Y_{n,1,1} - Y_1' Y_{n,1,1}'\right) + \frac{1}{n} \sum_{i \neq 1} Y_i (Y_{n,i,1} - Y_{n,i,1}').$$

One can check that $|Y_1 Y_{n,1,1} - Y_1' Y_{n,1,1}'| \leq 2L^2$. Introduce the following notations. Let $n[i]$ be the index of the first nearest neighbour of $\mathbf{X}_i$ from the set $\{\mathbf{X}_1, \mathbf{X}_2, \ldots, \mathbf{X}_n\} \setminus \{\mathbf{X}_i\}$. Similarly, let $n'[i]$ be the index of the first nearest neighbour of $\mathbf{X}_i$ from the set $\{\mathbf{X}_1', \mathbf{X}_2, \ldots, \mathbf{X}_n\} \setminus \{\mathbf{X}_i\}$. For fixed $i \neq 1$, notice

$$\{Y_{n,i,1} - Y_{n,i,1}' \neq 0\} \subset \{n[i] = 1\} \cup \{n'[i] = 1\}.$$

Thus

$$\left|\sum_{i \neq 1} Y_i (Y_{n,i,1} - Y_{n,i,1}')\right| \leq L \sum_{i \neq 1} |Y_{n,i,1} - Y_{n,i,1}'|$$

$$\leq 2L^2 \left(\sum_{i \neq 1} \mathbf{1}_{n[i]=1} + \sum_{i \neq 1} \mathbf{1}_{n'[i]=1}\right) \leq 2L^2 (\gamma_d + \gamma_d) = 4L^2 \gamma_d$$

a.s., where in the last step we applied Lemma 14.1. Summarizing these bounds we get that

$$\mathbf{Var}(R_n) \leq \frac{n}{2}\left(\frac{1}{n} 2L^2 + \frac{1}{n} 4L^2 \gamma_d\right)^2 = \frac{2(1 + 2\gamma_d)^2 L^4}{n}$$

a.s., and the proof of (14.20) is complete. For the bias term $\mathbf{E}\{\tilde{L}_n\} - L^*$, notice that

$$\mathbf{E}\{\tilde{L}_n\} - L^* = \mathbf{E}\{m(\mathbf{X}_1) m(\mathbf{X}_{n,1,1})\} - \mathbf{E}\{m(\mathbf{X}_1)^2\}.$$

Because of

$$m(\mathbf{X}_1)m(\mathbf{X}_{n,1,1}) - m(\mathbf{X}_1)^2 = \frac{1}{2}(m(\mathbf{X}_{n,1,1})^2 - m(\mathbf{X}_1)^2) - \frac{1}{2}(m(\mathbf{X}_{n,1,1}) - m(\mathbf{X}_1))^2,$$

the Lipschitz condition (14.1) implies that

$$|\mathbf{E}\{m(\mathbf{X}_1)m(\mathbf{X}_{n,1,1})\} - \mathbf{E}\{m(\mathbf{X}_1)^2\}|$$
$$\leq \frac{1}{2}|\mathbf{E}\{m(\mathbf{X}_{n,1,1})^2\} - \mathbf{E}\{m(\mathbf{X}_1)^2\}| + \frac{1}{2}\mathbf{E}\{(m(\mathbf{X}_{n,1,1}) - m(\mathbf{X}_1))^2\}$$
$$\leq \frac{1}{2}|\mathbf{E}\{m(\mathbf{X}_{n,1,1})^2\} - \mathbf{E}\{m(\mathbf{X}_1)^2\}| + \frac{C}{2}\mathbf{E}\{\|\mathbf{X}_{n,1,1} - \mathbf{X}_1\|^2\},$$

where C is the Lipschitz constant in (14.1). For $d \geq 3$, Lemma 6.4 in Györfi et al. [7], and for $d \geq 2$, Theorem 3.2 in Liitiäinen et al. [11] say that

$$\mathbf{E}\{\|\mathbf{X}_{n,1,1} - \mathbf{X}_1\|^2\} \leq c_3 n^{-2/d}.$$

Therefore

$$|\mathbf{E}\{m(\mathbf{X}_1)m(\mathbf{X}_{n,1,1})\} - \mathbf{E}\{m(\mathbf{X}_1)^2\}| \leq \frac{1}{2}|\mathbf{E}\{m(\mathbf{X}_{n,1,1})^2\} - \mathbf{E}\{m(\mathbf{X}_1)^2\}| + c_4 n^{-2/d},$$

and so we have to prove that

$$|\mathbf{E}\{m(\mathbf{X}_{n,1,1})^2\} - \mathbf{E}\{m(\mathbf{X}_1)^2\}| \leq c_4 n^{-1/2} + c_5 n^{-2/d}. \quad (14.21)$$

In order to show (14.21), let's calculate the density f_n of $(\mathbf{X}_1, \mathbf{X}_{n,1,1})$ with respect to $\mu \times \mu$. We have that

$$\mathbf{P}\{\mathbf{X}_1 \in A, \mathbf{X}_{n,1,1} \in B\}$$
$$= \sum_{i=2}^{n} \mathbf{P}\{\mathbf{X}_1 \in A, \mathbf{X}_i \in B, \mathbf{X}_{n,1,1} = \mathbf{X}_i\}$$
$$= (n-1)\mathbf{P}\{\mathbf{X}_1 \in A, \mathbf{X}_2 \in B, \mathbf{X}_{n,1,1} = \mathbf{X}_2\}$$
$$= (n-1)\mathbf{E}\{\mathbf{P}\{\mathbf{X}_1 \in A, \mathbf{X}_2 \in B, \mathbf{X}_{n,1,1} = \mathbf{X}_2 \mid \mathbf{X}_1, \mathbf{X}_2\}\}$$
$$= (n-1)\mathbf{E}\{\mathbf{P}\{\cap_{i=3}^{n}\{\|\mathbf{X}_1 - \mathbf{X}_i\| \geq \|\mathbf{X}_1 - \mathbf{X}_2\|\} \mid \mathbf{X}_1, \mathbf{X}_2\} \mathbf{I}_{\{\mathbf{X}_1 \in A, \mathbf{X}_2 \in B\}}\}$$
$$= (n-1)\mathbf{E}\left\{(1 - \mu(S_{\mathbf{X}_1, \|\mathbf{X}_1 - \mathbf{X}_2\|}))^{n-2} \mathbf{I}_{\{\mathbf{X}_1 \in A, \mathbf{X}_2 \in B\}}\right\}.$$

Therefore

$$f_n(\mathbf{x}_1, \mathbf{x}_2) = (n-1)\left(1 - \mu(S_{\mathbf{x}_1, \|\mathbf{x}_1 - \mathbf{x}_2\|})\right)^{n-2}.$$

It implies that

$$\mathbf{E}\{m(\mathbf{X}_1)^2\} = \mathbf{E}\{m(\mathbf{X}_1)^2 f_n(\mathbf{X}_1, \mathbf{X}_2)\}$$

and

$$\mathbf{E}\{m(\mathbf{X}_{n,1,1})^2\} = \mathbf{E}\{m(\mathbf{X}_2)^2 f_n(\mathbf{X}_1, \mathbf{X}_2)\} = \mathbf{E}\{m(\mathbf{X}_1)^2 f_n(\mathbf{X}_2, \mathbf{X}_1)\}.$$

Thus,

$$\mathbf{E}\{m(\mathbf{X}_{n,1,1})^2\} - \mathbf{E}\{m(\mathbf{X}_1)^2\} = \mathbf{E}\{m(\mathbf{X}_1)^2 (f_n(\mathbf{X}_2, \mathbf{X}_1) - f_n(\mathbf{X}_1, \mathbf{X}_2))\},$$

and interchanging $\mathbf{X}_1$ and $\mathbf{X}_2$ we get that

$$\mathbf{E}\{m(\mathbf{X}_{n,1,1})^2\} - \mathbf{E}\{m(\mathbf{X}_1)^2\} = -\mathbf{E}\{m(\mathbf{X}_2)^2 (f_n(\mathbf{X}_2, \mathbf{X}_1) - f_n(\mathbf{X}_1, \mathbf{X}_2))\},$$

and so

$$\mathbf{E}\{m(\mathbf{X}_{n,1,1})^2\} - \mathbf{E}\{m(\mathbf{X}_1)^2\} = \frac{1}{2}\mathbf{E}\{(m(\mathbf{X}_1)^2 - m(\mathbf{X}_2)^2)(f_n(\mathbf{X}_2, \mathbf{X}_1) - f_n(\mathbf{X}_1, \mathbf{X}_2))\}. \tag{14.22}$$

m satisfies the Lipschitz condition (14.1). Therefore

$$|m(\mathbf{x})^2 - m(\mathbf{z})^2| \leq |m(\mathbf{x}) - m(\mathbf{z})|(|m(\mathbf{x})| + |m(\mathbf{z})|) \leq 2LC\|\mathbf{x} - \mathbf{z}\|,$$

and so (14.22) implies that

$$|\mathbf{E}\{m(\mathbf{X}_{n,1,1})^2\} - \mathbf{E}\{m(\mathbf{X}_1)^2\}| \leq LC\mathbf{E}\{\|\mathbf{X}_1 - \mathbf{X}_2\| \cdot |f_n(\mathbf{X}_2, \mathbf{X}_1) - f_n(\mathbf{X}_1, \mathbf{X}_2)|\}.$$

For any $0 < a < b < 1$, we have the inequality

$$0 < (1-a)^n - (1-b)^n < n(b-a)(1-a)^{n-1}.$$

Therefore

$$|\mathbf{E}\{m(\mathbf{X}_{n,1,1})^2\} - \mathbf{E}\{m(\mathbf{X}_1)^2\}|$$
$$\leq LCn^2 \mathbf{E}\Big\{\|\mathbf{X}_1 - \mathbf{X}_2\| \cdot \big|\mu(S_{\mathbf{X}_1, \|\mathbf{X}_1 - \mathbf{X}_2\|}) - \mu(S_{\mathbf{X}_2, \|\mathbf{X}_1 - \mathbf{X}_2\|})\big|$$
$$\cdot \big(e^{-(n-2)\mu(S_{\mathbf{X}_1, \|\mathbf{X}_1 - \mathbf{X}_2\|})} + e^{-(n-2)\mu(S_{\mathbf{X}_2, \|\mathbf{X}_1 - \mathbf{X}_2\|})}\big)\Big\}.$$

If $c_d := Vol(S_{\mathbf{0},1})$ then condition (i) implies that

$$\big|\mu(S_{\mathbf{X}_1, \|\mathbf{X}_1 - \mathbf{X}_2\|}) - \mu(S_{\mathbf{X}_2, \|\mathbf{X}_1 - \mathbf{X}_2\|})\big| \leq c_d \|\mathbf{X}_1 - \mathbf{X}_2\|^d \max_{\|\mathbf{x} - \mathbf{z}\| \leq 2\|\mathbf{X}_1 - \mathbf{X}_2\|} |f(\mathbf{x}) - f(\mathbf{z})|$$

$$\le c_9 \|\mathbf{X}_1 - \mathbf{X}_2\|^{d+1}.$$

Because of condition (ii), both $e^{-(n-2)\mu(S_{\mathbf{X}_1,\|\mathbf{X}_1-\mathbf{X}_2\|})}$ and $e^{-(n-2)\mu(S_{\mathbf{X}_2,\|\mathbf{X}_1-\mathbf{X}_2\|})}$ are upper bounded by $e^{-(n-1)\gamma\|\mathbf{X}_1-\mathbf{X}_2\|^d}$. Therefore

$$|\mathbf{E}\{m(\mathbf{X}_{n,1,1})^2\} - \mathbf{E}\{m(\mathbf{X}_1)^2\}| \le c_{10} n^2 \mathbf{E}\left\{\|\mathbf{X}_1 - \mathbf{X}_2\|^{d+2} e^{-(n-1)\gamma\|\mathbf{X}_1-\mathbf{X}_2\|^d}\right\}.$$

Note that the random variable $R := \|\mathbf{X}_1 - \mathbf{X}_2\|$ has a density on $[0, 2K]$ bounded above by $c_{11} r^{d-1}$. Therefore

$$|\mathbf{E}\{m(\mathbf{X}_{n,1,1})^2\} - \mathbf{E}\{m(\mathbf{X}_1)^2\}| \le c_{12} n^2 \int_0^{2K} r^{d+2} e^{-n\gamma r^d} r^{d-1} dr$$

$$\le c_{12} n^{-2/d} \int_0^\infty r^{1+2/d} e^{-\gamma r} dr.$$

□

Acknowledgements This work was partially supported by the European Union and the European Social Fund through project FuturICT.hu (grant no.: TAMOP-4.2.2.C-11/1/KONV-2012-0013).

References

1. Chow, Y.S.: Local convergence of martingales and the law of large numbers. Ann. Math. Stat. **36**, 552–558 (1965)
2. Devroye, L., Györfi, L., Krzyżak, A., Lugosi, G.: On the strong universal consistency of nearest neighbor regression function estimation. Ann. Stat. **22**, 1371–1385 (1994)
3. Devroye, L., Schäfer, D., Györfi, L., Walk, H.: The estimation problem of minimum mean squared error. Stat. Decis. **21**, 15–28 (2003)
4. Dudoit, S., van der Laan, M.: Asymptotics of cross-validated risk estimation in estimator selection and performance assessment. Stat. Methodol. **2**, 131–154 (2005)
5. Efron, B., Stein, C.: The jackknife estimate of variance. Ann. Stat. **9**, 586–596 (1981)
6. Ferrario, P.G., Walk, H.: Nonparametric partitioning estimation of residual and local variance based on first and second nearest neighbors. J. Nonparametric Stat. **24**, 1019–1039 (2012)
7. Györfi, L., Kohler, M., Krzyżak, A., Walk, H.: A Distribution-Free Theory of Nonparametric Regression. Springer, New York (2002)
8. Landau, E.: Über die Bedeutung einiger neuer Grenzwertsätze der Herren Hardy und Axer. Prace mat.-fiz. **21**, 91–177 (1910)
9. Liitiäinen, E., Corona, F., Lendasse, A.: On nonparametric residual variance estimation. Neural Process. Lett. **28**, 155–167 (2008)
10. Liitiäinen, E., Verleysen, M., Corona, F., Lendasse, A.: Residual variance estimation in machine learning. Neurocomputing **72**, 3692–3703 (2009)
11. Liitiäinen, E., Corona, F., Lendasse, A.: Residual variance estimation using a nearest neighbor statistic. J. Multivar. Anal. **101**, 811–823 (2010)
12. Müller, H.G., Stadtmüller, U.: Estimation of heteroscedasticity in regression analysis. Ann. Stat. **15**, 610–625 (1987)

13. Müller, U., Schick, A., Wefelmeyer, W.: Estimating the error variance in nonparametric regression by a covariate-matched U-statistic. Statistics **37**, 179–188 (2003)
14. Neumann, M.H.: Fully data-driven nonparametric variance estimators. Statistics **25**, 189–212 (1994)
15. Stadtmüller, U., Tsybakov, A.: Nonparametric recursive variance estimation. Statistics **27**, 55–63 (1995)
16. Walk, H.: Strong laws of large numbers by elementary Tauberian arguments. Monatsh. Math. **144**, 329–346 (2005)

Chapter 15
The Median Hypothesis

Ran Gilad-Bachrach and Chris J.C. Burges

Abstract The classification task uses observations and prior knowledge to select a hypothesis that will predict class assignments well. In this work we ask the question: what is the best hypothesis to select from a given hypothesis class? To address this question we adopt a PAC-Bayesian approach. According to this viewpoint, the observations and prior knowledge are combined to form a belief probability over the hypothesis class. Therefore, we focus on the next part of the learning process, in which one has to choose the hypothesis to be used given the belief. We call this problem the hypothesis selection problem. Based on recent findings in PAC-Bayesian analysis, we suggest that a good hypothesis has to be close to the Bayesian optimal hypothesis. We define a measure of "depth" for hypotheses to measure their proximity to the Bayesian optimal hypothesis and we show that deeper hypotheses have stronger generalization bounds. Therefore, we propose algorithms to find the deepest hypothesis. Following the definitions of depth in multivariate statistics, we refer to the deepest hypothesis as the median hypothesis. We show that similarly to the univariate and multivariate medians, the median hypothesis has good stability properties in terms of the breakdown point. Moreover, we show that the Tukey median is a special case of the median hypothesis. Therefore, the algorithms proposed here also provide a polynomial time approximation for the Tukey median. This algorithm makes the mildest assumptions compared to other efficient approximation algorithms for the Tukey median.

R. Gilad-Bachrach (✉) · C.J.C. Burges
Microsoft Research, 1 Microsoft Way, Redmond, WA, USA
e-mail: rang@microsoft.com; cburges@microsoft.com

15.1 Introduction

According to the PAC-Bayesian point of view, learning can be split into three phases. First, a prior belief is introduced. Then, observations are used to transform the prior belief into a posterior belief. Finally, a hypothesis is selected. In this study, we concentrate on the last step. This allows us to propose methods that are independent of the first two phases. For example, the observations used to form the posterior belief can be supervised, unsupervised, semi-supervised, or something entirely different. The most commonly used method for selecting a hypothesis is to select the maximum a posteriori (MAP) hypothesis. For example, many learning algorithms use the following evaluation function (energy function):

$$E(f) = \sum_{i=1}^{n} l(f(x_i), y_i) + r(f) , \qquad (15.1)$$

where l is a convex loss function, $\{(x_i, y_i)\}_{i=1}^{n}$ are the observations and r is a convex regularization term. This can be viewed as a prior P over the hypothesis class with density $p(f) = \frac{1}{Z_p} e^{-r(f)}$ and a posterior belief Q with density $q(f) = \frac{1}{Z_q} e^{-E[f]}$. The common practice is then to select the hypothesis that minimizes the evaluation function, i.e., the MAP hypothesis. However, this choice has two significant drawbacks. First, since it considers only the maximal point, it misses much of the information encoded in the posterior belief. As a result it is straightforward to construct pathological examples: in [14] we give an example where the MAP classifier solution disagrees with the Bayes optimal hypothesis on every point, and where the Bayes optimal hypothesis[1] in fact *minimizes* the posterior probability. Second, the MAP framework is sensitive to perturbations in the posterior belief. That is, if we think of the MAP hypothesis as a statistic of the posterior, it has a low breakdown point [17]: in fact, its breakdown point is 0 [14].

This motivates us to study the problem of selecting the best hypothesis, given the posterior belief. The goal is to select a hypothesis that will generalize well. Two well-known methods for achieving this are the Bayes optimal hypothesis, and the Gibbs hypothesis, which selects a random classifier according to the posterior belief. However, the Gibbs hypothesis is non-deterministic, and in most cases the Bayes optimal hypothesis is not a member of the hypothesis class; these drawbacks are often shared by other hypothesis selection methods. This restricts the usability of these approaches. For example, in some cases, due to practical constraints, only a hypothesis from a given class can be used. Furthermore stochasticity in the predictive model can lead to adverse results in terms of consistency of the model and the ability to debug it. Therefore, in this work we limit the discussion to the following question: given a hypothesis class $\mathcal{F}$ and a posterior belief Q, how can one

---

[1]The Bayes optimal hypothesis is also known as the Bayes optimal classifier. It performs a weighted majority vote on each prediction according to the posterior.

15 The Median Hypothesis

select a hypothesis $f \in \mathcal{F}$ that will generalize well? We further limit the discussion to the binary classification setting.[2]

To answer this question we extend the notions of depth and the multivariate median, which are commonly used in multivariate statistics [21], to the classification setting. The depth function measures the centrality of a point in a sample or a distribution. For example, if Q is a probability measure over $\mathbb{R}^d$, the Tukey depth for a point $x \in \mathbb{R}^d$, also known as the half-space depth [26], is defined as

$$D(x|Q) = \inf_{H \text{ s.t. } x \in H \text{ and } H \text{ is halfspace}} Q(H) . \tag{15.2}$$

That is, the depth of a point x is the minimal measure of a half-space that contains it.[3] The Tukey depth also has a minimum entropy interpretation: each hyperplane containing x defines a Bernoulli distribution by splitting the distribution Q in two. Choose that hyperplane whose Bernoulli distribution has minimum entropy. The Tukey depth is then the probability mass on the side of that hyperplane with the lowest mass.

The depth function is thus a measure of centrality. The median is then simply defined as the deepest point. It is easy to verify that in the univariate case, the Tukey median is indeed the standard median. In this work we extend Tukey's definition beyond half-spaces and define depth for any hypothesis class. We show that the generalization error of a hypothesis is inversely proportional to its depth. Hence, the median hypothesis has the best generalization guaranty. We present algorithms for approximating the depth and the median. Since the Tukey depth is a special case of the hypothesis depth, our algorithms provide polynomial approximations to the Tukey depth and Tukey median as well. We analyze the stability of the median hypothesis and also discuss the case where a convex evaluation function $E(f)$ (see Eq. (15.1)) is used to form the posterior belief. We show that in this special case, the average hypothesis has a depth of at least $1/e$, independent of the dimension. Hence, it enjoys good generalization bounds.

After first introducing the notion of hypothesis depth, we address the issues of approximating depth and of approximating the median. Since we show that the Tukey median is a special case of the median hypothesis (see Sect. 15.3.1), the algorithms presented in Sect. 15.4 can be used to measure the Tukey depth and find the Tukey median efficiently. While some polynomial time algorithms for these problems do exist (see Sect. 15.2), they require restrictive assumptions that our algorithms avoid.

Regarding notation: we denote the sample space by $\mathcal{X}$, with instances $x \in \mathcal{X}$, measure μ, and a sample $S = \{x_1, \ldots, x_u\}$. We let $\mathcal{F}$ denote a function class with functions $f \in \mathcal{F}$, $f : \mathcal{X} \mapsto \pm 1$, measures P, Q, Q' and a sample

---

[2] Due to space constraints we omit proofs, all of which can be found in [14].
[3] Note that we can restrict the half-spaces in (15.2) to those half-spaces for which x lies on the boundary.

$T = \{f_1, \ldots, f_n\}$. $D_Q(f \mid x)$ is the depth of function f on the instance x with respect to measure Q, and $D_Q^{\delta,\mu}(f)$ is the δ-insensitive depth of f with respect to Q and μ. $\hat{D}_T(f \mid x)$ is the *empirical depth* of f on the instance x with respect to the sample T, and $\hat{D}_T^S(f)$ is the empirical depth of f with respect to the samples T and S. ν is a probability measure over $\mathcal{X} \times \{\pm 1\}$ and $\mathcal{S}$ is a sample $\{(x_i, y_i)\}_{i=1}^m$ from $(\mathcal{X} \times \{\pm 1\})^m$. Finally, the test error of f is $R_\nu(f) = \Pr_{(x,y) \sim \nu} [f(x) \neq y]$ and the empirical error of f is $R_\mathcal{S}(f) = \Pr_{(x,y) \sim \mathcal{S}} [f(x) \neq y]$.

15.2 Relation to Previous Work

In this section we survey the literature as it relates to our work.

Depth for functional data: Fraiman and Muniz [10] introduces an extension of univariate depth to function spaces. For a real function f, the depth of f is defined to be $E_x [D(f(x))]$ where $D(\cdot)$ is the univariate depth function. If we choose the rank function such that the rank of a value is the probability that a function will assign this value, we arrive at a similar definition to the one we propose. Whereas in [10] the depth is an average over all x's, we instead take the infimum, which plays a key role in the analysis. López-Pintado and Romo [22] also studies depth for functions. The definition of depth used therein is closer in spirit to the simplicial depth in the multivariate case [20]. As a consequence it is defined only for the case where the measure over the function space is an empirical measure over a finite set of functions. Zuo [27] studies a projection-based depth. The functional depth we present in this work can in fact be derived from this quantity [14].

Classification depth: Ghosh and Chaudhuri [13] uses depth for classification purposes. Given samples from the different classes, authors create depth functions for each of the classes, and at inference time, the algorithm associates an instance x with the class in which x is deepest. Ghosh and Chaudhuri [13] proves generalization bounds for elliptic class distributions. Cuevas et al. [8] uses a similar approach and compares the performance of different depth functions empirically. Jörnsten [19] uses a similar approach with an L_1-based depth function. Billor et al. [2] proposes another variant of this technique.

Ghosh [12] introduces two variants of depth functions to be used for learning linear classifiers. Authors show that as the sample size goes to infinity, the maximal depth classifier is the optimal linear classifier. However, since maximizing this quantity is known to be hard, the authors suggest using the logistic function as a surrogate to the indicator function. These methods are therefore very close (and in some cases identical) to logistic regression.

Gilad-Bachrach [15] uses the Tukey depth to analyze the generalization performance of the Bayes Point Machine [18]. Our work uses depth in a similar fashion, although the approach in [15] is limited to the linear classification case, and its analysis is restricted to the realizable case where there exists a classifier which correctly classifies the entire data set. We do not make these assumptions.

Regression depth: Rousseeuw and Hubert [25] introduces the notion of regression depth. Its authors discuss linear regression but their definition can be extended to general function classes in the following way: Let $\mathcal{F} = \{f : \mathcal{X} \mapsto \mathbb{R}\}$ be a function class and let $S = \{(x_i, y_i)\}_{i=1}^n$ be a sample such that $x_i \in \mathcal{X}$ and $y_i \in \mathbb{R}$. We say that the function $f \in \mathcal{F}$ has depth 0 ("non-fit" in [25]) if there exists $g \in \mathcal{F}$ that is strictly better than f on every point in S. That is, for every point (x_i, y_i), either $f(x_i) < g(x_i) \leq y_i$ or $f(x_i) > g(x_i) \geq y_i$. We say that a function $f \in \mathcal{F}$ has a depth d if d is the minimal number of points that should be removed from S to make f a non-fit. For example, if f is a perfect match, that is, $\forall i, \ f(x_i) = y_i$, then f will become a non-fit only if all the points in S are removed, and hence f has a depth n which is the largest value possible. Christmann [5] then applies the regression depth to the classification task, using the logit function to convert the classification task to a regression problem. Its authors show that in this setting the regression depth is closely related to logistic regression and to the well-known risk minimization technique.

Methods for computing the Tukey median: We propose algorithms for computing the hypothesis depth and for approximating the median hypothesis. Since the Tukey depth is a special case of the hypothesis depth, here we survey the literature on computing the Tukey median. Chan [4] presents a randomized algorithm that can find the Tukey median for a sample of n points with expected computational complexity of $O(n \log n)$ when the data is in $\mathbb{R}^2$ and $O(n^{d-1})$ when the data is in $\mathbb{R}^d$ for $d > 2$. Its authors conjecture that these results are optimal for finding the exact median. Massé [23] analyzes the asymptotic behavior of the empirical Tukey depth, which is the Tukey depth function when it is applied to an empirical measure. Massé [23] shows that the empirical depth converges uniformly to the true depth with probability 1, and that the empirical median converges to the true median at a rate that scales as $1/\sqrt{n}$. Cuesta-Albertos and Nieto-Reyes [7] proposes picking k random directions and computing the univariate depth of each candidate point for each direction. Its authors define the random Tukey depth for a given point to be the minimum univariate depth of the point with respect to the k random directions, and they search for the number of directions needed to obtain a good empirical approximation of the depth.

Note that the empirical depth of [23] and the random Tukey depth of [7] are different quantities. In the case of empirical depth, when evaluating the depth of a point x, one considers every possible hyperplane and evaluates the measure of the corresponding half-space using only a sample. For the random depth, one evaluates only k different hyperplanes, but for each hyperplane it is assumed that the true probability of the half-space is computable. Therefore, each of these approaches solves one of the problems involved in computing the Tukey depth, and in order to solve both problems simultaneously, an additional approximation is needed.

The solution we present addresses both issues and proves the convergence of the outcome to the Tukey depth, as well as giving the rate of convergence.

Since finding the deepest point is hard, some studies focus on just finding a deep point. Clarkson et al. [6] present an algorithm for finding a point with depth $\Omega\left(1/d^2\right)$ in polynomial time. However when the distribution is log-concave, it is known that there exists a point with depth $1/e$, independent of the dimension [3], and for any distribution there is a point with a depth of at least $1/d+1$.

15.3 The Hypothesis Depth: Definitions and Properties

In this study, unlike Tukey, who used the depth function on the instance space, we view the depth function as operating on the space of classification functions. Moreover, the definition here extends beyond the linear case to any function class. The depth function measures the agreement of the function f with the weighted majority vote on x. A deep function is a function that will always have a large agreement with its prediction over the class $\mathcal{F}$.

Definition 15.1. Let $\mathcal{F}$ be a function class and let Q be a probability measure over $\mathcal{F}$. The depth of f on the instance $x \in \mathcal{X}$ with respect to Q is defined as

$$D_Q(f \mid x) = \Pr_{g \sim Q}[g(x) = f(x)] \;.$$

The depth of f with respect to Q is defined as

$$D_Q(f) = \inf_{x \in \mathcal{X}} D_Q(f \mid x) = \inf_{x \in \mathcal{X}} \Pr_{g \sim Q}[g(x) = f(x)] \;.$$

The Tukey depth is a special case of this definition, as discussed in Sect. 15.3.1. The depth $D_Q(f)$ is defined as the infimum over all points $x \in \mathcal{X}$. We relax this by defining the δ-insensitive depth:

Definition 15.2. Let $\mathcal{F}$ be a function class and let Q be a probability measure over $\mathcal{F}$. Let μ be a probability measure over $\mathcal{X}$ and let $\delta \geq 0$. The δ-insensitive depth of f with respect to Q and μ is defined as

$$D_Q^{\delta, \mu}(f) = \sup_{\mathcal{X}' \subseteq \mathcal{X}, \mu(\mathcal{X}') \leq \delta} \inf_{x \in \mathcal{X} \setminus \mathcal{X}'} D_Q(f \mid x) \;.$$

Thus instead of requiring that the function f always have a large agreement in the class $\mathcal{F}$, the δ-insensitive depth makes this requirement on all but a set of instances with probability mass δ.

With these definitions in hand, we next provide generalization bounds for deep hypotheses. The first theorem shows that the error of a deep function is close to the error of the Gibbs classifier.

15 The Median Hypothesis

Theorem 15.1 (Deep vs. Gibbs). *Let Q be a measure on $\mathcal{F}$. Let v be a measure on $\mathcal{X} \times \{\pm 1\}$ with the marginal μ on $\mathcal{X}$. For every $f \in F$ the following holds:*

$$R_v(f) \leq \frac{1}{D_Q(f)} E_{g \sim Q} [R_v(g)]$$

and

$$R_v(f) \leq \frac{1}{D_Q^{\delta,\mu}(f)} E_{g \sim Q} [R_v(g)] + \delta .$$

Note that the term $E_{g \sim Q}[R_v(g)]$ is the expected error of the Gibbs classifier.[4] Thus we see that the generalization error of a deep hypothesis cannot be large, provided that the expected error of the Gibbs classifier is not large.

Theorem 15.1 bounds the ratio of the generalization error of the Gibbs classifier to the generalization error of a given classifier as a function of the depth of that classifier. For example, consider the Bayes optimal classifier. By definition, the depth of this classifier is at least 1/2; thus Theorem 15.1 recovers the well-known result that the generalization error of the Bayes optimal classifier is at most twice as large as the generalization error of the Gibbs classifier.

Next, we combine Theorem 15.1 with PAC-Bayesian bounds [24] to bound the difference between the training error and the test error. We use the version of the PAC-Bayesian bounds in Theorem 3.1 of [11].

Theorem 15.2 (Generalization Bounds). *Let v be a probability measure on $\mathcal{X} \times \{\pm 1\}$, let P be a probability measure of $\mathcal{F}$ and let $\delta, \kappa > 0$. With a probability greater than $1 - \delta$ over the sample S sampled from v^m:*

$$\forall Q, \forall f, R_v(f) \leq$$

$$\frac{1}{(1-e^{-\kappa}) D_Q(f)} \left(\kappa E_{g \sim Q} [R_S(g)] + \frac{1}{m} \left[KL(Q||P) + \ln \frac{1}{\delta} \right] \right) .$$

Furthermore, for every $\delta' > 0$, the following holds with a probability greater than $1 - \delta$ over the sample S sampled from v^m:

$$\forall Q, \forall f, R_v(f) \leq$$

$$\frac{1}{(1-e^{-\kappa}) D_Q^{\delta',\mu}(f)} \left(\kappa E_{g \sim Q} [R_S(g)] + \frac{1}{m} \left[KL(Q||P) + \ln \frac{1}{\delta} \right] \right) + \delta'$$

where μ is the marginal of v on $\mathcal{X}$.

---

[4] Note that this is not necessarily the same as the expected error of the Bayes optimal hypothesis.

Theorem 15.2 shows that if a deep function exists, then it is expected to generalize well, provided that the PAC-Bayes bound for Q is sufficiently smaller than the depth of f. This justifies our pursuit to find the deepest function, that is, the median. However, the question remains: are there any deep functions? We next show that for linear classifiers, indeed there are.

15.3.1 Depth for Linear Classifiers

We wish to show that deep functions exist for linear classifiers and that the Tukey depth is a special case of the hypothesis depth. To that end we use a variant of linear classifiers called *linear threshold functions*. In this setting $\mathcal{F} = \mathbb{R}^d$ and $\mathcal{X} = \mathbb{R}^d \times \mathbb{R}$ such that $f \in \mathcal{F}$ operates on $x = (x_v, x_\theta) \in \mathcal{X}$ by $f(x) = \text{sign}(f \cdot x_v - x_\theta)$.

Theorem 15.3. *Let $\mathcal{X} = \mathbb{R}^d \times \mathbb{R}$ and $\mathcal{F}$ be the class of linear threshold functions over $\mathcal{X}$. Let Q be a probability measure over $\mathcal{F}$ with density function $q(f)$ such that $q(f) = \frac{1}{Z} \exp(-E(f))$ where $E(f)$ is a convex function. Then there exists a function $f^* \in \mathcal{F}$ such that $D_Q(f^*) \geq 1/e$. Moreover, $f^* = E_{f \sim Q}[f]$.*

As noted above, many learning algorithms use convex energy functions of the form presented in (15.1). Hence, in all these cases, the median has a depth of at least $1/e$. This leads to the following conclusion:

Conclusion 1. *In the setting of Theorem 15.3, let $f^* = E_{f \sim Q}[f]$. Let ν be a probability measure on $\mathcal{X} \times \{\pm 1\}$, let P be a probability measure of $\mathcal{F}$ and let $\delta, \kappa > 0$. With a probability greater than $1 - \delta$ over the sample $\mathcal{S}$ sampled from ν^m:*

$$R_\nu(f^*) \leq \frac{e}{(1-e^{-\kappa})} \left(\kappa E_{g \sim Q}[R_\mathcal{S}(g)] + \frac{1}{m}\left[KL(Q\|P) + \ln\frac{1}{\delta} \right] \right) .$$

We now show that the Tukey depth is a special case of the hypothesis depth. Again, we will use the class of linear threshold functions. Since $\mathcal{F} = \mathbb{R}^d$ we can treat $f \in \mathcal{F}$ both as a function and as a point. Therefore, a probability measure Q over $\mathcal{F}$ is also considered as a probability measure over $\mathbb{R}^d$. The following theorem shows that for any $f \in \mathcal{F}$, the Tukey depth of f and the hypothesis depth of f are the same.

Theorem 15.4. *If $\mathcal{F}$ is the class of threshold functions then for every $f \in \mathcal{F}$:*

$$D_Q(f) = \text{Tukey depth}_Q(f) .$$

15.3.2 Breakdown Point

We now discuss another important property of the hypothesis selection: the breakdown point. Any solution to the hypothesis selection problem may be viewed as a statistic of the posterior Q. An important property of any such statistic is its stability [17], as quantified by its breakdown point:

Definition 15.3. Let **Est** be a function that maps probability measures to $\mathcal{F}$. For two probability measures Q and Q' let $\delta(Q, Q')$ be the total variation distance: $\delta(Q, Q') = \sup\{|Q(A) - Q'(A)| : A \text{ is measurable}\}$. For every function $f \in \mathcal{F}$ let $d(\mathbf{Est}, Q, f)$ be the distance to the closest Q' such that $\mathbf{Est}(Q') = f$: $d(\mathbf{Est}, Q, f) = \inf\{\delta(Q, Q') : \mathbf{Est}(Q') = f\}$. The breakdown **Est** at Q is defined to be the distance to the furthest function: $\mathbf{breakdown}(\mathbf{Est}, Q) = \sup_{f \in \mathcal{F}} d(\mathbf{Est}, Q, f)$.

This definition may be interpreted as follows: if $s = \mathbf{breakdown}(\mathbf{Est}, Q)$ then for every $f \in \mathcal{F}$, we can force the estimator **Est** to use f as its estimate by changing Q by at most s in terms of total variation distance. Therefore, the larger the breakdown point of an estimator, the more stable it is with respect to perturbations in Q. The following theorem lower bounds the stability of the median estimator as a function of its depth.

Theorem 15.5. *Let Q be a posterior over $\mathcal{F}$. Let*

$$X' = \{x \in \mathcal{X} \text{ s.t. } \forall f_1, f_2 \in \mathcal{F}, \ f_1(x) = f_2(x)\} \text{ and}$$

$$p = \inf_{x \in \mathcal{X} \setminus X', y \in \pm 1} Q\{f : f(x) = y\} \ .$$

If d is the depth of the median for Q then $\mathbf{breakdown}(\mathbf{median}, Q) \geq \frac{d-p}{2}$.

15.4 Measuring Depth

So far, we have motivated the use of depth as a criterion for selecting a hypothesis. However, finding the deepest function, even in the case of linear functions, can be hard. In this section we focus on algorithms that measure the depth of functions. The main results are an efficient algorithm for approximating the depth uniformly over the entire function class, and an algorithm for approximating the median.

The depth estimation algorithm (Algorithm 1) takes as inputs two samples. One sample, $S = \{x_1, \ldots, x_u\}$, is a sample of points from the domain $\mathcal{X}$. The other sample, $T = \{f_1, \ldots, f_n\}$, is a sample of functions from $\mathcal{F}$. Given a function f for which we would like to compute the depth, the algorithm first estimates its depth on the points $x_1, \ldots, x_u$ and then uses the minimal value as an estimate of the global depth. The depth on a point x_i is estimated by counting the fraction of the functions $f_1, \ldots, f_n$ that make the same prediction as f on the point x_i. Since samples are used to estimate depth, we call the value returned by this algorithm, $\hat{D}_T^S(f)$, the empirical depth of f.

Despite its simplicity, the depth estimation algorithm can provide good estimates of the true depth, which in fact are uniformly good over all the functions $f \in \mathcal{F}$,

Algorithm 1 Depth estimation algorithm

Inputs:

- A sample $S = \{x_1, \ldots, x_u\}$ such that $x_i \in \mathcal{X}$
- A sample $T = \{f_1, \ldots, f_n\}$ such that $f_j \in \mathcal{F}$
- A function f

Output:

- $\hat{D}_T^S(f)$ – an approximation for the depth of f

Algorithm:

1. For $i = 1, \ldots, u$ compute $\hat{D}_T(f \mid x_i) = \frac{1}{n} \sum_j 1_{f_j(x_i) = f(x_i)}$
2. Return $\hat{D}_T^S(f) = \min_i \hat{D}(f \mid x_i)$

---

as the following theorem shows. This will be an essential building block when we seek to find the median in Sect. 15.4.1.

Theorem 15.6 (Uniform convergence of depth). *Let Q be a probability measure on $\mathcal{F}$ and let μ be a probability measure on $\mathcal{X}$. Let $\epsilon, \delta > 0$. For every $f \in \mathcal{F}$ let the function f_δ be such that $f_\delta(x) = 1$ if $D_Q(f \mid x) \leq D_Q^{\delta,\mu}(f)$ and $f_\delta(x) = -1$ otherwise. Let $\mathcal{F}_\delta = \{f_\delta\}_{f \in \mathcal{F}}$. Assume $\mathcal{F}_\delta$ has a finite VC dimension $d < \infty$ and define $\phi(d, k) = \sum_{i=0}^d \binom{k}{i}$ if $d < k$, $\phi(d, k) = 2^k$ otherwise. If S and T are chosen at random from μ^u and Q^n respectively such that $u \geq 8/\delta$, then with probability*

$$1 - u \exp\left(-2n\epsilon^2\right) - \phi(d, 2u) 2^{1 - \delta u/2}$$

the following holds:

$$\forall f \in \mathcal{F}, \quad D_Q(f) - \epsilon \leq D_Q^{0,\mu}(f) - \epsilon \leq \hat{D}_T^S(f) \leq D_Q^{\delta,\mu}(f) + \epsilon$$

where $\hat{D}_T^S(f)$ is the empirical depth computed by the depth measure algorithm.

Theorem 15.6 shows that the estimated depth converges uniformly to the true depth. However, since we are interested in deep hypotheses, it is suffices that the estimate is accurate for these hypotheses, as long as "shallow" hypotheses are distinguishable from deep ones. This is the motivation for our next theorem:

Theorem 15.7. *Let Q be a probability measure on $\mathcal{F}$ and let μ be a probability measure on $\mathcal{X}$. Let $\epsilon, \delta > 0$. Assume $\mathcal{F}$ has a finite VC dimension $d < \infty$ and define $\phi(d, k)$ as before. Let $D = \sup_{f \in \mathcal{F}} D_Q(f)$. If S and T are chosen at random from μ^u and Q^n respectively such that $u \geq 8/\delta$ then with probability $1 - u \exp\left(-2n\epsilon^2\right) - \phi(d, 2u) 2^{1 - \delta u/2}$ the following hold:*

1. For every f such that $D_Q^{\delta,\mu}(f) < D$ we have that $\hat{D}_S^T(f) \le D_Q^{\delta,\mu}(f) + \epsilon$
2. For every f we have that $\hat{D}_S^T(f) \ge D_Q^{0,\mu}(f) - \epsilon \ge D_Q(f) - \epsilon$

where $\hat{D}_T^S(f)$ is the empirical depth computed by the depth measure algorithm.

15.4.1 Finding the Median

We have seen that if the samples S and T are large enough, then with high probability the estimated depth is accurate uniformly for all functions $f \in \mathcal{F}$. We now use these findings to present an algorithm which approximates the median. Recall that the median f maximizes the depth, that is, $f = \arg\max_{f \in \mathcal{F}} D_Q(f)$. As an approximation, we will present an algorithm which finds a function f that maximizes the empirical depth, that is, $f = \arg\max_{f \in \mathcal{F}} \hat{D}_T^S(f)$.

The intuition behind the algorithm is simple. Let $S = \{x_i\}_{i=1}^u$. A function that has large empirical depth will agree with the majority vote on these points. However, such a function may not exist. If we are forced to find a hypothesis that does not agree with the majority on some instances, the empirical depth will be higher if these points are such that the majority vote on them wins by a small margin. Therefore, we take a sample $T = \{f_j\}_{j=1}^n$ of functions and use them to compute the majority vote on every x_i and the fraction q_i of functions which disagree with the majority vote. A viable strategy will first try to find a function that agrees with the majority votes on all the points in S. If such a function does not exist, we remove the point for which q_i is the largest and try to find a function that agrees with the majority vote on the remaining points. This process continues until a consistent function[5] is found. This function is the maximizer of $\hat{D}_T^S(f)$. In the Median Approximation algorithm, this process is accelerated by using binary search. Assuming that the consistency algorithm requires $O(u^c)$ when working on a sample of size u, the linear search described above requires $O(nu + u\log(u) + u^{c+1})$ operations, while using binary search reduces the complexity to $O(nu + u\log(u) + u^c \log(u))$.

The Median Approximation (MA) algorithm is presented in Algorithm 2. One of the key advantages of the MA algorithm is that it uses a consistency oracle instead of an oracle that minimizes the empirical error. Minimizing the empirical error is hard in many cases, and even hard to approximate [1]. In contrast, the MA algorithm requires only access to an oracle that is capable of finding a consistent hypothesis if one exists. For example, in the case of a linear classifier, finding a consistent hypothesis can be achieved in polynomial time by linear programming while finding a hypothesis which approximates the one with minimal empirical error is NP-hard. The following theorem lists key properties of the MA algorithm.

---

[5] A function is defined to be consistent with a labelled sample if it labels correctly all the instances in the sample.

Algorithm 2 Median Approximation (MA)

Inputs:

- A sample $S = \{x_1, \ldots, x_u\} \in \mathcal{X}^u$ and a sample $T = \{f_1, \ldots, f_n\} \in \mathcal{F}^n$.
- A learning algorithm $\mathcal{A}$ that given a sample returns a function consistent with it if such a function exists.

Outputs:

- A function $f \in \mathcal{F}$ and its depth estimation $\hat{D}_T^S(f)$

Details:

1. For each $i = 1, \ldots, u$ compute $p_i^+ = \frac{1}{n} |\{j : f_j(x_i) = 1\}|$ and $q_i = \min \{p_i^+, 1 - p_i^+\}$.
2. Sort $x_1, \ldots, x_u$ such that $q_1 \geq q_2 \geq \ldots \geq q_m$
3. For each $i = 1, \ldots, u$ let $y_i = 1$ if $p_i^+ \geq 0.5$; otherwise, let $y_i = -1$
4. Use binary search to find i^*, the smallest i for which $\mathcal{A}$ can find a consistent function f with the sample $S^i = \{(x_k, y_k)\}_{k=i}^u$
5. If $i^* \equiv 1$ return f and depth $\hat{D} = 1 - q_1$, else return f and depth $\hat{D} = q_{i^*-1}$

Theorem 15.8 (The MA Theorem). *The MA algorithm (Algorithm 2) has the following properties:*

1. *The algorithm will always terminate and return a function $f \in \mathcal{F}$ and an empirical depth $\hat{D}$.*
2. *If f and $\hat{D}$ are the outputs of the MA algorithm then $\hat{D} = \hat{D}_T^S(f)$.*
3. *If f is the function returned by the MA algorithm then $f = \arg\max_{f \in \mathcal{F}} \hat{D}_T^S(f)$.*
4. *Let $\epsilon, \delta > 0$. If the sample S is taken from μ^u such that $u \geq 8/\delta$, and the sample T is taken from Q^n, then with probability at least*

$$1 - u \exp\left(-2n\epsilon^2\right) - \phi(d, 2u)\, 2^{1-\delta u/2}$$

the f returned by the MA algorithm is such that

$$D_Q^{\delta,\mu}(f) \geq \sup_{g \in F} D_Q^{0,\mu}(g) - 2\epsilon \geq \sup_{g \in F} D_Q(g) - 2\epsilon$$

where d is the minimum between the VC dimension of $\mathcal{F}$ and the VC dimension of the class $\mathcal{F}_\delta$ defined in Theorem 15.6.

15.4.2 Implementation Issues

The MA algorithm requires access to three oracles that provide: (1) a sample of unlabelled instances $x_1, \ldots, x_u$ from μ, (2) a sample of hypotheses $f_1, \ldots, f_n$ from the belief distribution Q, and (3) a learning algorithm $\mathcal{A}$ that returns a hypothesis consistent with the sample of instances (if such a hypothesis exists).

The first requirement is usually trivial. In a sense, the MA algorithm converts the consistency algorithm $\mathcal{A}$ to a semi-supervised learning algorithm by using this sample. The third requirement is not too restrictive. Many learning algorithms would be much simpler if they required a hypothesis which is consistent with the entire sample as opposed to a hypothesis which minimizes the number of mistakes (see, for example, [1]). The second requirement, that is, sampling hypotheses, is challenging.

Sampling hypotheses is hard even in very restrictive cases. For example, even if Q is uniform over a convex body, sampling from it is challenging but theoretically possible [9]. A closer look at the MA algorithm and the depth estimation algorithm reveals that they use the sample of functions in order to estimate the marginal $Q[Y = 1|X = x] = \Pr_{g \sim Q}[g(x) = 1]$. In some cases, it is possible to directly estimate this value. For example, many learning algorithms output a real value such that the sign of the output is the predicted label and the amplitude is the margin. Using a sigmoid function, this can be viewed as an estimate of $Q[Y = 1|X = x]$. This can be used directly in the above algorithms. Moreover, the results of Theorems 15.6 and 15.8 apply with $\epsilon = 0$. Note that the algorithm that is used for computing the probabilities might be infeasible for run-time applications but can still be used in the process of finding the median.

Another option is to sample from a distribution Q' that approximates Q [16]. The way to use a sample from Q' is to reweigh the functions when computing $\hat{D}_T(f | x)$. Note that computing $\hat{D}_T(f | x)$ such that it is close to $D_Q(f | x)$ is sufficient for estimating the depth using the depth measure algorithm (Algorithm 1) and for finding the approximated median using the MA algorithm (Algorithm 2). Therefore, in this section we will focus only on computing the empirical conditional depth $\hat{D}_T(f | x)$. The following definition provides the estimate for $D_Q(f | x)$ given a sample T sampled from Q':

Definition 15.4. Given a sample T and the relative density function $\frac{dQ}{dQ'}$, we define

$$\hat{D}_{T, \frac{dQ}{dQ'}}(f) = \frac{1}{n} \sum_j \frac{dQ(f_j)}{dQ'(f_j)} 1_{f_j(x) = f(x)}.$$

To see the intuition behind this definition, recall that

$$D_Q(f | x) = \Pr_{g \sim Q}[g(x)]$$

and

$$\hat{D}_T(f | x) = \frac{1}{n} \sum_j 1_{f_j(x) = f(x)}$$

where $T = \{f_j\}_{j=1}^n$. If T is sampled from Q^n we have that

$$E_{T\sim Q^n}\left[\hat{D}_T(f\mid x)\right] = \frac{1}{n}\sum_j E\left[1_{f_j(x)=f(x)}\right]$$
$$= \frac{1}{n}\sum_j \Pr[f_j(x) = f(x)] = D_Q(f\mid x).$$

Therefore, we will show that $\hat{D}_{T,\frac{dQ}{dQ'}}(f)$ is an unbiased estimate of $D_Q(f\mid x)$ and that it is concentrated around its expected value.

Theorem 15.9. *Let Q and Q' be probability measures over $\mathcal{F}$. Then:*

1. *For every f, $E_{T\sim Q'^n}\left[\hat{D}_{T,\frac{dQ}{dQ'}}(f)\right] = D_Q(f\mid x)$.*
2. *If $\frac{dQ}{dQ'}$ is bounded such that $\frac{dQ}{dQ'} \leq c$ then*

$$\Pr_{T\sim Q'^n}\left[\left|\hat{D}_{T,\frac{dQ}{dQ'}}(f) - D_Q(f\mid x)\right| > \epsilon\right] < 2\exp\left(-\frac{2n\epsilon^2}{c^2}\right).$$

15.5 Discussion

We studied the problem of selecting the best hypothesis, given a posterior belief over the hypothesis class. We defined a depth function for classifiers and showed that the generalization of a classifier is tied to its depth, which suggested that the deepest classifier, the median, is a good hypothesis to select. We analyzed the breakdown properties of the median and showed that it is related to the depth as well. We discussed the algorithmic aspects of our proposed solution and presented efficient algorithms for uniformly measuring the depth and for finding the median. Since the Tukey depth is a special case of the depth presented here, it follows that the Tukey depth and the Tukey median can be approximated in polynomial time by our algorithms.

Our discussion was limited to the binary classification case. It will be interesting to see if this work can be extended to other scenarios, for example, regression, multi-class classification, and ranking.

References

1. Ben-David, S., Eiron, N., Long, P.M.: On the difficulty of approximately maximizing agreements. J. Comput. Syst. Sci. **66**(3), 496–514 (2003)
2. Billor, N., Abebe, A., Turkmen, A., Nudurupati, S.V.: Classification based on depth transvariations. J. Classif. **25**(2), 249–260 (2008)

3. Caplin, A., Nalebuff, B.: Aggregation and social choice: a mean voter theorem. Econometrica **59**(1), 1–23 (1991)
4. Chan, T.M.: An optimal randomized algorithm for maximum Tukey depth. In: SODA, New Orleans, pp. 430–436 (2004)
5. Christmann, A.: Regression depth and support vector machine. DIMACS Ser. Discret. Math. Theor. Comput. Sci. **72**, 71 (2006)
6. Clarkson, K.L., Eppstein, D., Miller, G.L., Sturtivant, C., Teng, S.H.: Approximating center points with iterative Radon points. Int. J. Comput. Geom. Appl. **6**(3), 357–377 (1996)
7. Cuesta-Albertos, J.A., Nieto-Reyes, A.: The random Tukey depth. Comput. Stat. Data Anal. **52**, 4979–4988 (2008)
8. Cuevas, A., Febrero, M., Fraiman, R.: Robust estimation and classification for functional data via projection-based depth notions. Comput. Stat. **22**, 481–496 (2007)
9. Fine, S., Gilad-Bachrach, R., Shamir, E.: Query by committee, linear separation and random walks. Theor. Comput. Sci. **284**(1), 25–51 (2002)
10. Fraiman, R., Muniz, G.: Trimmed means for functional data. Test **10**(2), 419–440 (2001)
11. Germain, P., Lacasse, A., Laviolette, F., Marchand, M., Shanian, S.: From PAC-Bayes bounds to KL regularization. In: NIPS, Vancouver (2009)
12. Ghosh, A., Chaudhuri, P.: On data depth and distribution-free discriminant analysis using separating surfaces. Bernoulli **11**(1), 1–27 (2005)
13. Ghosh, A.K., Chaudhuri, P.: On maximum depth and related classifiers. Scand. J. Stat. **32**(2), 327–350 (2005)
14. Gilad-Bachrach, R., Burges, C.J.C.: Classifier selection using the predicate depth. Technical Report MSR-TR-2013-8, Microsoft Research (2013)
15. Gilad-Bachrach, R., Navot, A., Tishby, N.: Bayes and Tukey meet at the center point. In: Proceedings of the Conference on Learning Theory (COLT), Banff, pp. 549–563. Springer (2004)
16. Gilad-Bachrach, R., Navot, A., Tishby, N.: Query by committee made real. In: NIPS, Vancouver (2005)
17. Hampel, F.R.: A general qualitative definition of robustness. Ann. Math. Stat. **42**(6), 1887–1896 (1971)
18. Herbrich, R., Graepel, T., Campbell, C.: Bayes point machines. J. Mach. Learn. Res. **1**, 245–279 (2001)
19. Jörnsten, R.: Clustering and classification based on the L_1 data depth. J. Multivar. Anal. **90**, 67–89 (2004)
20. Liu, R.Y.: On a notion of data depth based on random simplices. Ann. Stat. **18**(1), 405–414 (1990)
21. Liu, R.Y., Parelius, J.M., Singh, K.: Multivariate analysis by data depth: descriptive statistics, graphics and inference (with discussion and a rejoinder by Liu and Singh). Ann. Stat. **27**(3), 783–858 (1999)
22. López-Pintado, S., Romo, J.: On the concept of depth for functional data. J. Am. Stat. Assoc. **104**, 718–734 (2009)
23. Massé, J.C.: Asymptotics for the Tukey median. J. Multivar. Anal. **81**, 286–300 (2002)
24. McAllester, D.A.: Some PAC-Bayesian theorems. Mach. Learn. **37**(3), 355–363 (1999)
25. Rousseeuw, P.J., Hubert, M.: Regression depth. J. Am. Stat. Assoc. **94**(446), 388–402 (1999)
26. Tukey, J.: Mathematics and the picturing of data. In: Proceedings of the International Congress of Mathematicians, Vancouver (1975)
27. Zuo, Y.: Projection-based depth functions and associated medians. Ann. Stat. **31**, 1460–1490 (2003)

Chapter 16
Efficient Transductive Online Learning via Randomized Rounding

Nicolò Cesa-Bianchi and Ohad Shamir

Abstract Most traditional online learning algorithms are based on variants of mirror descent or follow-the-leader. In this chapter, we present an online algorithm based on a completely different approach, tailored for transductive settings, which combines "random playout" and randomized rounding of loss subgradients. As an application of our approach, we present the first computationally efficient online algorithm for collaborative filtering with trace-norm constrained matrices. As a second application, we solve an open question linking batch learning and transductive online learning.

16.1 Introduction

Online learning algorithms, which have received much attention in recent years, enjoy an attractive combination of computational efficiency, lack of distributional assumptions, and strong theoretical guarantees. Informally speaking, online learning is framed as a sequential game between a *learner*, who provides predictions, and an all-powerful *adversary*, who chooses the outcomes on which the learner's predictions are tested. The learner's goal is to attain low regret—that is, low excess loss—with respect to a comparison class of experts or predictors (see Sect. 16.2 for a more precise statement). Using standard online-to-batch techniques (e.g. [10]), one can convert online learning methods into simple and effective batch learning algorithms in a stochastic setting, where training and test examples are sampled from a distribution.

N. Cesa-Bianchi (✉)
Dipartimento di Informatica, Università degli Studi di Milano, 20135 Milano, Italy
e-mail: nicolo.cesa-bianchi@unimi.it

O. Shamir
Microsoft Research, 1 Memorial Drive, Cambridge, MA 02142, USA
e-mail: ohadsh@microsoft.com

In this work, we focus on transductive online learning, where the predictions of the experts/predictors can all be computed in advance. For example, consider the case where a sequence of unlabelled instances $\{x_t\}$ are given, and the learner needs to predict the corresponding labels $\{y_t\}$ which are sequentially chosen and revealed by the adversary. Thus, for a given fixed predictor h, we can compute its predictions $\{h(x_t)\}$ beforehand. This is a natural online analogue of the transductive learning framework introduced by Vapnik in a statistical batch setting [27], where the test instances on which one needs to predict are known in advance.

Despite the effectiveness of online learning methods, it is probably fair to say that, at their core, most of them are based on the same small set of fundamental techniques, in particular mirror descent and regularized follow-the-leader (see, for instance, [14, 23]). In this work we revisit, and significantly extend, an algorithm which uses a completely different approach. This algorithm, known as the *Minimax Forecaster*, was introduced in [9,11] for the setting of prediction with static experts. The Forecaster computes minimax predictions in the case of a fixed horizon, binary outcomes, and absolute loss. Although the original version is computationally expensive, it can easily be made efficient through randomization.

We extend the analysis of [9] to the case of non-binary outcomes, unknown horizons, and arbitrary convex and Lipschitz loss functions. The new algorithm is based on a combination of "random playout" and randomized rounding, which assigns random binary labels to future unseen instances, in a way depending on the loss subgradients. Our resulting *Randomized Rounding (R^2) Forecaster* has a parameter trading off regret performance for computational complexity, and runs in polynomial time. The idea of "random playout", in the context of online learning, has also been used in [1, 16], but we apply this idea in a different way.

Interestingly, our work, which focuses on online learning, has close links to methods and concepts from statistical learning, and thus can be seen as bridging the two fields. For example, the R^2 Forecaster uses empirical risk minimization—a standard statistical learning method—as a subroutine. Moreover, the regret of the R^2 Forecaster is determined by the Rademacher complexity of the comparison class, which is a measure of the generalization performance of the class in a statistical setting. The connection between online learnability and Rademacher complexity has also been explored in [2, 19]. Recently, [20] provided a significant generalization of these ideas, implying new algorithms and extending in a sense the work presented here.

As an application of our results, we describe how the R^2 Forecaster can be used to design the first efficient online learning algorithm for collaborative filtering with trace-norm constrained matrices. While this is a well-known setting, a straightforward application of standard online learning approaches, such as mirror descent, appear to give only trivial performance guarantees. Moreover, our regret bound matches the best known sample complexity bound in the batch distribution-free setting [24].

As a different application, we consider general reductions between batch learning and transductive online learning. The relationship between these two settings was analyzed in [16], in the context of binary prediction with respect to classes of bounded VC dimension. Their main result was that efficient learning in a statistical setting implies efficient learning in the transductive online setting, but at an inferior rate of $T^{3/4}$ (where T is the number of rounds). The main open question posed by that chapter is whether a better rate can be obtained. Using the R^2 Forecaster, we improve on those results, and provide an efficient algorithm with the optimal $\sqrt{T}$ rate, for a wide class of losses. This shows that efficient batch learning not only implies efficient transductive online learning (the main thesis of [16]), but also that the same rates can be obtained, and for possibly non-binary prediction problems as well.

We emphasize that the R^2 Forecaster requires computing many empirical risk minimizers (ERMs) at each round, which might be prohibitive in practice. Thus, while it does run in polynomial time whenever an ERM can be efficiently computed, we make no claim that it is a practical algorithm. Nevertheless, it seems to be a useful tool in showing that *efficient* online learnability is possible in various settings, often working in cases where more standard techniques appear to fail. Moreover, we hope the techniques we employ might prove useful in deriving practical online algorithms in other contexts.

16.2 The Minimax Forecaster

We start by formally introducing our online learning setting, known as prediction with expert advice (see [8]). The game is played between a forecaster and an adversary, and is specified by an outcome space $\mathcal{Y}$, a prediction space $\mathcal{P}$, a nonnegative loss function $\ell : \mathcal{P} \times \mathcal{Y} \to \mathbb{R}$, which measures the discrepancy between the forecaster's prediction and the outcome, and an expert class $\mathcal{F}$. Here we focus on classes $\mathcal{F}$ of *static experts*, whose prediction at each round t does not depend on the outcomes in previous rounds. Therefore, we think of each $\mathbf{f} \in \mathcal{F}$ simply as a sequence $\mathbf{f} = (f_1, f_2, \dots)$ where each $f_t \in \mathcal{P}$. At each step $t = 1, 2, \dots$ of the game, the forecaster outputs a prediction $p_t \in \mathcal{P}$ and simultaneously the adversary reveals an outcome $y_t \in \mathcal{Y}$. The forecaster's goal is to predict the outcome sequence almost as well as the best expert in the class $\mathcal{F}$, irrespective of the outcome sequence $\mathbf{y} = (y_1, y_2, \dots)$. The performance of a forecasting strategy A is measured by the worst-case regret

$$V_T(A, \mathcal{F}) = \sup_{\mathbf{y} \in \mathcal{Y}^T} \left(\sum_{t=1}^{T} \ell(p_t, y_t) - \inf_{\mathbf{f} \in \mathcal{F}} \sum_{t=1}^{T} \ell(f_t, y_t) \right)$$

viewed as a function of the horizon (number of rounds) T.

Consider now the special case where the horizon T is fixed and known in advance, the outcome space is $\mathcal{Y} = \{-1, +1\}$, the prediction space is $\mathcal{P} = [-1, +1]$, and the loss is the absolute loss $\ell(p, y) = |p - y|$. To simplify notation, let $L(\mathbf{f}, \mathbf{y}) = \sum_{t=1}^{T} |f_t - y_t|$. We will denote the regret in this special case as $V_T^{\text{abs}}(A, \mathcal{F})$.

The Minimax Forecaster—which is based on work presented in [9] and [11] (see also [8] for an exposition)—is derived by an explicit analysis of the minimax regret $\inf_A V_T^{\text{abs}}(A, \mathcal{F})$, where the infimum is over all forecasters A producing at round t a prediction p_t as a function of $p_1, y_1, \ldots p_{t-1}, y_{t-1}$. For general online learning problems, the analysis of this quantity is intractable. However, for the specific setting we focus on (absolute loss and binary outcomes), one can get both an explicit expression for the minimax regret, as well as an explicit algorithm, provided $\inf_{f \in \mathcal{F}} \sum_{t=1}^{T} \ell(f_t, y_t)$ can be efficiently computed for any sequence $y_1, \ldots, y_T$. This procedure is akin to performing empirical risk minimization (ERM) in statistical learning. A full development of the analysis is out of scope of this chapter, but is outlined in section "Appendix: Derivation of the Minimax Forecaster". In a nutshell, the idea is to begin by calculating the optimal prediction in the last round T, and then work backwards, calculating the optimal prediction at round $T-1, T-2$, and so on. Remarkably, the value of $\inf_A V_T^{\text{abs}}(A, \mathcal{F})$ is *exactly* the Rademacher complexity $\mathcal{R}_T(\mathcal{F})$ of the class $\mathcal{F}$, which is known to play a crucial role in controlling the sample complexity in statistical learning [4]. In this chapter, we define it as:

$$\mathcal{R}_T(\mathcal{F}) = \mathbb{E}\left[\sup_{\mathbf{f} \in \mathcal{F}} \sum_{t=1}^{T} \sigma_t f_t \right] \quad (16.1)$$

where $\sigma_1, \ldots, \sigma_T$ are i.i.d. Rademacher random variables, taking values $-1, +1$ with equal probability. When $\mathcal{R}_T(\mathcal{F}) = o(T)$, we get a minimax regret $\inf_A V_T^{\text{abs}}(A, \mathcal{F}) = o(T)$ which implies a vanishing per-round regret.

In terms of an explicit algorithm, the optimal prediction p_t at round t is given by a complicated-looking recursive expression, involving exponentially many terms. Indeed, for general online learning problems, this is the most one seems able to hope for. However, an apparently little-known fact is that when one deals with a class $\mathcal{F}$ of fixed binary sequences as discussed above, then one can write the optimal prediction p_t in a much simpler way. Letting $Y_1, \ldots, Y_T$ be i.i.d. Rademacher random variables, the optimal prediction at round t can be written as

$$p_t = \mathbb{E}\left[\inf_{\mathbf{f} \in \mathcal{F}} L\left(\mathbf{f}, y_1 \cdots y_{t-1}(-1) Y_{t+1} \cdots Y_T\right) \right.$$
$$\left. - \inf_{\mathbf{f} \in \mathcal{F}} L\left(\mathbf{f}, y_1 \cdots y_{t-1} 1 Y_{t+1} \cdots Y_T\right) \right]. \quad (16.2)$$

In words, the prediction is simply the expected difference between the minimal cumulative loss over $\mathcal{F}$, when the adversary plays -1 at round t and random values afterwards, and the minimal cumulative loss over $\mathcal{F}$, when the adversary plays $+1$ at round t, and the same random values afterwards. Again, we refer the reader to section "Appendix: Derivation of the Minimax Forecaster" for how this is derived. We call this optimal strategy (for absolute loss and binary outcomes) the Minimax Forecaster (MF); see Algorithm 3. The relevant guarantee for MF is summarized in the following theorem.

Algorithm 3 Minimax Forecaster (MF)

for $t = 1$ to T **do**
  Predict p_t as defined in (16.2)
  Receive outcome y_t and suffer loss $|p_t - y_t|$
end for

Algorithm 4 Minimax Forecaster with efficient implementation (MF*)

for $t = 1$ to T **do**
  For $i = t + 1, \ldots, T$, let Y_i be a Rademacher random variable
  Let

$$p_t := \inf_{\mathbf{f} \in \mathcal{F}} L\left(\mathbf{f}, y_1 \ldots y_{t-1} (-1) Y_{t+1} \ldots Y_T\right)$$

$$- \inf_{\mathbf{f} \in \mathcal{F}} L\left(\mathbf{f}, y_1 \ldots y_{t-1} 1 Y_{t+1} \ldots Y_T\right)$$

  Predict p_t, receive outcome y_t and suffer loss $|p_t - y_t|$
end for

Theorem 16.1. *For any class $\mathcal{F} \subseteq [-1, +1]^T$ of static experts, the regret of the Minimax Forecaster (Algorithm 3) satisfies $\mathcal{V}_T^{\text{abs}}(\text{MF}, \mathcal{F}) = \mathcal{R}_T(\mathcal{F})$.*

The Minimax Forecaster described above is not computationally efficient, as the computation of p_t requires averaging over exponentially many ERMs. However, by a martingale argument, it is not hard to show that it is in fact sufficient to compute only two ERMs per round.

Theorem 16.2. *For any class $\mathcal{F} \subseteq [-1, +1]^T$ of static experts, the regret of the randomized forecasting strategy MF* (Algorithm 4) satisfies*

$$\mathcal{V}_T^{\text{abs}}(\text{MF*}, \mathcal{F}) \le \mathcal{R}_T(\mathcal{F}) + \sqrt{2T \ln(1/\delta)}$$

with probability at least $1 - \delta$. Moreover, if the predictions $\mathbf{p} = (p_1, \ldots, p_T)$ are computed reusing the random values $Y_1, \ldots, Y_T$ computed at the first iteration of the algorithm, rather than by drawing fresh values at each iteration, then it holds that

$$\mathbb{E}\left[L(\mathbf{p}, \mathbf{y}) - \inf_{\mathbf{f} \in \mathcal{F}} L(\mathbf{f}, \mathbf{y})\right] \le \mathcal{R}_T(\mathcal{F}) \quad \text{for all } \mathbf{y} \in \{-1, +1\}^T.$$

Proof sketch. To prove the second statement, note that $\left|\mathbb{E}[p_t] - y_t\right| = \mathbb{E}\left[|p_t - y_t|\right]$ for any fixed $y_t \in \{-1, +1\}$ and p_t bounded in $[-1, +1]$, and use Theorem 16.1. To prove the first statement, note that $|p_t - y_t| - \left|\mathbb{E}_{p_t}[p_t] - y_t\right|$ for $t = 1, \ldots, T$ is a martingale difference sequence with respect to $p_1, \ldots, p_T$, and apply Azuma's inequality. □

The second statement in the theorem bounds the regret only in expectation and is thus weaker than the first one. On the other hand, it might have algorithmic benefits. Indeed, if we reuse the same values for $Y_1, \ldots, Y_T$, then the computation of the infima over **f** in MF* are with respect to an outcome sequence which changes only at one point in each round. Depending on the specific learning problem, it might be easier to re-compute the infimum after changing a single point in the outcome sequence, as opposed to computing the infimum over a different outcome sequence in each round.

16.3 The R^2 Forecaster

The Minimax Forecaster presented above is very specific to the absolute loss $\ell(f, y) = |f - y|$ and for binary outcomes $\mathcal{Y} = \{-1, +1\}$, which limits its applicability. We note that, extending the forecaster to other losses or different outcome spaces is not trivial: indeed, the recursive unwinding of the minimax regret term, leading to an explicit expression and an explicit algorithm, does not work as is for other cases. Nevertheless, we will now show how one can deal with general (convex, Lipschitz) loss functions and outcomes belonging to any real interval $[-b, b]$.

The algorithm we propose essentially uses the Minimax Forecaster as a subroutine, by feeding it with a carefully chosen sequence of binary values z_t, and using predictions f_t which are scaled to lie in the interval $[-1, +1]$. The values of z_t are based on a randomized rounding of values in $[-1, +1]$, which depend in turn on the loss subgradient. Thus, we denote the algorithm as the Randomized Rounding (R^2) Forecaster.

To describe the algorithm, we introduce some notation. For any scalar $f \in [-b, b]$, define $\tilde{f} = f/b$ to be the scaled versions of f in the range $[-1, +1]$. For vectors **f**, define $\tilde{\mathbf{f}} = (1/b)\mathbf{f}$. Also, we let $\partial_{p_t}\ell(p_t, y_t)$ denote any subgradient of the loss function ℓ with respect to the prediction p_t. As before, we define $L(\tilde{\mathbf{f}}, \mathbf{y}) = \sum_{t=1}^T |\tilde{f}_t - y_t|$. The pseudocode of the R^2 Forecaster is presented as Algorithm 5 below, and its regret guarantee is summarized in Theorem 16.3.

Theorem 16.3. *Suppose ℓ is convex and ρ-Lipschitz in its first argument. For any $\mathcal{F} \subseteq [-b, b]^T$, with probability at least $1 - \delta$ the regret of the R^2 Forecaster (Algorithm 5) satisfies*

$$\mathcal{V}_T(R^2, \mathcal{F}) \leq \rho \mathcal{R}_T(\mathcal{F}) + \rho b \left(\sqrt{\frac{1}{\eta}} + 2\right) \sqrt{2T \ln\left(\frac{2T}{\delta}\right)}.$$

Proof. Let $Y(t)$ denote the set of Bernoulli random variables chosen at round t. Let $\mathbb{E}_{z_t}$ denote expectation with respect to z_t, conditioned on $z_1, Y(1), \ldots, z_{t-1}, Y(t-1)$ as well as $Y(t)$. Let $\mathbb{E}_{Y(t)}$ denote the expectation with respect to the random drawing of $Y(t)$, conditioned on $z_1, Y(1), \ldots, z_{t-1}, Y(t-1)$.

Algorithm 5 The R^2 Forecaster

Input: Upper bound b on $|f_t|, |y_t|$ for all $t = 1, \ldots, T$ and $\mathbf{f} \in \mathcal{F}$; upper bound ρ on $\sup_{p,y \in [-b,b]} |\partial_p \ell(p, y)|$; precision parameter $\eta \geq \frac{1}{T}$.

for $t = 1$ to T **do**
  $p_t := 0$
  for $j = 1$ to ηT **do**
    For $i = t, \ldots, T$, let Y_i be a Rademacher random variable
    Draw

$$\Delta := \inf_{\mathbf{f} \in \mathcal{F}} L\left(\tilde{\mathbf{f}}, z_1 \ldots z_{t-1} (-1) Y_{t+1} \ldots Y_T\right)$$

$$- \inf_{\mathbf{f} \in \mathcal{F}} L\left(\tilde{\mathbf{f}}, z_1 \ldots z_{t-1} 1 Y_{t+1} \ldots Y_T\right)$$

    Let $p_t := p_t + \frac{b}{\eta T} \Delta$
  end for
  Predict p_t
  Receive outcome y_t and suffer loss $\ell(p_t, y_t)$
  Let $r_t := \frac{1}{2}\left(1 - \frac{1}{\rho}\partial_{p_t}\ell(p_t, y_t)\right) \in [0, 1]$
  Let $z_t := 1$ with probability r_t, and $z_t := -1$ with probability $1 - r_t$
end for

We will need two simple observations. First, by convexity of the loss function, we have that for any $p_t, f_t, y_t, \ell(p_t, y_t) - \ell(f_t, y_t) \leq (p_t - f_t) \partial_{p_t} \ell(p_t, y_t)$. Second, by definition of r_t and z_t, we have that for any fixed p_t, f_t,

$$\frac{1}{\rho b}(p_t - f_t) \partial_{p_t} \ell(p_t, y_t) = \frac{1}{b}(p_t - f_t)(1 - 2r_t)$$

$$= \frac{1}{b} r_t(f_t - p_t) + \frac{1}{b}(1 - r_t)(p_t - f_t)$$

$$= r_t(\tilde{f}_t - \tilde{p}_t) + (1 - r_t)(\tilde{p}_t - \tilde{f}_t)$$

$$= r_t\left((1 - \tilde{p}_t) - \left(1 - \tilde{f}_t\right)\right) + (1 - r_t)\left((\tilde{p}_t + 1) - \left(\tilde{f}_t + 1\right)\right)$$

$$= \mathbb{E}_{z_t}\left[|\tilde{p}_t - z_t| - \left|\tilde{f}_t - z_t\right|\right] .$$

The last transition uses the fact that $\tilde{p}_t, \tilde{f}_t \in [-1, +1]$. By these two observations, we have

$$\sum_{t=1}^{T} (\ell(p_t, y_t) - \ell(f_t, y_t)) \leq \sum_{t=1}^{T} (p_t - f_t) \partial_{p_t} \ell(p_t, y_t)$$

$$= \rho b \sum_{t=1}^{T} \mathbb{E}_{z_t}\left[|\tilde{p}_t - z_t| - \left|\tilde{f}_t - z_t\right|\right] . \quad (16.3)$$

Now, note that $|\tilde{p}_t - z_t| - |\tilde{f}_t - z_t| - \mathbb{E}_{z_t}\left[|\tilde{p}_t - z_t| - |\tilde{f}_t - z_t|\right]$ for $t = 1, \ldots, T$ is a martingale difference sequence: for any values of $z_1, Y(1), \ldots, z_{t-1}, Y(t-1), Y(t)$ (which fixes $\tilde{p}_t$), the conditional expectation of this expression over z_t is 0. Using Azuma's inequality, we can upper bound (16.3) with probability at least $1 - \delta/2$ by

$$\rho b \sum_{t=1}^{T} \left(|\tilde{p}_t - z_t| - |\tilde{f}_t - z_t|\right) + \rho b \sqrt{8T \ln(2/\delta)}. \tag{16.4}$$

The next step is to relate (16.4) to $\rho b \sum_{t=1}^{T}\left(\left|\mathbb{E}_{Y(t)}[\tilde{p}_t] - z_t\right| - |\tilde{f}_t - z_t|\right)$. It might be tempting to appeal to Azuma's inequality again. Unfortunately, there is no martingale difference sequence here, since z_t is itself a random variable whose distribution is influenced by $Y(t)$. Thus, we need to turn to coarser methods. Equation (16.4) can be upper bounded by

$$\rho b \sum_{t=1}^{T} \left(\left|\mathbb{E}_{Y(t)}[\tilde{p}_t] - z_t\right| - |\tilde{f}_t - z_t|\right) + \rho b \sum_{t=1}^{T} \left|\tilde{p}_t - \mathbb{E}_{Y(t)}[\tilde{p}_t]\right| + \rho b \sqrt{8T \ln(2/\delta)}. \tag{16.5}$$

Recall that $\tilde{p}_t$ is an average over ηT i.i.d. random variables, with expectation $\mathbb{E}_{Y(t)}[\tilde{p}_t]$. By Hoeffding's inequality, this implies that for any $t = 1, \ldots, T$, with probability at least $1 - \delta/2T$ over the choice of $Y(t)$, $\left|\tilde{p}_t - \mathbb{E}_{Y(t)}[\tilde{p}_t]\right| \leq \sqrt{2\ln(2T/\delta)/(\eta T)}$. By a union bound, it follows that with probability at least $1 - \delta/2$ over the choice of $Y(1), \ldots, Y(T)$,

$$\sum_{t=1}^{T} \left|\tilde{p}_t - \mathbb{E}_{Y(t)}[\tilde{p}_t]\right| \leq \sqrt{\frac{2T \ln(2T/\delta)}{\eta}}.$$

Combining this with (16.5), we get that with probability at least $1 - \delta$,

$$\rho b \sum_{t=1}^{T} \left(\left|\mathbb{E}_{Y(t)}[\tilde{p}_t] - z_t\right| - |\tilde{f}_t - z_t|\right) + \rho b \sqrt{\frac{2T \ln(2T/\delta)}{\eta}} + \rho b \sqrt{8T \ln(2/\delta)}. \tag{16.6}$$

Finally, by definition of $\tilde{p}_t = p_t/b$, we have that $\mathbb{E}_{Y(t)}[\tilde{p}_t]$ equals

$$\mathbb{E}_{Y(t)}\left[\inf_{f \in \mathcal{F}} L\left(\tilde{f}, z_1 \ldots z_{t-1}(-1) Y_{t+1} \ldots Y_T\right)\right.$$

$$\left. - \inf_{f \in \mathcal{F}} L\left(\tilde{f}, z_1 \ldots z_{t-1} 1 Y_{t+1} \ldots Y_T\right)\right].$$

This is exactly the Minimax Forecaster's prediction at round t, with respect to the sequence of outcomes $z_1, \ldots, z_{t-1} \in \{-1, +1\}$, and the class $\tilde{\mathcal{F}} := \{\tilde{\mathbf{f}} : \mathbf{f} \in \mathcal{F}\} \subseteq [-1, 1]^T$. Therefore, using Theorem 16.1, we can upper bound (16.6) by

$$\rho b \, \mathcal{R}_T(\tilde{\mathcal{F}}) + \rho b \sqrt{\frac{2T \ln(2T/\delta)}{\eta}} + \rho b \sqrt{8T \ln(2/\delta)}.$$

By definition of $\tilde{\mathcal{F}}$ and the Rademacher complexity, it is straightforward to verify that $\mathcal{R}_T(\tilde{\mathcal{F}}) = \frac{1}{b}\mathcal{R}_T(\mathcal{F})$. Using that to rewrite the bound, and slightly simplifying for readability, the result stated in the theorem follows. □

The computed prediction p_t is an empirical approximation to

$$b \, \mathbb{E}_{Y_{t+1}, \ldots, Y_T} \left[\inf_{\mathbf{f} \in \mathcal{F}} L\left(\tilde{\mathbf{f}}, z_1 \ldots z_{t-1} \, 0 \, Y_{t+1} \ldots Y_T\right) \right.$$

$$\left. - \inf_{\mathbf{f} \in \mathcal{F}} L\left(\tilde{\mathbf{f}}, z_1 \cdots z_{t-1} \, 1 \, Y_{t+1} \cdots Y_T\right) \right]$$

by repeatedly drawing independent values of $Y_{t+1}, \ldots, Y_T$ and averaging. The accuracy of the approximation is reflected in the precision parameter η. A larger value of η improves the regret bound, but also increases the runtime of the algorithm. Thus, η provides a trade-off between the computational complexity of the algorithm and its regret guarantee. We note that even when η is taken to be a constant fraction, the resulting algorithm still runs in polynomial time $\mathcal{O}(T^2 c)$, where c is the time to compute a single ERM. In subsequent results pertaining to this Forecaster, we will assume that η is taken to be a constant fraction.

The R^2 forecaster, as presented so far, assumes that the horizon T is known in advance. We now turn to describe how it can be readily extended to the case where it is unknown. The standard generic method to achieve this is known as the "doubling" trick (see [8]), and is based on guessing the value of T (initially $T = 1$), and running the algorithm with this guess. If the game did not end after T rounds, the guess is doubled and the algorithm is restarted with this new value. If the actual horizon T equals $2^0 + 2^1 + 2^2 + \ldots + 2^r$ for some integer r, then it is easy to show that our algorithm enjoys the same regret bound as before, plus a moderate multiplicative factor.[1] The only case we need to worry about is when T is not of this form, i.e., that the game ends in the middle of the algorithm's run. In that case, it is enough to ensure that the algorithm's regret bound, designed for horizon T, also bounds the regret after a smaller number $t < T$ of rounds. This can be shown to hold quite generically, given a very mild assumption on the loss function:

---

[1] Specifically, we divide the rounds into r consecutive epochs, such that epoch i consists of 2^i rounds, and use Theorem 16.3 with confidence $\delta' = \delta/2^{i+1}$, and a union bound, to get a regret bound of $\mathcal{O}(\mathcal{R}_{2^i}(\mathcal{F}) + \sqrt{(i + \log(1/\delta)) \, 2^i})$ over any epoch i. In the typical case where $\mathcal{R}_T(\mathcal{F}) = \mathcal{O}(\sqrt{T})$, summing over $i = 1, \ldots, r$ where $r = \log_2(T + 1) - 1$ yields a total regret bound of order $\mathcal{O}(\sqrt{\log(T/\delta) T})$. Up to log factors, this is the same bound as if T were known in advance.

Lemma 16.1. *Consider a (possibly randomized) forecaster A for a class $\mathcal{F}$ whose regret after T steps satisfies $V_T(A, \mathcal{F}) \leq G$ with probability at least $1 - \delta > \frac{1}{2}$. Furthermore, suppose the loss function is such that*

$$\inf_{p' \in \mathcal{P}} \sup_{y \in \mathcal{Y}} \inf_{p \in \mathcal{P}} \left(\ell(p, y) - \ell(p', y) \right) \geq 0.$$

Then

$$\max_{t=1,\ldots,T} V_t(A, \mathcal{F}) \leq G \qquad \text{with probability at least } 1 - \delta.$$

Note that for the assumption on the loss to hold, a simple sufficient condition is that $\mathcal{P} = \mathcal{Y}$ and $\ell(p, y) \geq \ell(y, y)$ for all $p, y \in \mathcal{P}$.

Proof. The proof assumes that the infimum and supremum of certain functions over $\mathcal{Y}, \mathcal{F}$ are attainable. If not, the proof can be easily adapted by finding attainable values which are ϵ-close to the infimum or supremum, and then taking $\epsilon \to 0$.

For the purpose of contradiction, suppose there exists a strategy for the adversary and a round $r \leq T$ such that at the end of round r, the forecaster suffers a regret $G' > G$ with probability larger than δ. Consider the following modified strategy for the adversary: the adversary plays according to the aforementioned strategy until round r. It then computes

$$f^* = \operatorname*{argmin}_{f \in \mathcal{F}} \sum_{t=1}^{r} \ell(f_t, y_t) \,.$$

At all subsequent rounds $t = r+1, r+2, \ldots, T$, the adversary chooses

$$y_t^* = \operatorname*{argmax}_{y \in \mathcal{Y}} \inf_{p \in \mathcal{P}} \left(\ell(p, y) - \ell(f_t^*, y) \right).$$

By the assumption on the loss function,

$$\ell(p_t, y_t^*) - \ell(f_t^*, y_t^*) \geq \inf_{p \in \mathcal{P}} \left(\ell(p, y_t^*) - \ell(f_t^*, y_t^*) \right)$$

$$= \sup_{y \in \mathcal{Y}} \inf_{p \in \mathcal{P}} \left(\ell(p, y) - \ell(f_t^*, y) \right) \geq 0 \,.$$

Thus, the regret over all T rounds, with respect to f^*, is

$$\sum_{t=1}^{r} \left(\ell(p_t, y_t) - \ell(f_t^*, y_t) \right) + \sum_{t=r+1}^{T} \left(\ell(p_t, y_t^*) - \ell(f_t^*, y_t^*) \right)$$

$$\geq \sum_{t=1}^{r} \ell(p_t, y_t) - \inf_{f \in \mathcal{F}} \sum_{t=1}^{r} \ell(f_t, y_t)$$

which is at least G' with probability larger than δ. On the other hand, we know that the learner's regret is at most G with probability at least $1 - \delta$. Thus we have a contradiction and the proof is concluded. □

We end this section with a remark that plays an important role in what follows.

Remark 16.1. The predictions of our forecasting strategies do not depend on the ordering of the predictions of the experts in $\mathcal{F}$. In other words, all the results proven so far also hold in a setting where the elements of $\mathcal{F}$ are functions $f : \{1, \ldots, T\} \to \mathcal{P}$, and the adversary has control on the permutation $\pi_1, \ldots, \pi_T$ of $\{1, \ldots, T\}$ that is used to define the prediction $f(\pi_t)$ of expert f at time t.[2] Also, Theorem 16.1 implies that the value of $\mathcal{V}_T^{\text{abs}}(\mathcal{F})$ remains unchanged irrespective of the permutation chosen by the adversary.

16.4 Application 1: Transductive Online Learning

The first application we consider is a rather straightforward one, in the context of transductive online learning [5]. In this model, we have an arbitrary sequence of labelled examples $(x_1, y_1), \ldots, (x_T, y_T)$, where only the set $\{x_1, \ldots, x_T\}$ of unlabelled instances is known to the learner in advance. At each round t, the learner must provide a prediction p_t for the label of y_t. The true label y_t is then revealed, and the learner incurs a loss $\ell(p_t, y_t)$. The learner's goal is to minimize the transductive online regret $\sum_{t=1}^{T}(\ell(p_t, y_t) - \inf_{f \in \mathcal{F}} \ell(f(x_t), y_t))$ with respect to a fixed class of predictors $\mathcal{F}$ of the form $\{x \mapsto f(x)\}$.

The work [16] considers the binary classification case with 0–1 loss. Their main result is that if a class $\mathcal{F}$ of binary functions has bounded VC dimension d, and there exists an efficient algorithm to perform empirical risk minimization, then one can construct an efficient randomized algorithm for transductive online learning, whose regret is at most $\mathcal{O}(T^{3/4}\sqrt{d \ln(T)})$ in expectation. The significance of this result is that efficient batch learning (via empirical risk minimization) implies efficient learning in the transductive online setting. This is an important result, as online learning can be computationally harder than batch learning—see, e.g., [7] for an example in the context of Boolean learning.

A major open question posed by Kakade and Kalai [16] was whether one can achieve the optimal rate $\mathcal{O}(\sqrt{dT})$, matching the rate of a batch learning algorithm in the statistical setting. Using the R^2 Forecaster, we can easily achieve the above result, as well as similar results in a strictly more general setting. This shows that efficient batch learning not only implies efficient transductive online learning (the main thesis of [16]), but also that the same rates can be obtained, and for possibly non-binary prediction problems as well.

---

[2]Formally, at each step t: (1) the adversary chooses and reveals the next element π_t of the permutation; (2) the forecaster chooses $p_t \in \mathcal{P}$ and simultaneously the adversary chooses $y_t \in \mathcal{Y}$.

Theorem 16.4. *Suppose we have a computationally efficient algorithm for empirical risk minimization (with respect to the 0–1 loss) over a class $\mathcal{F}$ of $\{0, 1\}$-valued functions with VC dimension d. Then, in the transductive online model, the efficient randomized forecaster* MF* *achieves an expected regret of $\mathcal{O}(\sqrt{dT})$ with respect to the 0–1 loss.*

Moreover, for an arbitrary class $\mathcal{F}$ of $[-b, b]$-valued functions with Rademacher complexity $\mathcal{R}_T(\mathcal{F})$, and any convex ρ-Lipschitz loss function, if there exists a computationally efficient algorithm for empirical risk minimization, then the R^2 Forecaster is computationally efficient and achieves, in the transductive online model, a regret of $\rho \mathcal{R}_T(\mathcal{F}) + \mathcal{O}(\rho b \sqrt{T \ln(T/\delta)})$ with probability at least $1 - \delta$.

Proof. Since the set $\{x_1, \ldots, x_T\}$ of unlabelled examples is known, we reduce the online transductive model to prediction with expert advice in the setting of Remark 16.1. This is done by mapping each function $f \in \mathcal{F}$ to a function $f : \{1, \ldots, T\} \to \mathcal{P}$ by $t \mapsto f(x_t)$, which is equivalent to an expert in the setting of Remark 16.1. When $\mathcal{F}$ maps to $\{0, 1\}$, and we care about the 0–1 loss, we can use the forecaster MF* to compute randomized predictions and apply Theorem 16.2 to bound the expected transductive online regret with $\mathcal{R}_T(\mathcal{F})$. For a class with VC dimension d, $\mathcal{R}_T(\mathcal{F}) \leq \mathcal{O}(\sqrt{dT})$ for some constant $c > 0$, using Dudley's chaining method [12], and this concludes the proof of the first part of the theorem. The second part is an immediate corollary of Theorem 16.3. □

We close this section by contrasting our results for online transductive learning with those of [6] for standard online learning. If $\mathcal{F}$ contains $\{0, 1\}$-valued functions, then the optimal regret bound for online learning is of order $\sqrt{d'T}$, where d' is the Littlestone dimension of $\mathcal{F}$. Since the Littlestone dimension of a class is never smaller than its VC dimension, we conclude that online learning is a harder setting than online transductive learning.

16.5 Application 2: Online Collaborative Filtering

We now turn to discuss the application of our results in the context of collaborative filtering with trace-norm constrained matrices, presenting the first computationally efficient online algorithm for this problem.

In collaborative filtering, the learning problem is to predict entries of an unknown $m \times n$ matrix based on a subset of its observed entries. A common approach is norm regularization, where we seek a low-norm matrix which matches the observed entries as best as possible. The norm is often taken to be the trace-norm [3, 21, 26], although other norms have also been considered, such as the max-norm [18] and the weighted trace-norm [13, 22].

Previous theoretical treatments of this problem assumed a stochastic setting, where the observed entries are picked according to some underlying distribution (e.g., [24, 25]). However, even when the guarantees are distribution-free, assuming a fixed distribution fails to capture important aspects of collaborative filtering in practice, such as non-stationarity [17]. Thus, an online adversarial setting, where no distributional assumptions whatsoever are required, seems to be particularly well suited to this problem domain.

In an online setting, at each round t the adversary reveals an index pair (i_t, j_t) and secretly chooses a value y_t for the corresponding matrix entry. After that, the learner selects a prediction p_t for that entry. Then y_t is revealed and the learner suffers a loss $\ell(p_t, y_t)$. Hence, the goal of a learner is to minimize the regret with respect to a fixed class $\mathcal{W}$ of prediction matrices, $\sum_{t=1}^{T} \ell(p_t, y_t) - \inf_{W \in \mathcal{W}} \sum_{t=1}^{T} \ell(W_{i_t, j_t}, y_t)$. Following reality, we will assume that the adversary picks a different entry in each round. When the learner's performance is measured by the regret after all $T = mn$ entries have been predicted, the online collaborative filtering setting reduces to prediction with expert advice, as discussed in Remark 16.1.

As mentioned previously, $\mathcal{W}$ is often taken to be a convex class of matrices with bounded trace-norm. Many convex learning problems, such as linear and kernel-based predictors, as well as matrix-based predictors, can be learned efficiently both in a stochastic and an online setting, using mirror descent or regularized follow-the-leader methods. However, for reasonable choices of $\mathcal{W}$, a straightforward application of these techniques leads to algorithms with trivial bounds. In particular, in the case of $\mathcal{W}$ consisting of $m \times n$ matrices with trace-norm at most r, standard online regret bounds would scale as $\mathcal{O}(r\sqrt{T})$. Since for this norm one typically has $r = \mathcal{O}(\sqrt{mn})$, we get a per-round regret guarantee of $\mathcal{O}(\sqrt{mn/T})$. This is a trivial bound, since it becomes "meaningful" (smaller than a constant) only after all $T = mn$ entries have been predicted. In this section, we show how to obtain a computationally efficient algorithm for this problem, using the R^2 Forecaster. We note that following our work, other efficient algorithms were proposed in [15, 20].

Consider first the transductive online setting, where the set of indices to be predicted is known in advance, and the adversary may only choose the order and values of the entries. It is readily seen that the R^2 Forecaster can be applied in this setting, using any convex class $\mathcal{W}$ of fixed matrices with bounded entries to compete against, and any convex Lipschitz loss function. To do so, we let $\{i_k, j_k\}_{k=1}^{T}$ be the set of entries, and run the R^2 Forecaster with respect to $\mathcal{F} = \{t \mapsto W_{i_t, j_t} : W \in \mathcal{W}\}$, which corresponds to a class of experts as discussed in Remark 16.1.

What is perhaps more surprising is that the R^2 Forecaster can also be applied in a *non-transductive* setting, where the indices to be predicted are not known in advance. Moreover, the Forecaster doesn't need to know the horizon T in advance. The key idea is to utilize the non-asymptotic nature of the learning problem—namely, that the game is played over a finite $m \times n$ matrix, so the time horizon is necessarily bounded.

The algorithm we propose is very simple: we apply the R^2 Forecaster as if we are in a setting with time horizon $T = mn$, which is played over *all* entries of the $m \times n$ matrix. By Remark 16.1, the R^2 Forecaster does not need to know the order in which these $m \times n$ entries are going to be revealed. Whenever $\mathcal{W}$ is convex and ℓ is a convex function, we can find an ERM in polynomial time by solving a convex problem. Hence, we can implement the R^2 Forecaster efficiently.

Using Lemma 16.1, the following theorem exemplifies how we can obtain a regret guarantee for our algorithm, in the case of $\mathcal{W}$ consisting of the convex set of matrices with bounded trace-norm and bounded entries. For the sake of clarity, we will consider $n \times n$ square matrices.

Theorem 16.5. *Let ℓ be a loss function which satisfies the conditions of Lemma 16.1. Also, let $\mathcal{W}$ consist of $n \times n$ matrices with trace-norm at most $r = \mathcal{O}(n)$ and entries at most $b = \mathcal{O}(1)$; suppose we apply the R^2 Forecaster over time horizon n^2 and all entries of the matrix. Then with probability at least $1 - \delta$, after T rounds, the algorithm achieves an average per-round regret of at most*

$$\mathcal{O}\left(\frac{n^{3/2} + n\sqrt{\ln(n/\delta)}}{T}\right) \quad \text{uniformly over } T = 1, \ldots, n^2.$$

Proof. In our setting, where the adversary chooses a different entry at each round, [24, Theorem 6] implies that for the class $\mathcal{W}'$ of all matrices with trace-norm at most $r = \mathcal{O}(n)$, it holds that $\mathcal{R}_T(\mathcal{W}')/T \leq \mathcal{O}(n^{3/2}/T)$. Therefore, $\mathcal{R}_{n^2}(\mathcal{W}') \leq \mathcal{O}(n^{3/2})$. Since $\mathcal{W} \subseteq \mathcal{W}'$, we get by definition of the Rademacher complexity that $\mathcal{R}_{n^2}(\mathcal{W}) = \mathcal{O}(n^{3/2})$ as well. By Theorem 16.3, the regret after n^2 rounds is $\mathcal{O}(n^{3/2} + n\sqrt{\ln(n/\delta)})$ with probability at least $1 - \delta$. Applying Lemma 16.1, we get that the cumulative regret at the end of any round $T = 1, \ldots, n^2$ is at most $\mathcal{O}(n^{3/2} + n\sqrt{\ln(n/\delta)})$, as required. □

This bound becomes non-trivial after $n^{3/2}$ entries are revealed, which is still a vanishingly small portion of all n^2 entries. While the regret might seem unusual compared to standard regret bounds (which usually have rates of $1/\sqrt{T}$ for general losses), it is a natural outcome of the non-asymptotic nature of our setting, where T can never be larger than n^2. In fact, this is the same rate one would obtain in a batch setting, where the entries are drawn from an arbitrary distribution.

As mentioned in the introduction, other online learning algorithms for this problem have been published since this work appeared [15, 20], using other techniques and assumptions.

Appendix: Derivation of the Minimax Forecaster

In this appendix, we outline how the Minimax Forecaster is derived, as well as its associated guarantees. This outline closely follows the exposition in [8, Chap. 8], to which we refer the reader for some of the technical derivations.

First, we note that the Minimax Forecaster as presented in [8] actually refers to a slightly different setup than ours, where the outcome space is $\mathcal{Y} = \{0, 1\}$ and the prediction space is $\mathcal{P} = [0, 1]$, rather than $\mathcal{Y} = \{-1, +1\}$ and $\mathcal{P} = [-1, +1]$. We will first derive the forecaster for the first setting, and then show how to convert it to the second setting.

Our goal is to find a predictor which minimizes the worst-case regret,

$$\max_{\mathbf{y} \in \{0,1\}^T} \left(L(\mathbf{p}, \mathbf{y}) - \inf_{\mathbf{f} \in \mathcal{F}} L(\mathbf{f}, \mathbf{y}) \right)$$

where $\mathbf{p} = (p_1, \ldots, p_T)$ is the prediction sequence.

For convenience, in the following we sometimes use the notation $\mathbf{y}^t$ to denote a vector in $\{0, 1\}^t$. The idea of the derivation is to work backwards, starting with computing the optimal prediction at the last round T, then deriving the optimal prediction at round $T - 1$, and so on. In the last round T, the first $T - 1$ outcomes $\mathbf{y}^{T-1}$ have been revealed, and we want to find the optimal prediction p_T. Since our goal is to minimize the worst-case regret with respect to the absolute loss, we just need to compute p_T, which minimizes

$$L(\mathbf{p}^{T-1}, \mathbf{y}^{T-1}) + \max\left\{ p_T - \inf_{\mathbf{f} \in \mathcal{F}} L(\mathbf{f}, \mathbf{y}^{T-1} 0), (1 - p_T) - \inf_{\mathbf{f} \in \mathcal{F}} L(\mathbf{f}, \mathbf{y}^{T-1} 1) \right\}.$$

In our setting, it is not hard to show that

$$\left| \inf_{\mathbf{f} \in \mathcal{F}} L(\mathbf{f}, \mathbf{y}^{t-1} 0) - \inf_{\mathbf{f} \in \mathcal{F}} L(\mathbf{f}, \mathbf{y}^{t-1} 1) \right| \leq 1$$

(see [8, Lemma 8.1]). Using this, we can compute the optimal p_T to be

$$p_T = \frac{1}{2} \left(A_T(\mathbf{y}^{T-1} 1) - A_T(\mathbf{y}^{T-1} 0) + 1 \right) \tag{16.7}$$

where $A_T(\mathbf{y}^T) = -\inf_{\mathbf{f} \in \mathcal{F}} L(\mathbf{f}, \mathbf{y}^T)$.

Having determined p_T, we can continue to the previous prediction p_{T-1}. This is equivalent to minimizing

$$L(\mathbf{p}^{T-2}, \mathbf{y}^{T-2}) + \max\left\{ p_{T-1} + A_{T-1}(\mathbf{y}^{T-2} 0), (1 - p_{T-1}) + A_{T-1}(\mathbf{y}^{T-2} 1) \right\}$$

where

$$A_{T-1}(\mathbf{y}^{T-1}) =$$

$$\min_{p_T \in [0,1]} \max \left\{ p_T - \inf_{\mathbf{f} \in \mathcal{F}} L(\mathbf{f}, \mathbf{y}^{T-1} 0), (1 - p_T) - \inf_{\mathbf{f} \in \mathcal{F}} L(\mathbf{f}, \mathbf{y}^{T-1} 1) \right\}. \tag{16.8}$$

Note that, by plugging in the value of p_T from (16.7), we also get the following equivalent formulation for $A_{T-1}(\mathbf{y}^{T-1})$:

$$A_{T-1}(\mathbf{y}^{T-1}) = \frac{1}{2} \left(A_T(\mathbf{y}^{T-1} 0) + A_T(\mathbf{y}^{T-1} 1) + 1 \right).$$

Again, it is possible to show that the optimal value of p_{T-1} is

$$p_{T-1} = \frac{1}{2}\Big(A_{T-1}(\mathbf{y}^{T-2}1) - A_T(\mathbf{y}^{T-2}0) + 1\Big).$$

Repeating this procedure, one can show that at any round t, the minimax optimal prediction is

$$p_t = \frac{1}{2}\Big(A_t(\mathbf{y}^{t-1}1) - A_t(\mathbf{y}^{t-1}0) + 1\Big) \qquad (16.9)$$

where A_t is defined recursively as $A_T(\mathbf{y}^T) = -\inf_{\mathbf{f}\in\mathcal{F}} L(\mathbf{f}, \mathbf{y}^T)$ and, for all t,

$$A_{t-1}(\mathbf{y}^{t-1}) = \frac{1}{2}\Big(A_t(\mathbf{y}^{t-1}0) + A_t(\mathbf{y}^{t-1}1) + 1\Big). \qquad (16.10)$$

At first glance, computing p_t from (16.9) might seem tricky, since it requires computing $A_t(\mathbf{y}^t)$, whose recursive expansion in (16.10) involves exponentially many terms. Luckily, the recursive expansion has a simple structure, and it is not hard to show that

$$\begin{aligned}
A_t(\mathbf{y}^t) &= \frac{T-t}{2} - \frac{1}{2^T}\sum_{\mathbf{y}\in\{0,1\}^T}\Big(\inf_{\mathbf{f}\in\mathcal{F}} L(\mathbf{f}, \mathbf{y}^t Y^{T-t})\Big) \\
&= \frac{T-t}{2} - \mathbb{E}\Big[\inf_{\mathbf{f}\in\mathcal{F}} L(\mathbf{f}, \mathbf{y}^t Y^{T-t})\Big]
\end{aligned} \qquad (16.11)$$

where Y^{T-t} is a sequence of $T-t$ i.i.d. Bernoulli random variables, which take values in $\{0, 1\}$ with equal probability. Plugging this into the formula for the minimax prediction in (16.9), we get that[3]

$$p_t = \frac{1}{2}\Big(\mathbb{E}\Big[\inf_{\mathbf{f}\in\mathcal{F}} L(\mathbf{f}, \mathbf{y}^{t-1}0 Y^{T-t}) - \inf_{\mathbf{f}\in\mathcal{F}} L(\mathbf{f}, \mathbf{y}^{t-1}1 Y^{T-t})\Big] + 1\Big). \qquad (16.12)$$

This prediction rule constitutes the Minimax Forecaster as presented in [8].

After deriving the algorithm, we turn to analyze its regret performance. To do so, we just need to note that A_0 equals the worst-case regret—see the recursive definition (16.8). Using the alternative explicit definition in (16.11), we get that the worst-case regret equals

$$\begin{aligned}
\frac{T}{2} - \mathbb{E}\Big[\inf_{\mathbf{f}\in\mathcal{F}}\sum_{t=1}^T |f_t - Y_t|\Big] &= \mathbb{E}\Big[\sup_{\mathbf{f}\in\mathcal{F}}\sum_{t=1}^T \Big(\frac{1}{2} - |f_t - Y_t|\Big)\Big] \\
&= \mathbb{E}\Big[\sup_{\mathbf{f}\in\mathcal{F}}\sum_{t=1}^T \Big(f_t - \frac{1}{2}\Big)\sigma_t\Big]
\end{aligned}$$

---

[3]This fact appears in an implicit form in [9]; see also [8, Exercise 8.4].

where σ_t are i.i.d. Rademacher random variables (taking values of -1 and $+1$ with equal probability). Recalling the definition of Rademacher complexity, (16.1), we get that the regret is bounded by the Rademacher complexity of the shifted class, which is obtained from $\mathcal{F}$ by taking every $\mathbf{f} \in \mathcal{F}$ and replacing every coordinate f_t by $f_t - 1/2$.

Finally, it remains to show how to convert the forecaster and analysis above to the setting discussed in this chapter, where the outcomes are in $\{-1, +1\}$ rather than $\{0, 1\}$ and the predictions are in $[-1, +1]$ rather than $[0, 1]$. To do so, consider a learning problem in this new setting, with some class $\mathcal{F}$. For any vector $\mathbf{y}$, define $\tilde{\mathbf{y}}$ to be the shifted vector $(\mathbf{y} + \mathbf{1})/2$, where $\mathbf{1} = (1, \ldots, 1)$ is the all-1s vector. Also, define $\tilde{\mathcal{F}}$ to be the shifted class $\tilde{\mathcal{F}} = \{(\mathbf{f} + \mathbf{1})/2 \, : \, \mathbf{f} \in \mathcal{F}\}$. It is easily seen that $L(\mathbf{f}, \mathbf{y}) = 2L(\tilde{\mathbf{f}}, \tilde{\mathbf{y}})$ for any $\mathbf{f}, \mathbf{y}$. As a result, if we look at the prediction p_t given by our forecaster in (16.2), then $\tilde{p}_t = (p_t + 1)/2$ is the minimax optimal prediction given by (16.12) with respect to the class $\tilde{\mathcal{F}}$ and the outcomes $\tilde{\mathbf{y}}^T$. So our analysis above applies, and we get that

$$\max_{\mathbf{y} \in \{-1,+1\}^T} \left(L(\mathbf{p}, \mathbf{y}) - \inf_{\mathbf{f} \in \mathcal{F}} L(\mathbf{f}, \mathbf{y}) \right) = \max_{\tilde{\mathbf{y}} \in [0,1]^T} 2 \left(L(\tilde{\mathbf{p}}, \tilde{\mathbf{y}}) - \inf_{\tilde{\mathbf{f}} \in \tilde{\mathcal{F}}} L(\tilde{\mathbf{f}}, \tilde{\mathbf{y}}) \right)$$

$$= 2\mathbb{E}\left[\sup_{\tilde{\mathbf{f}} \in \tilde{\mathcal{F}}} \sum_{t=1}^T \left(\tilde{f}_t - \frac{1}{2} \right) \sigma_t \right] = \mathbb{E}\left[\sup_{\mathbf{f} \in \mathcal{F}} \sum_{t=1}^T \sigma_t f_t \right].$$

which is exactly the Rademacher complexity of the class $\mathcal{F}$.

Acknowledgements The first author acknowledges partial support by the PASCAL2 NoE under EC grant FP7-216886.

References

1. Abernethy, J., Warmuth, M.: Repeated games against budgeted adversaries. In: NIPS, Vancouver (2010)
2. Abernethy, J., Bartlett, P., Rakhlin, A., Tewari, A.: Optimal strategies and minimax lower bounds for online convex games. In: COLT, Montreal (2009)
3. Bach, F.: Consistency of trace-norm minimization. J. Mach. Learn. Res. **9**, 1019–1048 (2008)
4. Bartlett, P., Mendelson, S.: Rademacher and Gaussian complexities: risk bounds and structural results. In: COLT, Amsterdam (2001)
5. Ben-David, S., Kushilevitz, E., Mansour, Y.: Online learning versus offline learning. Mach. Learn. **29**(1), 45–63 (1997)
6. Ben-David, S., Pál, D., Shalev-Shwartz, S.: Agnostic online learning. In: COLT, Montreal (2009)
7. Blum, A.: Separating distribution-free and mistake-bound learning models over the Boolean domain. SIAM J. Comput. **23**(5), 990–1000 (1994)
8. Cesa-Bianchi, N., Lugosi, G.: Prediction, Learning, and Games. Cambridge University Press, New York (2006)

9. Cesa-Bianchi, N., Freund, Y., Haussler, D., Helmbold, D., Schapire, R., Warmuth, M.: How to use expert advice. J. ACM **44**(3), 427–485 (1997)
10. Cesa-Bianchi, N., Conconi, A., Gentile, C.: On the generalization ability of on-line learning algorithms. IEEE Trans. Inf. Theory **50**(9), 2050–2057 (2004)
11. Chung, T.: Approximate methods for sequential decision making using expert advice. In: COLT, New Brunswick (1994)
12. Dudley, R.M.: A Course on Empirical Processes, École de Probabilités de St. Flour, 1982. Lecture Notes in Mathematics, vol. 1097. Springer, Berlin (1984)
13. Foygel, R., Salakhutdinov, R., Shamir, O., Srebro, N.: Learning with the weighted trace-norm under arbitrary sampling distributions. In: NIPS, Granada (2011)
14. Hazan, E.: The convex optimization approach to regret minimization. In: Nowozin, S., Sra, S., Wright, S. (eds.) Optimization for Machine Learning. MIT, Cambridge (2012)
15. Hazan, E., Kale, S., Shalev-Shwartz, S.: Near-optimal algorithms for online matrix prediction. In: COLT, Edinburgh (2012)
16. Kakade, S., Kalai, A.: From batch to transductive online learning. In: NIPS, Vancouver (2005)
17. Koren, Y.: Collaborative filtering with temporal dynamics. In: KDD, Paris (2009)
18. Lee, J., Recht, B., Salakhutdinov, R., Srebro, N., Tropp, J.: Practical large-scale optimization for max-norm regularization. In: NIPS, Vancouver (2010)
19. Rakhlin, A., Sridharan, K., Tewari, A.: Online learning: random averages, combinatorial parameters, and learnability. In: NIPS, Vancouver (2010)
20. Rakhlin, A., Shamir, O., Sridharan, K.: Relax and localize: from value to algorithms. CoRR abs/1204.0870 (2012)
21. Salakhutdinov, R., Mnih, A.: Probabilistic matrix factorization. In: NIPS, Vancouver (2007)
22. Salakhutdinov, R., Srebro, N.: Collaborative filtering in a non-uniform world: learning with the weighted trace norm. In: NIPS, Vancouver (2010)
23. Shalev-Shwartz, S.: Online learning and online convex optimization. Found. Trends Mach. Learn. **4**(2), 107–194 (2012)
24. Shamir, O., Shalev-Shwartz, S.: Collaborative filtering with the trace norm: learning, bounding, and transducing. In: COLT, Budapest (2011)
25. Srebro, N., Shraibman, A.: Rank, trace-norm and max-norm. In: COLT, Bertinoro (2005)
26. Srebro, N., Rennie, J., Jaakkola, T.: Maximum-margin matrix factorization. In: NIPS, Vancouver (2004)
27. Vapnik, V.: Statistical Learning Theory. Wiley, New York (1998)

Chapter 17
Pivotal Estimation in High-Dimensional Regression via Linear Programming

Eric Gautier and Alexandre B. Tsybakov

Abstract We propose a new method of estimation in high-dimensional linear regression models. It allows for very weak distributional assumptions, including heteroscedasticity, and does not require knowledge of the variance of random errors. The method is based on linear programming only, so that its numerical implementation is faster than for previously known techniques using conic programs, and it allows one to deal with higher-dimensional models. We provide upper bounds for estimation and prediction errors of the proposed estimator, showing that it achieves the same rate as in the more restrictive situation of fixed design and i.i.d. Gaussian errors with known variance. Following Gautier and Tsybakov (High-dimensional instrumental variables regression and confidence sets. ArXiv e-prints 1105.2454, 2011), we obtain the results under weaker sensitivity assumptions than the restricted eigenvalue or assimilated conditions.

17.1 Introduction

In this chapter, we consider the linear regression model

$$y_i = x_i^T \beta^* + u_i, \quad i = 1, \ldots, n,$$

where x_i are random vectors of explanatory variables in $\mathbb{R}^p$, and $u_i \in \mathbb{R}$ is a random error. The aim is to estimate the vector $\beta^* \in \mathbb{R}^p$ from n independent, not necessarily identically distributed realizations $(y_i, x_i^T), i = 1, \ldots, n$. We are mainly interested in high-dimensional models where p can be much larger than n under the sparsity scenario, where only few components β_k^* of β^* are non-zero (β^* is sparse).

E. Gautier (✉) · A.B. Tsybakov
CREST-ENSAE, 3, avenue Pierre Larousse, 92245 Malakoff Cedex, France
e-mail: eric.gautier@ensae.fr; alexandre.tsybakov@ensae.fr

The most studied techniques for high-dimensional regression under the sparsity scenario are the Lasso, the Dantzig selector (see, e.g., Candès and Tao [8], Bickel et al. [6]; more references can be found in Bühlmann and van de Geer [7] and Koltchinskii [14]), and aggregation by exponential weighting (see Dalalyan and Tsybakov [10], Rigollet and Tsybakov [12, 16] and the references cited therein). Most of the literature on high-dimensional regression assumes that the random errors are Gaussian or subgaussian with known variance (or noise level). However, quite recently several methods have been proposed which are independent of the noise level (see, e.g., Städler et al. [17], Antoniadis [1], Belloni et al. [3], Gautier and Tsybakov [2], Sun and Zhang [12], Belloni et al. [18], Dalalyan [4,9]). Among these, the methods of Belloni et al. [2], Belloni et al. [4], and Gautier and Tsybakov [12] allow us to handle non-identically distributed errors u_i and are *pivotal*, i.e., rely on very weak distributional assumptions. In Gautier and Tsybakov [12], the regressors x_i can be correlated with the errors u_i, and an estimator is suggested that makes use of instrumental variables, called the *STIV* (Self-Tuned Instrumental Variables) estimator. In a particular instance, the *STIV* estimator can be applied in classical linear regression models where all regressors are uncorrelated with the errors. This yields a pivotal extension of the Dantzig selector based on conic programming. Gautier and Tsybakov [12] also present a method to obtain finite sample confidence sets that are robust to non-Gaussian and heteroscedastic errors.

Another important issue is to relax the assumptions on the model under which the validity of the Lasso type methods is proved, such as the restricted eigenvalue condition of Bickel et al. [6] and its various analogues. Belloni et al. [2] obtain fast rates for prediction for the Square-root Lasso under a relaxed version of the restricted eigenvalue condition. In the context of known noise variance, Ye and Zhang [19] introduce cone invertibility factors instead of restricted eigenvalues. For pivotal estimation, an approach based on the sensitivities and sparsity certificates is introduced in Gautier and Tsybakov [12]; see more details below. Finally, note that aggregation by exponential weighting [10, 15, 16] does not require any condition on the model, but its numerical realization is based on MCMC algorithms in high dimensions whose convergence rates are hard to assess theoretically.

In this chapter, we introduce a new pivotal estimator, called the Self-tuned Dantzig estimator. It is defined as a linear program, so from the numerical point of view it is simpler than the previously known pivotal estimators based on conic programming. We obtain upper bounds on its estimation and prediction errors under weak assumptions on the model and on the distribution of the errors, showing that it achieves the same rate as in the more restrictive situation of fixed design and i.i.d. Gaussian errors with known variance. The model assumptions are based on the sensitivity analysis from Gautier and Tsybakov [12]. Distributional assumptions allow for dependence between x_i and u_i. When x_i's are independent from u_i's, it is enough to assume, for example, that the errors u_i are symmetric and have a finite second moment.

17.2 Notation

We set $\mathbf{Y} = (y_1, \ldots, y_n)^T$ and $\mathbf{U} = (u_1, \ldots, u_n)^T$, and we denote by $\mathbf{X}$ the matrix of dimension $n \times p$ with rows x_i^T, $i = 1, \ldots, n$. We denote by $\mathbf{D}$ the $p \times p$ diagonal normalizing matrix with diagonal entries $d_{kk} > 0$, $k = 1, \ldots, p$. Typical examples are: $d_{kk} \equiv 1$ or

$$d_{kk} = \left(\frac{1}{n}\sum_{i=1}^n x_{ki}^2\right)^{-1/2}, \quad \text{and} \quad d_{kk} = \left(\max_{i=1,\ldots,n} |x_{ki}|\right)^{-1}$$

where x_{ki} is the kth component of x_i. For a vector $\beta \in \mathbb{R}^p$, let $J(\beta) = \{k \in \{1, \ldots, p\} : \beta_k \neq 0\}$ be its support, i.e., the set of indices corresponding to its non-zero components β_k. We denote by $|J|$ the cardinality of a set $J \subseteq \{1, \ldots, p\}$ and by J^c its complement: $J^c = \{1, \ldots, p\} \setminus J$. The ℓ_p norm of a vector Δ is denoted by $|\Delta|_p$, $1 \leq p \leq \infty$. For $\Delta = (\Delta_1, \ldots \Delta_p)^T \in \mathbb{R}^p$ and a set of indices $J \subseteq \{1, \ldots, p\}$, we consider $\Delta_J \triangleq (\Delta_1 \mathbf{1}_{\{1 \in J\}}, \ldots, \Delta_p \mathbf{1}_{\{p \in J\}})^T$, where $\mathbf{1}_{\{\cdot\}}$ is the indicator function. For $a \in \mathbb{R}$, we set $a_+ \triangleq \max(0, a)$, $a_+^{-1} \triangleq (a_+)^{-1}$.

17.3 The Estimator

We say that a pair $(\beta, \sigma) \in \mathbb{R}^p \times \mathbb{R}^+$ satisfies the *Self-tuned Dantzig constraint* if it belongs to the set

$$\hat{\mathcal{D}} \triangleq \left\{(\beta, \sigma) \; \beta \in \mathbb{R}^p, \; \sigma > 0, \; \left|\frac{1}{n}\mathbf{D}\mathbf{X}^T(\mathbf{Y} - \mathbf{X}\beta)\right|_\infty \leq \sigma r\right\}$$

for some $r > 0$ (specified below).

Definition 17.1. We call the *Self-tuned Dantzig* estimator any solution $(\hat{\beta}, \hat{\sigma})$ of the following minimization problem:

$$\min_{(\beta,\sigma) \in \hat{\mathcal{D}}} \left(|\mathbf{D}^{-1}\beta|_1 + c\sigma\right), \qquad (17.1)$$

for some positive constant c.

Finding the Self-tuned Dantzig estimator is a linear program. The term $c\sigma$ is included in the criterion to prevent us from choosing σ arbitrarily large. The choice of the constant c will be discussed later.

17.4 Sensitivity Characteristics

The sensitivity characteristics are defined by the action of the matrix

$$\Psi_n \triangleq \frac{1}{n}\mathbf{D}\mathbf{X}^T\mathbf{X}\mathbf{D}$$

on the so-called *cone of dominant coordinates*

$$C_J^{(\gamma)} \triangleq \{\Delta \in \mathbb{R}^p : |\Delta_{J^c}|_1 \leq (1+\gamma)|\Delta_J|_1\},$$

for some $\gamma > 0$. It is straightforward that for $\delta \in C_J^{(\gamma)}$,

$$|\Delta|_1 \leq (2+\gamma)|\Delta_J|_1 \leq (2+\gamma)|J|^{1-1/q}|\Delta_J|_q, \quad \forall 1 \leq q \leq \infty. \quad (17.2)$$

We now recall some definitions from Gautier and Tsybakov [12]. For $q \in [1, \infty]$, we define the ℓ_q *sensitivity* as the following random variable:

$$\kappa_{q,J}^{(\gamma)} \triangleq \inf_{\Delta \in C_J^{(\gamma)}: |\Delta|_q = 1} |\Psi_n \Delta|_\infty.$$

Given a subset $J_0 \subset \{1, \ldots, p\}$ and $q \in [1, \infty]$, we define the $\ell_q - J_0$ *block sensitivity* as

$$\kappa_{q,J_0,J}^{(\gamma)} \triangleq \inf_{\Delta \in C_J^{(\gamma)}: |\Delta_{J_0}|_q = 1} |\Psi_n \Delta|_\infty.$$

By convention, we set $\kappa_{q,\emptyset,J}^{(\gamma)} = \infty$. Also, recall that the restricted eigenvalue of Bickel et al. [6] is defined by

$$\kappa_{RE,J}^{(\gamma)} \triangleq \inf_{\Delta \in \mathbb{R}^p \setminus \{0\}: \Delta \in C_J^{(\gamma)}} \frac{|\Delta^T \Psi_n \Delta|}{|\Delta_J|_2^2}$$

and a closely related quantity is

$$\kappa_{RE,J}^{'(\gamma)} \triangleq \inf_{\Delta \in \mathbb{R}^p \setminus \{0\}: \Delta \in C_J^{(\gamma)}} \frac{|J||\Delta^T \Psi_n \Delta|}{|\Delta_J|_1^2}.$$

The next result establishes a relation between restricted eigenvalues and sensitivities. It follows directly from the Cauchy-Schwarz inequality and (17.2).

Proposition 17.1.

$$\kappa_{RE,J}^{(\gamma)} \leq \kappa_{RE,J}^{'(\gamma)} \leq (2+\gamma)|J|\kappa_{1,J,J}^{(\gamma)} \leq (2+\gamma)^2|J|\kappa_{1,J}^{(\gamma)}.$$

17 Pivotal Estimation in High-Dimensional Regression via Linear Programming

The following proposition gives a useful lower bound on the sensitivity.

Proposition 17.2. *If* $|J| \leq s$,

$$\kappa^{(\gamma)}_{1,J,J} \geq \frac{1}{s} \min_{k=1,\ldots,p} \left\{ \min_{\Delta_k=1,\, |\Delta|_1 \leq (2+\gamma)s} |\Psi_n \Delta|_\infty \right\} \triangleq \kappa^{(\gamma)}_{1,0}(s). \qquad (17.3)$$

Proof. We have

$$\kappa^{(\gamma)}_{1,J,J} = \inf_{\Delta:\, |\Delta_J|_1=1,\, |\Delta_{J^c}|_1 \leq 1+\gamma} |\Psi_n \Delta|_\infty$$

$$\geq \inf_{\Delta:\, |\Delta|_\infty \geq \frac{1}{s},\, |\Delta|_1 \leq 2+\gamma} |\Psi_n \Delta|_\infty$$

$$= \frac{1}{s} \inf_{\Delta:\, |\Delta|_\infty \geq 1,\, |\Delta|_1 \leq (2+\gamma)s} |\Psi_n \Delta|_\infty \quad \text{(by homogeneity)}$$

$$= \frac{1}{s} \inf_{\Delta:\, |\Delta|_\infty \geq 1,\, |\Delta|_1 \leq (2+\gamma)s} |\Delta|_\infty \frac{|\Psi_n \Delta|_\infty}{|\Delta|_\infty}$$

$$\geq \frac{1}{s} \inf_{\Delta:\, |\Delta|_\infty = 1,\, |\Delta|_1 \leq (2+\gamma)s|\Delta|_\infty} |\Psi_n \Delta|_\infty \quad \text{(by homogeneity)}$$

$$= \frac{1}{s} \inf_{\Delta:\, |\Delta|_\infty = 1,\, |\Delta|_1 \leq (2+\gamma)s} |\Psi_n \Delta|_\infty$$

$$= \frac{1}{s} \min_{k=1,\ldots,p} \left\{ \inf_{\Delta:\, \Delta_k=1,\, |\Delta|_1 \leq (2+\gamma)s} |\Psi_n \Delta|_\infty \right\}. \qquad \square$$

Note that the random variable $\kappa^{(\gamma)}_{1,0}(s)$ depends only on the observed data. It is not difficult to see that it can be obtained by solving p linear programs. For more details and further results on the sensitivity characteristics, see Gautier and Tsybakov [12].

17.5 Bounds on the Estimation and Prediction Errors

In this section, we use the notation $\Delta \triangleq \mathbf{D}^{-1}(\hat{\beta} - \beta)$. Let $0 < \alpha < 1$ be a given constant. We choose the tuning parameter r in the definition of $\hat{\mathcal{D}}$ as follows:

$$r = \sqrt{\frac{2\log(4p/\alpha)}{n}}.$$

Theorem 17.1. *Let, for all* $i = 1,\ldots,n$, *and* $k = 1,\ldots,p$, *the random variables* $x_{ki}u_i$ *be symmetric. Let* $Q^* > 0$ *be a constant such that*

$$\mathbb{P}\left(\max_{k=1,\dots,p}\frac{d_{kk}^2}{n}\sum_{i=1}^n x_{ki}^2 u_i^2 > Q^*\right) \leq \alpha/2. \tag{17.4}$$

Assume that $|J(\beta^*)| \leq s$, and set in (17.1)

$$c = \frac{(2\gamma+1)r}{\kappa_{1,0}^{(\gamma)}(s)}, \tag{17.5}$$

where γ is a positive number. Then, with probability at least $1 - \alpha$, for any $\gamma > 0$ and any $\hat{\beta}$ such that $(\hat{\beta}, \hat{\sigma})$ is a solution of the minimization problem (17.1) with c defined in (17.5), we have the following bounds on the ℓ_1 estimation error and on the prediction error:

$$|\Delta|_1 \leq \left(\frac{(\gamma+2)(2\gamma+1)\sqrt{Q^*}}{\gamma \kappa_{1,0}^{(\gamma)}(s)}\right) r, \tag{17.6}$$

$$\Delta^T \Psi_n \Delta \leq \left(\frac{(\gamma+2)(2\gamma+1)^2 Q^*}{\gamma^2 \kappa_{1,0}^{(\gamma)}(s)}\right) r^2. \tag{17.7}$$

Proof. Set

$$\hat{Q}(\beta) \triangleq \max_{k=1,\dots,p}\frac{d_{kk}^2}{n}\sum_{i=1}^n x_{ki}^2 (y_i - x_i^T\beta)^2,$$

and define the event

$$\mathcal{G} = \left\{\left\|\frac{1}{n}\mathbf{D}\mathbf{X}^T\mathbf{U}\right\|_\infty \leq r\sqrt{\hat{Q}(\beta^*)}\right\}$$

$$= \left\{\left|\frac{d_{kk}}{n}\sum_{i=1}^n x_{ki}u_i\right| \leq r\sqrt{\hat{Q}(\beta^*)},\ k=1,\dots,p\right\}.$$

Then

$$\mathcal{G}^c \subset \bigcup_{k=1,\dots,p}\left\{\left|\frac{\sum_{i=1}^n x_{ki}u_i}{\sqrt{\sum_{i=1}^n (x_{ki}u_i)^2}}\right| \geq \sqrt{n}r\right\}$$

and the union bound yields

$$\mathbb{P}(\mathcal{G}^c) \leq \sum_{k=1}^p \mathbb{P}\left(\left|\frac{\sum_{i=1}^n x_{ki}u_i}{\sqrt{\sum_{i=1}^n (x_{ki}u_i)^2}}\right| \geq \sqrt{n}r\right). \tag{17.8}$$

We now use the following result on deviations of self-normalized sums due to Efron [11].

Lemma 17.1. *If $\eta_1, \ldots, \eta_n$ are independent symmetric random variables, then*

$$\mathbb{P}\left(\frac{\left|\frac{1}{n}\sum_{i=1}^{n}\eta_i\right|}{\sqrt{\frac{1}{n}\sum_{i=1}^{n}\eta_i^2}} \geq t\right) \leq 2\exp\left(-\frac{nt^2}{2}\right), \quad \forall\, t > 0.$$

For each of the probabilities on the right-hand side of (17.8), we apply Lemma 17.1 with $\eta_i = x_{ki}u_i$. This and the definition of r yield $\mathbb{P}(\mathcal{G}^c) \leq \alpha/2$. Thus, the event $\mathcal{G}$ holds with probability at least $1 - \alpha/2$. On the event $\mathcal{G}$ we have

$$\begin{aligned}
|\Psi_n \Delta|_\infty &\leq \left|\frac{1}{n}\mathbf{DX}^T(\mathbf{Y}-\mathbf{X}\hat{\beta})\right|_\infty + \left|\frac{1}{n}\mathbf{DX}^T(\mathbf{Y}-\mathbf{X}\beta^*)\right|_\infty \\
&\leq r\hat{\sigma} + \left|\frac{1}{n}\mathbf{DX}^T\mathbf{U}\right|_\infty \qquad (17.9) \\
&\leq r\left(\hat{\sigma} + \sqrt{\hat{Q}(\beta^*)}\right) \\
&\leq r\left[2\sqrt{\hat{Q}(\beta^*)} + \left(\hat{\sigma} - \sqrt{\hat{Q}(\beta^*)}\right)\right].
\end{aligned}$$

Inequality (17.9) holds because $(\hat{\beta}, \hat{\sigma})$ belongs to the set $\hat{\mathcal{D}}$ by definition. Notice that, on the event $\mathcal{G}$, $\left(\beta^*, \sqrt{\hat{Q}(\beta^*)}\right)$ belongs to the set $\hat{\mathcal{D}}$. On the other hand, $(\hat{\beta}, \hat{\sigma})$ minimizes the criterion $\left|\mathbf{D}^{-1}\beta\right|_1 + c\sigma$ on the same set $\hat{\mathcal{D}}$. Thus, on the event $\mathcal{G}$,

$$\left|\mathbf{D}^{-1}\hat{\beta}\right|_1 + c\hat{\sigma} \leq |\mathbf{D}^{-1}\beta^*|_1 + c\sqrt{\hat{Q}(\beta^*)}. \qquad (17.10)$$

This implies, again on the event $\mathcal{G}$,

$$\begin{aligned}
|\Psi_n \Delta|_\infty &\leq r\left[2\sqrt{\hat{Q}(\beta^*)} + \frac{1}{c}\sum_{k \in J(\beta^*)}\left(\left|d_{kk}^{-1}\beta_k^*\right| - \left|d_{kk}^{-1}\hat{\beta}_k\right|\right) - \frac{1}{c}\sum_{k \in J(\beta^*)^c}\left|d_{kk}^{-1}\hat{\beta}_k\right|\right] \\
&\leq r\left(2\sqrt{\hat{Q}(\beta^*)} + \frac{1}{c}|\Delta_{J(\beta^*)}|_1\right) \qquad (17.11)
\end{aligned}$$

where $\beta_k^*, \hat{\beta}_k$ are the kth components of $\beta^*, \hat{\beta}$. Similarly, (17.10) implies that, on the event $\mathcal{G}$,

$$|\Delta_{J(\beta^*)^c}|_1 = \sum_{k \in J(\beta^*)^c} \left|d_{kk}^{-1} \hat{\beta}_k\right|$$

$$\leq \sum_{k \in J(\beta^*)} \left(|d_{kk}^{-1} \beta_k^*| - \left|d_{kk}^{-1} \hat{\beta}_k\right|\right) + c\left(\sqrt{\hat{Q}(\beta^*)} - \hat{\sigma}\right)$$

$$\leq |\Delta_{J(\beta^*)}|_1 + c\sqrt{\hat{Q}(\beta^*)}. \tag{17.12}$$

We now distinguish between the following two cases.

Case 1: $c\sqrt{\hat{Q}(\beta^*)} \leq \gamma |\Delta_{J(\beta^*)}|_1$. In this case (17.12) implies

$$|\Delta_{J(\beta^*)^c}|_1 \leq (1+\gamma)|\Delta_{J(\beta^*)}|_1.$$

Thus, $\Delta \in C_{J(\beta^*)}^{(\gamma)}$ on the event $\mathcal{G}$. By definition of $\kappa_{1,J(\beta^*),J(\beta^*)}^{(\gamma)}$ and (17.3),

$$|\Delta_{J(\beta^*)}|_1 \leq \frac{|\Psi_n \Delta|_\infty}{\kappa_{1,J(\beta^*),J(\beta^*)}^{(\gamma)}} \leq \frac{|\Psi_n \Delta|_\infty}{\kappa_{1,0}^{(\gamma)}(s)}.$$

This and (17.11) yield

$$|\Delta_{J(\beta^*)}|_1 \leq \frac{2r\sqrt{\hat{Q}(\beta^*)}}{\kappa_{1,0}^{(\gamma)}(s)}\left(1 - \frac{r}{c\kappa_{1,0}^{(\gamma)}(s)}\right)_+^{-1}.$$

Case 2: $c\sqrt{\hat{Q}(\beta^*)} > \gamma |\Delta_{J(\beta^*)}|_1$. Then, obviously, $|\Delta_{J(\beta^*)}|_1 < \frac{c}{\gamma}\sqrt{\hat{Q}(\beta^*)}$.
Combining the two cases we obtain, on the event $\mathcal{G}$,

$$|\Delta_{J(\beta^*)}|_1 \leq \sqrt{\hat{Q}(\beta^*)} \max\left\{\frac{2r}{\kappa_{1,0}^{(\gamma)}(s)}\left(1 - \frac{r}{c\kappa_{1,0}^{(\gamma)}(s)}\right)_+^{-1}, \frac{c}{\gamma}\right\}. \tag{17.13}$$

In this argument, $c > 0$ and $\gamma > 0$ were arbitrary. The value of c given in (17.5) is the minimizer of the right-hand side of (17.13). Plugging it in (17.13) we find that, with probability at least $1 - \alpha/2$,

$$|\Delta|_1 \leq \frac{(\gamma+2)(2\gamma+1)r}{\gamma \kappa_{1,0}^{(\gamma)}(s)}\sqrt{\hat{Q}(\beta^*)},$$

where we have used (17.12). Now, by (17.4), $\hat{Q}(\beta^*) \leq Q^*$ with probability at least $1 - \alpha/2$. Thus, we get that (17.6) holds with probability at least $1 - \alpha$. Next, using (17.11) we obtain that, on the same event of probability at least $1 - \alpha$,

$$|\Psi_n \Delta|_\infty \leq \frac{(2\gamma + 1)r}{\gamma}\sqrt{Q^*}.$$

Combining this inequality with (17.6) yields (17.7). □

Discussion of Theorem 17.1.

1. In view of Proposition 17.1, $\kappa^{(\gamma)}_{1,J(\beta^*),J(\beta^*)} \geq (2+\gamma)^{-2}\kappa^{(\gamma)}_{RE,J(\beta^*)}/s$. Also, it is easy to see from Proposition 17.2 that $\kappa^{(\gamma)}_{1,0}(s)$ is of the order $1/s$ when Ψ_n is the identity matrix and $p \gg s$ (this is preserved for Ψ_n that are small perturbations of the identity). Thus, the bounds (17.6) and (17.7) take the form

$$|\Delta|_1 \leq C\left(s\sqrt{\frac{\log p}{n}}\right), \quad \Delta^T \Psi_n \Delta \leq C\left(\frac{s \log p}{n}\right),$$

for some constant C, and we recover the usual rates for the ℓ_1 estimation and for the prediction error respectively; cf. Bickel et al. [6].

2. Theorem 17.1 does not assume that the x_{ki} are independent from the u_i. The only assumption is the symmetry of $x_{ki}u_i$. However, if the x_{ki} are independent from u_i, then by conditioning on x_{ki} in the bound for $\mathbb{P}(\mathcal{G})$, it is enough to assume the symmetry of the u_i. Furthermore, while we have chosen the symmetry since it makes the conditions of Theorem 17.1 simple and transparent, it is not essential for our argument to be applied. The only point in the proof where we use the symmetry is the bound for the probability of deviations of self-normalized sums $\mathbb{P}(\mathcal{G})$. This probability can be bounded in many other ways without the symmetry assumption; cf., e.g., Gautier and Tsybakov [12]. It is enough to have $\mathbb{E}[x_{ki}u_i] = 0$ and a control, uniformly in k, of the ratio

$$\frac{(\sum_{i=1}^n \mathbb{E}[x_{ki}^2 u_i^2])^{1/2}}{(\sum_{i=1}^n \mathbb{E}[|x_{ki}u_i|^{2+\delta}])^{1/(2+\delta)}}$$

for some $\delta > 0$, cf. [13] or [5].

3. The quantity Q^* is not present in the definition of the estimator and is needed only to assess the rate of convergence. It is not hard to find Q^* in various situations. The simplest case is when $d_{kk} \equiv 1$ and the random variables x_{ki} and u_i are bounded uniformly in k, i by a constant L. Then we can take $Q^* = L^4$. If only x_{ki} are bounded uniformly in k by L, condition (17.4) holds when $\mathbb{P}\left(\frac{1}{n}\sum_{i=1}^n u_i^2 > Q^*/L^2\right) \leq \alpha/2$, and then for Q^* to be bounded it is enough to assume that the u_i have a finite second moment. The same remark applies when $d_{kk} = (\max_{i=1,\ldots,n} |x_{ki}|)^{-1}$, with an advantage that in this case we guarantee that Q^* is bounded under no assumption on x_{ki}.

4. The bounds in Theorem 17.1 depend on $\gamma > 0$ that can be optimized. Indeed, the functions of γ on the right-hand sides of (17.6) and (17.7) are data-driven and

can be minimized on a grid of values of γ. Thus, we obtain an optimal (random) value $\gamma = \hat{\gamma}$, for which (17.6) and (17.7) remain valid, since these results hold for any $\gamma > 0$.

References

1. Antoniadis, A.: Comments on: l^1-penalization for mixture regression models (with discussion). Test **19**, 257–258 (2010)
2. Belloni, A., Chernozhukov, V., Wang, L.: Pivotal estimation of nonparametric functions via square-root Lasso. arXiv e-prints 1105.1475 (2011)
3. Belloni, A., Chernozhukov, V., Wang, L.: Square-root Lasso: pivotal recovery of sparse signals via conic programming. Biometrika **98**, 791–806 (2011)
4. Belloni, A., Chen, D., Chernozhukov, V., Hansen, C.: Sparse models and methods for optimal instruments with an application to eminent domain. Econometrica **80**, 2369–2430 (2012)
5. Bertail, P., Gauthérat, E., Harari-Kermadec, H.: Exponential inequalities for self normalized sums. Electron. Commun. Probab. **13**, 628–640 (2009)
6. Bickel, P., Ritov, J.Y., Tsybakov, A.B.: Simultaneous analysis of Lasso and Dantzig selector. Ann. Stat. **37**, 1705–1732 (2009)
7. Bühlmann, P., van de Geer, S.: Statistics for High-Dimensional Data. Springer, New York (2011)
8. Candès, E., Tao, T.: The Dantzig selector: statistical estimation when p is much larger than n. Ann. Stat. **35**, 2313–2351 (2007)
9. Dalalyan, A.: SOCP based variance free Dantzig selector with application to robust estimation. C. R. Math. Acad. Sci. Paris **350**, 785–788 (2012)
10. Dalalyan, A., Tsybakov, A.: Aggregation by exponential weighting, sharp PAC-Bayesian bounds and sparsity. J. Mach. Learn. Res. **72**, 39–61 (2008)
11. Efron, B.: Student's t-test under symmetry conditions. J. Am. Stat. Assoc. **64**, 1278–1302 (1969)
12. Gautier, E., Tsybakov, A.: High-dimensional instrumental variables regression and confidence sets. arXiv e-prints 1105.2454 (2011)
13. Jing, B.Y., Shao, Q.M., Wang, Q.: Self-normalized Cramér-type large deviations for independent random variables. Ann. Probab. **31**, 2167–2215 (2003)
14. Koltchinskii, V.: Oracle Inequalities for Empirical Risk Minimization and Sparse Recovery Problems. Lecture Notes in Mathematics, vol. 2033. Springer, New York (2011)
15. Rigollet, P., Tsybakov, A.: Exponential screening and optimal rates of sparse estimation. Ann. Stat. **39**, 731–771 (2011)
16. Rigollet, P., Tsybakov, A.: Sparse estimation by exponential weighting. Stat. Sci. **27**, 558–575 (2012)
17. Städler, N., Bühlmann, P., van de Geer, S.: l^1-penalization for mixture regression models. Test **19**, 209–256 (2010)
18. Sun, T., Zhang, C.H.: Scaled sparse linear regression. arXiv e-prints 1104.4595 (2011)
19. Ye, F., Zhang, C.H.: Rate minimaxity of the Lasso and Dantzig selector for the l_q loss in l_r balls. J. Mach. Learn. Res. **11**, 3519–3540 (2010)

Chapter 18
On Sparsity Inducing Regularization Methods for Machine Learning

Andreas Argyriou, Luca Baldassarre, Charles A. Micchelli, and Massimiliano Pontil

Dedicated to Vladimir Vapnik with esteem and gratitude for his fundamental contribution to Machine Learning.

Abstract During the past few years there has been an explosion of interest in learning methods based on sparsity regularization. In this chapter, we discuss a general class of such methods, in which the regularizer can be expressed as the composition of a convex function ω with a linear function. This setting includes several methods such as the Group Lasso, the Fused Lasso, multi-task learning and many more. We present a general approach for solving regularization problems of this kind, under the assumption that the proximity operator of the function ω is available. Furthermore, we comment on the application of this approach to support vector machines, a technique pioneered by the groundbreaking work of Vladimir Vapnik.

A. Argyriou (✉)
Ecole Centrale de Paris, Grande Voie des Vignes, 92 295 Chatenay-Malabry, France
e-mail: andreas.argyriou@ecp.fr

L. Baldassarre
Laboratory for Information and Inference Systems, EPFL, ELD 243, Station 11 CH-1015 Lausanne, Switzerland
e-mail: luca.baldassarre@epfl.ch

C.A. Micchelli
Department of Mathematics and Statistics, University at Albany, Earth Science 110, Albany, NY 12222, USA
e-mail: charles_micchelli@gotmail.com

M. Pontil
Department of Computer Science, University College London, Malet Place, London WC1E 6BT, UK
e-mail: m.pontil@cs.ucl.ac.uk

18.1 Introduction

In this chapter, we address supervised learning methods which are based on the optimization problem

$$\min_{x \in \mathbb{R}^d} \{f(x) + g(x)\}, \tag{18.1}$$

where the function f measures the fit of a vector x (linear predictor) to available training data and g is a penalty term or regularizer which encourages certain types of solutions. Specifically, we let $f(x) = E(y, Ax)$, where $E : \mathbb{R}^s \times \mathbb{R}^s \to [0, \infty)$ is an error function, $y \in \mathbb{R}^s$ is a vector of measurements and $A \in \mathbb{R}^{s \times d}$ is a matrix whose rows are the input vectors. This class of regularization methods arise in machine learning, signal processing and statistics and have a wide range of applications.

Different choices of the error function and the penalty function correspond to specific techniques. In this chapter, we are interested in solving problem (18.1) when f is a *strongly smooth convex* function (such as the square error $E(y, Ax) = \|y - Ax\|_2^2$) and the penalty function g is obtained as the composition of a "simple" function with a linear transformation B, that is,

$$g(x) = \omega(Bx), \tag{18.2}$$

where B is a prescribed $m \times d$ matrix and ω is a *nondifferentiable convex* function on $\mathbb{R}^d$. The class of regularizers (18.2) includes a variety of methods, depending on the choice of the function ω and of matrix B. Our motivation for studying this class of penalty functions arises from sparsity-inducing regularization methods which consider ω to be either the ℓ_1 norm or a mixed ℓ_1-ℓ_p norm. When B is the identity matrix and $p = 2$, the latter case corresponds to the well-known Group Lasso method [36], for which well-studied optimization techniques are available. Other choices of the matrix B give rise to different kinds of Group Lasso with overlapping groups [15, 38], which have proved to be effective in modelling structured sparse regression problems. Further examples can be obtained by considering composition with the ℓ_1 norm; for example, this includes the Fused Lasso penalty function [31] and the graph prediction problem of Herbster and Lever [13].

A common approach to solving many optimization problems of the general form (18.1) is via proximal-gradient methods. These are first-order iterative methods, whose computational cost per iteration is comparable to gradient descent. In some problems in which g has a simple expression, proximal-gradient methods can be combined with acceleration techniques [22, 25, 32] to yield significant gains in the number of iterations required to reach a certain approximation accuracy of the minimal value. The essential step of proximal-gradient methods requires the computation of the proximity operator of function g; see Definition 18.1 below. In certain cases of practical importance, this operator admits a closed form, which makes proximal-gradient methods appealing to use. However, in the general case (18.2) the proximity operator may not be easily computable.

We describe a general technique to compute the proximity operator of the composite regularizer (18.2) from the solution of a fixed-point problem, which depends on the proximity operator of the function ω and the matrix B. This problem can be solved by a simple and efficient iterative scheme when the proximity operator of ω has a closed form or can be computed in a finite number of steps. When f is a strongly smooth function, the above result can be used together with Nesterov's accelerated method [22, 25] to provide an efficient first-order method for solving the optimization problem (18.1).

The chapter is organized as follows. In Sect. 18.2, we review the notion of proximity operator, discuss useful facts from fixed point theory and present a convergent algorithm for the solution of problem (18.1) when f is a quadratic function followed by an algorithm to solve the associated optimization problem (18.1). In Sect. 18.3, we discuss some examples of composite functions of the form (18.2), which are valuable in applications. In Sect. 18.4 we apply our observations to support vector machines and obtain new algorithms for the solution of this problem. Finally, Sect. 18.5 contains concluding remarks.

18.2 Fixed Point Algorithms Based on Proximity Operators

In this section, we present an optimization approach which uses fixed point algorithms for nonsmooth problems of the form (18.1) under the assumption (18.2). We first recall some notation and then move on to present an approach to compute the proximity operator for composite regularizers.

18.2.1 Notation and Problem Formulation

We denote by $\langle \cdot, \cdot \rangle$ the Euclidean inner product on $\mathbb{R}^d$ and let $\|\cdot\|_2$ be the induced norm. If $v : \mathbb{R} \to \mathbb{R}$, for every $x \in \mathbb{R}^d$ we denote by $v(x)$ the vector $(v(x_i))_{i=1}^d$. For every $p \geq 1$, we define the ℓ_p norm of x as $\|x\|_p = (\sum_{i=1}^d |x_i|^p)^{\frac{1}{p}}$.

As the basic building block of our method, we consider the optimization problem (18.1) in the special case when f is a quadratic function and the regularization term g is obtained by the composition of a convex function with a linear function. That is, we consider the problem

$$\min \left\{ \frac{1}{2} y^\top Q y - x^\top y + \omega(By) : y \in \mathbb{R}^d \right\}, \tag{18.3}$$

where x is a given vector in $\mathbb{R}^d$ and Q a positive definite $d \times d$ matrix. The development of a convergent method for the solution of this problem requires the well-known concepts of proximity operator and the subdifferential of a convex function. Let us now review some of the salient features of these important notions which are needed for the analysis of problem (18.3).

The proximity operator on a Hilbert space was introduced by Moreau in [20].

Definition 18.1. Let ω be a real-valued convex function on $\mathbb{R}^d$. The proximity operator of ω is defined, for every $x \in \mathbb{R}^d$, by

$$\text{prox}_\omega(x) := \text{argmin}\left\{\frac{1}{2}\|y - x\|_2^2 + \omega(y) : y \in \mathbb{R}^d\right\}. \tag{18.4}$$

The proximity operator is well defined, because the above minimum exists and is unique.

Recall that the subdifferential of ω at x is defined as $\partial \omega(x) = \{u : u \in \mathbb{R}^d, \langle y - x, u \rangle + \omega(x) \leq \omega(y), \forall y \in \mathbb{R}^d\}$. The subdifferential is a nonempty compact and convex set. Moreover, if ω is differentiable at x then its subdifferential at x consists only of the gradient of ω at x.

The relationship between the proximity operator and the subdifferential of ω is essential for algorithmic development for the solution of (18.3); see [2, 9, 19, 21]. Generally the proximity operator is difficult to compute since it is expressed as the minimum of a convex optimization problem. However, there are some rare circumstances in which it can be obtained explicitly; for example, when $\omega(x)$ is a multiple of the ℓ_1 norm of x, the proximity operator relates to soft thresholding, and moreover a related formula allows for the explicit identification of the proximity operator for the ℓ_2 norm; see, for example, [2, 9, 19]. Our optimization problem (18.3) can be reduced to that of the identification of the proximity operator for the composition function $\omega \circ B$. Although the prox of ω may be readily available, it may still be a computational challenge to obtain the prox of $\omega \circ B$. We consider this essential issue in the next section.

18.2.2 Computation of a Generalized Proximity Operator with a Fixed Point Method

In this section we consider circumstances in which the proximity operator of ω can be explicitly computed in a finite number of steps and seek an algorithm for the solution of the optimization problem (18.3).

As we shall see, the method proposed here applies for any positive definite matrix Q. This will allow us in a future publication to provide a second-order method for solving (18.1). For the moment, we are content with focusing on (18.3) by providing a technique for the evaluation of $\text{prox}_{\omega \circ B}$.

First, we observe that the minimizer $\hat{y}$ of (18.3) exists and is *unique*. Indeed, this vector is characterised by the set inclusion

$$Q\hat{y} \in x - B^\top \partial \omega(B\hat{y}).$$

Algorithm 6 Proximal-gradient & fixed point algorithm.

$x_1 \leftarrow 0$
for t = 1,2,... **do**
    Compute $x_{t+1} \leftarrow \text{prox}_{\frac{\omega}{L} \circ B}\left(x_t - \frac{1}{L}\nabla f(x_t)\right)$
    by the Picard process.
end for

To make use of this observation, we introduce the affine transformation $A : \mathbb{R}^m \to \mathbb{R}^m$, defined, for fixed $x \in \mathbb{R}^d$, $\lambda > 0$, at $z \in \mathbb{R}^m$ by

$$Az := (I - \lambda B Q^{-1} B^\top)z + BQ^{-1}x,$$

and the nonlinear operator $H : \mathbb{R}^m \to \mathbb{R}^m$,

$$H := \left(I - \text{prox}_{\frac{\omega}{\lambda}}\right) \circ A. \tag{18.5}$$

The next theorem from [2] is a natural extension of an observation in [19], which only applies to the case $Q = I$.

Theorem 18.1. *If ω is a convex function on $\mathbb{R}^m$, $B \in \mathbb{R}^{m \times d}$, $x \in \mathbb{R}^d$, λ is a positive number, the operator H is defined as in (18.5), and $\hat{y}$ is the minimizer of (18.3) then*

$$\hat{y} = Q^{-1}(x - \lambda B^\top v)$$

if and only if $v \in \mathbb{R}^m$ is a fixed point of H.

This theorem provides us with a practical tool to solve problem (18.3) numerically by using the Picard iteration relative to the nonlinear mapping H. Under an additional hypothesis on the matrix $BQ^{-1}B^\top$, the mapping H is non-expansive; see [2]. Therefore, Opial's Theorem [37] allows us to conclude that the Picard iterate converges to the solution of (18.3); see [2, 19] for a discussion of this issue. Furthermore, under additional hypotheses the mapping H is a contraction. In that case, the Picard iterate converges linearly.

We may extend the range of applicability of our observations and provide a fixed point proximal-gradient method for solving problem (18.1) when the regularizer has the form (18.2) and the error f is a *strongly smooth* convex function, that is, the gradient of f, denoted by ∇f, is Lipschitz continuous with constant L. So far, the convergence of this extension has yet to be analyzed. The idea behind proximal-gradient methods—see [9, 25, 32] and references therein—is to update the current estimate of the solution x_t using the proximity operator of g and the gradient of f. This is equivalent to replacing f with its linear approximation around a point which is a function of the previous iterates of the algorithm. The simplest instance of this iterative algorithm is given in Algorithm 1–6. Extensions to acceleration schemes are described in [2].

18.2.3 Connection to the Forward-Backward Algorithm

In this section, we consider the special case $Q = I$ and interpret the Picard iteration of H in terms of a *forward-backward algorithm* in the dual; for a discussion of the forward-backward algorithm, see for example [9]

The Picard iteration is defined as

$$v_{t+1} \leftarrow (I - \mathrm{prox}_{\frac{\omega}{\lambda}})((I - \lambda BB^\top)v_t + Bx). \tag{18.6}$$

We first recall the Moreau decomposition—see, for example, [9] and references therein—which relates the proximity operators of a lower semicontinuous convex function $\varphi : \mathbb{R}^m \to \mathbb{R} \cup \{+\infty\}$ and its conjugate,

$$I = \mathrm{prox}_\varphi + \mathrm{prox}_{\varphi^*} . \tag{18.7}$$

Using Eq. (18.7), the iterative step (18.6) becomes

$$v_{t+1} \leftarrow \mathrm{prox}_{(\frac{\omega}{\lambda})^*} (v_t - (\lambda BB^\top v_t - Bx))$$

which is a forward-backward method. We can further simplify this iteration by introducing the vector $z_t := \lambda v_t$ and obtaining the iterative algorithm

$$z_{t+1} \leftarrow \lambda \, \mathrm{prox}_{(\frac{\omega}{\lambda})^*} \left(\frac{1}{\lambda} z_t - (BB^\top z_t - Bx) \right) .$$

Using the readily verified formulas

$$\frac{1}{\lambda} \mathrm{prox}_{\lambda g} \circ \lambda I = \mathrm{prox}_{\frac{1}{\lambda} g \circ \lambda I}$$

and

$$\left(\frac{\omega}{\lambda} \right)^* = \frac{1}{\lambda} \omega^* \circ \lambda I$$

(see, for example, [4]), we obtain the equivalent forward-backward iteration

$$z_{t+1} \leftarrow \mathrm{prox}_{\lambda \omega^*} (z_t - (\lambda BB^\top z_t - \lambda Bx)) .$$

This method is a forward-backward method of the type considered in [8, Algorithm 10.3] and solves the minimization problem

$$\min \left\{ \frac{1}{2} \| B^\top z - x \|^2 + \omega^*(z) : z \in \mathbb{R}^m \right\} .$$

This minimization problem in turn can be viewed as the dual of the primal problem

$$\min\left\{\frac{1}{2}\|u\|^2 - \langle x, u\rangle + \omega(Bu) : u \in \mathbb{R}^d\right\} \tag{18.8}$$

by using Fenchel's duality theorem; see, for example, [4]. Moreover, the primal and dual solutions are related through the conditions $-B^\top \hat{z} = \hat{u} - x$ and $\hat{z} \in \partial\omega(B\hat{u})$, the first of which implies that $x - \lambda B^\top \hat{v}$ equals the solution of the proximity problem (18.8), that is, equals $\text{prox}_{\omega \circ B}(x)$.

18.3 Examples of Composite Functions

In this section, we provide some examples of penalty functions which have appeared in the literature that fall within the class of linear composite functions (18.2).

We define for every $d \in \mathbb{N}$, $x \in \mathbb{R}^d$ and $J \subseteq \{1, \ldots, d\}$, the restriction of the vector x to the index set J as $x_{|J} = (x_i : i \in J)$. Our first example considers the Group Lasso penalty function, which is defined as

$$\omega_{\text{GL}}(x) = \sum_{\ell=1}^{k} \|x_{|J_\ell}\|_2, \tag{18.9}$$

where the J_ℓ are prescribed subsets of $\{1, \ldots, d\}$ (also called the "groups") such that $\cup_{\ell=1}^{k} J_\ell = \{1, \ldots, d\}$. The standard Group Lasso penalty—see, for example, [36]—corresponds to the case where the collection of groups $\{J_\ell : 1 \leq \ell \leq k\}$ forms a partition of the index set $\{1, \ldots, d\}$, that is, the groups do not overlap. In this case, the optimization problem (18.4) for $\omega = \omega_{\text{GL}}$ decomposes as the sum of separate problems, and the proximity operator is readily obtained by using the proximity operator of the ℓ_2-norm to each group separately. In many cases of interest, however, the groups overlap and the proximity operator cannot be easily computed.

Note that the function (18.9) is of the form (18.2). We let $d_\ell = |J_\ell|$, $m = \sum_{\ell=1}^{k} d_\ell$ and define, for every $z \in \mathbb{R}^m$, $\omega(z) = \sum_{\ell=1}^{k} \|z_\ell\|_2$, where, for every $\ell = 1, \ldots, k$ we let $z_\ell = (z_i : \sum_{j=1}^{\ell-1} d_j < i \leq \sum_{j=1}^{\ell} d_j)$. Moreover, we choose $B^\top = [B_1^\top, \ldots, B_k^\top]$, where B_ℓ is a $d_\ell \times d$ matrix defined as

$$(B_\ell)_{ij} = \begin{cases} 1 & \text{if } j = J_\ell[i] \\ 0 & \text{otherwise} \end{cases},$$

where for every $J \subseteq \{1, \ldots, d\}$ and $i \in \{1, \ldots, |J|\}$, we denote by $J[i]$ the i-th largest integer in J.

The second example concerns the Fused Lasso [31], which considers the penalty function $x \mapsto g(x) = \sum_{i=1}^{d-1} |x_i - x_{i+1}|$. This function falls into the class (18.2). Indeed, if we choose ω to be the ℓ_1 norm and B the first-order divided difference matrix

$$B = \begin{bmatrix} 1 & -1 & 0 & \dots & & \\ 0 & 1 & -1 & 0 & \dots & \\ \vdots & \ddots & \ddots & \ddots & \ddots & \end{bmatrix}$$

we get back g. The intuition behind the Fused Lasso is that it favors vectors which do not vary much across contiguous components. Further extensions of this case may be obtained by choosing B to be the incidence matrix of a graph, leading to the penalty $\sum_{(i,j)\in E}^{n} |x_i - x_j|$. This is a setting which is relevant, for example, in online learning over graphs [13, 14].

The next example considers composition with orthogonally invariant (OI) norms. Specifically, we choose a symmetric gauge function h, that is, a norm h which is both *absolute* and *invariant under permutations* [35], and define the function $\omega : \mathbb{R}^{d\times n} \to [0,\infty)$ at X by the formula $\omega(X) = h(\sigma(X))$, where $\sigma(X) \in [0,\infty)^r$ and $r = \min(d,n)$ is the vector formed by the singular values of matrix X in non-increasing order. An example of OI-norm is the Schatten p-norm, which corresponds to the case where ω is the ℓ_p-norm. The next proposition provides a formula for the proximity operator of an OI-norm. A proof can be found in [2].

Proposition 18.1. *With the above notation, it holds that*

$$\text{prox}_{h\circ\sigma}(X) = U \, \text{diag}\left(\text{prox}_h(\sigma(X))\right) V^\top$$

where $X = U\,\text{diag}(\sigma(X))V^\top$ and U and V are the matrices formed by the left and right singular vectors of X, respectively.

We can compose an OI-norm with a linear transformation B, this time between two spaces of matrices, obtaining yet another subclass of penalty functions of the form (18.2). This setting is relevant in the context of multi-task learning. For example, in [1] h is chosen to be the *trace* or *nuclear* norm and a specific linear transformation which models task relatedness is considered. Specifically, the regulariser is given by $g(X) = \left\|\sigma\left(X(I - \frac{1}{n}ee^\top)\right)\right\|_1$, where $e \in \mathbb{R}^d$ is the vector all of whose components are equal to 1.

18.4 Application to Support Vector Machines

In this section, we turn our attention to the important topic of support vector machines (SVMs), which are widely used in data analysis. SVMs were pioneered by the fundamental work of Vapnik [5, 10, 33] and inspired one of us to begin research in machine learning [11, 26, 27]. For that we are all very grateful to Vladimir Vapnik for his fundamental contributions to machine learning.

First, we recall the SVM primal and dual optimization problems; see [33]. To simplify the presentation we only consider the linear version of SVMs. A similar treatment using feature map representations is straightforward and so will not be discussed here, although this in an important extension of practical value. Moreover, we only consider SVMs for classification, but our approach can be applied to SVM regression and other variants of SVMs which have appeared in the literature.

The optimization problem of concern here is given by

$$\min\left\{C\sum_{i=1}^{m} V(y_i w^\top x_i) + \frac{1}{2}\|w\|^2 : w \in \mathbb{R}^d\right\} \quad (18.10)$$

where $V(z) = \max(0, 1 - z)$, $z \in \mathbb{R}$, is the hinge loss and C is a positive parameter balancing empirical error against margin maximization. We let $x_i \in \mathbb{R}^d$, $i \in \{1,\ldots,m\}$, be the input data and $y_i \in \{-1,+1\}$ be the class labels.

Problem (18.10) can be viewed as a proximity operator computation of the form (18.3), with $Q = I$, $x = 0$, $\omega(z) = C\sum_{i=1}^{m} V(z_i)$ and $B = [y_1 x_1 \ldots y_m x_m]^\top$. The proximity operator of the hinge loss is separable across the coordinates and simple to compute. In fact, for any $\zeta \in \mathbb{R}$ and $\mu > 0$, it is given by the formula

$$\text{prox}_{\mu V}(\zeta) = \min(\zeta + \mu, \max(\zeta, 1)).$$

Hence, we can solve problem (18.10) by the Picard iteration, namely

$$v_{t+1} \leftarrow \left(I - \text{prox}_{\frac{\omega}{\lambda}}\right)\left((I - \lambda BB^\top)v_t\right) \quad (18.11)$$

with λ satisfying $0 < \lambda < \frac{2}{\lambda_{\max}(BB^\top)}$, which ensures that the nonlinear mapping is strictly nonexpansive. Note that $v_t \in \mathbb{R}^m$ and that this iterative scheme maybe interpreted as acting on the SVM dual; see Sect. 18.2.3. In fact, there is a simple relation to the support vector coefficients given by the equation $v = \frac{-1}{\lambda}\alpha$. Consequently, this algorithmic approach is well suited when the sample size m is small compared to the dimensionality d. An estimate of the primal solution, if required, can be obtained by using the formula $w = -\lambda B^\top v$.

Recall that the dual problem of (18.10) is given [33]

$$\min\left\{\frac{1}{2}\|B^\top \alpha\|^2 - 1^\top \alpha : \alpha \in [0, C]^m\right\}. \quad (18.12)$$

This problem can be seen as the computation of a generalized proximity operator of the type (18.3). To explain what we have in mind we use the notation $\odot$ as the element-wise product between matrices of the same size (Schur product) and introduce the kernel matrix $K = [x_1,\ldots,x_m]^\top [x_1,\ldots,x_m]$.

Using this terminology, we conclude that problem (18.12) is of the form (18.3) with $Q = K \odot yy^\top$, $x = \mathbf{1}$ (the vector of all 1s), $B = I$ and $\omega = \omega_C$, where $\omega_C(\alpha) = 0$ if $\alpha \in [0, C]^m$ and $\omega(\alpha) = +\infty$ otherwise. Furthermore, the proximity operator for ω is given by the projection on the set $[0, C]^m$, that is, $\text{prox}_{\omega_C}(\alpha) =$

$\min(C, \max(0, \alpha))$. These observations yield the Picard iteration

$$v_{t+1} \leftarrow \left(I - \text{prox}_{\omega C}\right)\left((I - \lambda(K^{-1} \odot yy^\top))v_t + (K^{-1} \odot yy^\top)\mathbf{1}\right) \quad (18.13)$$

with $0 < \lambda < 2\lambda_{\min}(K)$. This iterative scheme requires that the kernel matrix K be invertible, which is frequently the case, for example, in the case of Gaussian kernels. Another requirement is that either K^{-1} be precomputed or a linear system involving K be solved at every iteration, which limits the scalability of this scheme to very large samples. In contrast, the iteration (18.11) can always be applied, even when K is not invertible. In fact, when K, and equivalently $BB^\top$, is invertible, both iterative methods (18.11), (18.13) converge linearly at a rate which depends on the condition number of K; see [2, 19].

Recall that algorithm (18.11) is equivalent to a forward-backward method in the dual; see Sect. 18.2.3. Thus, an accelerated variant akin to Nesterov's optimal method and FISTA [3] could also be used. However, in the case of an invertible kernel matrix, both versions converge linearly [25], and hence it is not clear whether there is any practical advantage from the Nesterov update. Furthermore, algorithm (18.13) could also be modified in a similar way.

On the other hand, if $m > d$, we would directly attempt to solve the primal problem. In this case, the Nesterov smoothing method can be employed, [24]. An advantage of such a method is that it only stores $O(d)$ variables, even though it needs $O(md)$ computations per iteration. The method described above, based on the Picard iteration, requires $\min(O(md), O(m^2))$ cost per iteration and stores $O(m)$ variables.

Let us finally remark that iterative methods similar to (18.11) or (18.13) can be applied to ℓ_2 regularization problems other than SVMs, provided that the proximity operator of the corresponding loss function is available. Common choices for the loss function, other than the hinge loss, are the logistic and square loss functions leading to logistic regression and least squares regression, respectively. In particular, in these two cases, the primal objective (18.10) is both smooth and strongly convex and hence a linearly convergent gradient descent or accelerated gradient descent method can be used [23], regardless of the conditioning of the kernel matrix.

18.5 Conclusion

We presented a general approach to solve a class of nonsmooth optimization problems, whose objective function is given by the sum of a smooth term and a nonsmooth term which is obtained by a linear function composition. The prototypical example covered by this setting is a linear regression regularization method, in which the smooth term is an error term and the nonsmooth term is a regularizer which favors certain desired parameter vectors. An important feature of our approach is that it can deal with a rich class of regularizers and, as shown numerically in [2], is competitive with the state-of-the-art methods. Using these ideas, we also provided a fixed-point scheme to solve support vector machines. Although numerical experiments have yet to be done, we believe this method is simple enough to deserve attention by practitioners.

We believe that the method presented here should be thoroughly investigated both in terms of convergence analysis, where ideas presented in [34] may be valuable, and numerical performance with other methods, such as alternate direction of multipliers (see, for example, [6]), block coordinate descent, alternate minimization and others. Finally, there are several other machine learning problems where ideas presented here apply. For example, in that regard we mention multiple kernel learning (see, for example, [17, 28–30] and references therein), some structured sparsity regularizers [16, 18] and multi-task learning (see, for example, [1, 7, 12]). We leave these tantalizing issues for future investigation.

Acknowledgements Part of this work was supported by EPSRC Grant EP/H027203/1, Royal Society International Joint Project Grant 2012/R2 and by the European Union Seventh Framework Programme (FP7 2007-2013) under grant agreement No. 246556.

References

1. Argyriou, A., Evgeniou, T., Pontil, M.: Convex multi-task feature learning. Mach. Learn. **73**(3), 243–272 (2008)
2. Argyriou, A., Micchelli, C., Pontil, M., Shen, L., Xu, Y.: Efficient first order methods for linear composite regularizers. CoRR **1104.1436** (2011)
3. Beck, A., Teboulle, M.: A fast iterative shrinkage-thresholding algorithm for linear inverse problems. SIAM J. Imaging Sci. **2**(1), 183–202 (2009)
4. Borwein, J., Lewis, A.: Convex Analysis and Nonlinear Optimization: Theory and Examples. CMS Books in Mathematics. Springer (2005)
5. Boser, B., Guyon, I., Vapnik, V.: A training algorithm for optimal margin classifiers. In: Proceedings of 5th Annual ACM Workshop on Computational Learning Theory, Pittsburgh, pp. 144–152 (1992)
6. Boyd, S., Parikh, N., Chu, E., Peleato, B., Eckstein, J.: Distributed optimization and statistical learning via the alternating direction method of multipliers. Found. Trends Mach. Learn. **3**(1), 1–122 (2011)
7. Cavallanti, G., Cesa-Bianchi, N., Gentile, C.: Linear algorithms for online multitask classification. J. Mach. Learn. Res. **11**, 2901–2934 (2010)
8. Combettes, P., Pesquet, J.C.: Proximal splitting methods in signal processing. In: Bauschke, H., Burachik, R., Combettes, P., Elser, V., Luke, D., Wolkowicz, H. (eds.) Fixed-Point Algorithms for Inverse Problems in Science and Engineering, pp. 185–212. Springer (2011)
9. Combettes, P., Wajs, V.: Signal recovery by proximal forward-backward splitting. Multiscale Model. Simul. **4**(4), 1168–1200 (2006)
10. Cortes, C., Vapnik, V.: Support-vector networks. Mach. Learn. **20**, 273–297 (1995)
11. Evgeniou, T., Pontil, M., Poggio, T.: Regularization networks and support vector machines. Adv. Comput. Math. **13**(1), 1–50 (2000)
12. Evgeniou, T., Pontil, M., Toubia, O.: A convex optimization approach to modeling heterogeneity in conjoint estimation. Mark. Sci. **26**, 805–818 (2007)
13. Herbster, M., Lever, G.: Predicting the labelling of a graph via minimum p-seminorm interpolation. In: Proceedings of the 22nd Conference on Learning Theory (COLT), Montreal (2009)

14. Herbster, M., Pontil, M.: Prediction on a Graph with the Perceptron. Advances in Neural Information Processing Systems 19, pp. 577–584. MIT (2007)
15. Jenatton, R., Audibert, J.Y., Bach, F.: Structured variable selection with sparsity-inducing norms. CoRR **0904.3523v2** (2009)
16. Maurer, A., Pontil, M.: Structured sparsity and generalization. J. Mach. Learn. Res. **13**, 671–690 (2012)
17. Micchelli, C., Pontil, M.: Feature space perspectives for learning the kernel. Mach. Learn. **66**, 297–319 (2007)
18. Micchelli, C., Morales, J., Pontil, M.: A family of penalty functions for structured sparsity. In: NIPS, Vancouver, pp. 1612–1623 (2010)
19. Micchelli, C., Shen, L., Xu, Y.: Proximity algorithms for image models: denoising. Inverse Probl. **27**(4) (2011)
20. Moreau, J.: Fonctions convexes duales et points proximaux dans un espace hilbertien. Acad. Sci. Paris Sér. A Math. **255**, 2897–2899 (1962)
21. Mosci, S., Rosasco, L., Santoro, M., Verri, A., Villa, S.: Solving structured sparsity regularization with proximal methods. In: Proceedings of European Conference on Machine Learning and Knowledge Discovery in Databases, Barcelona, pp. 418–433 (2010)
22. Nesterov, Y.: A method of solving a convex programming problem with convergence rate $O(1/k^2)$. Sov. Math. Dokl. **27**(2), 372–376 (1983)
23. Nesterov, Y.: Introductory Lectures on Convex Optimization: A Basic Course. Kluwer, Boston (2004)
24. Nesterov, Y.: Smooth minimization of non-smooth functions. Math. Program. **103**(1), 127–152 (2005)
25. Nesterov, Y.: Gradient methods for minimizing composite objective function. ECORE Discussion Paper 2007/76 (2007)
26. Pontil, M., Verri, A.: Properties of support vector machines. Neural Comput. **10**, 955–974 (1998)
27. Pontil, M., Rifkin, R., Evgeniou, T.: From regression to classification in support vector machines. In: Proceedings of 7th European Symposium on Artificial Neural Networks, Bruges, pp. 225–230 (1999)
28. Rakotomamonjy, A., Bach, F., Canu, S., Grandvalet, Y.: SimpleMKL. J. Mach. Learn. Res. **9**, 2491–2521 (2008)
29. Sonnenburg, S., Rätsch, G., Schäfer, C., Schölkopf, B.: Large scale multiple kernel learning. J. Mach. Learn. Res. **7**, 1531–1565 (2006)
30. Suzuki, T., Tomioka, R.: SpicyMKL: a fast algorithm for multiple kernel learning with thousands of kernels. Mach. Learn. **8**(1), 77–108 (2011)
31. Tibshirani, R., Saunders, M., Rosset, S., Zhu, J., Knight, K.: Sparsity and smoothness via the fused Lasso. J. R. Stat. Soc.: Ser. B (Stat. Methodol.) **67**(1), 91–108 (2005)
32. Tseng, P.: Approximation accuracy, gradient methods, and error bound for structured convex optimization. Math. Program. **125**(2), 263–295 (2010)
33. Vapnik, V.: The Nature of Statistical Learning Theory. Springer, New York (1999)
34. Villa, S., Salzo, S., Baldassarre, L., Verri, A.: Accelerated and inexact forward-backward splitting. SIAM J. Optim. **23**(3), 1607–1633 (2013)
35. von Neumann, J.: Some matrix-inequalities and metrization of matric-space. Mitt. Forsch.-Inst. Math. Mech. Univ. Tomsk **1**, 286–299 (1937)
36. Yuan, M., Lin, Y.: Model selection and estimation in regression with grouped variables. J. R. Stat. Soc. B **68**(1), 49–67 (2006)
37. Zălinescu, C.: Convex Analysis in General Vector Spaces. World Scientific, River Edge/London (2002)
38. Zhao, P., Rocha, G., Yu, B.: Grouped and hierarchical model selection through composite absolute penalties. Ann. Stat. **37**(6A), 3468–3497 (2009)

Chapter 19
Sharp Oracle Inequalities in Low Rank Estimation

Vladimir Koltchinskii

Abstract This chapter deals with the problem of penalized empirical risk minimization over a convex set of linear functionals on the space of Hermitian matrices with convex loss and nuclear norm penalty. Such penalization is often used in low rank matrix recovery in the cases when the target function can be well approximated by a linear functional generated by a Hermitian matrix of relatively small rank (comparing it with the size of the matrix). Our goal is to prove sharp low rank oracle inequalities that involve the excess risk (the approximation error) with constant equal to 1 and the random error term with correct dependence on the rank of the oracle.

19.1 Main Result

About 40 years ago, Vapnik and Chervonenkis pioneered the "structural risk minimization" method (see Vapnik [15] and references therein) that was later developed into penalized empirical risk minimization with more general complexity penalties. This method has become one of the most powerful estimation tools in modern statistics and machine learning. In this paper, a version of penalized empirical risk minimization in the problem of estimation of large matrices of relatively small rank, where the nuclear norm is used as a complexity penalty, will be discussed.

Let (X, Y) be a couple, where X is a random variable in the space $\mathbb{H}_m$ of $m \times m$ Hermitian matrices and Y is a random response variable with values in a Borel subset $T \subset \mathbb{R}$. Let P be the distribution of (X, Y) and let Π denote the marginal distribution of X. The goal is to predict Y based on an observation of X. More precisely, let $\ell : T \times \mathbb{R} \mapsto \mathbb{R}_+$ be a measurable loss function. We will assume in

---

V. Koltchinskii (✉)
School of Mathematics, Georgia Institute of Technology, Atlanta, GA 30332-0160, USA
e-mail: vlad@math.gatech.edu

what follows that, for all $y \in T$, $\ell(y;\cdot)$ is convex. Given a measurable function $f : \mathbb{H}_m \mapsto \mathbb{R}$ (a "prediction rule"), denote $(\ell \bullet f)(x, y) := \ell(y; f(x))$ and define the risk of f as

$$P(\ell \bullet f) = \mathbb{E}\ell(Y; f(X)).$$

Then, one can view the prediction problem as risk minimization: the goal is to find a function $f_* : \mathbb{H}_m \mapsto \mathbb{R}$ that minimizes the risk $P(\ell \bullet f)$ over the class of all measurable prediction rules $f : \mathbb{H}_m \mapsto \mathbb{R}$ (provided that such a function exists), or, more realistically, to find a reasonably good approximation of f_*. To this end, one wants to find a function f for which the excess risk $\mathcal{E}(f) := P(\ell \bullet f) - \inf_{g:\mathbb{H}_m \mapsto \mathbb{R}} P(\ell \bullet g)$ is small enough. Of course, the risk $P(\ell \bullet f)$ depends on the distribution P of (X, Y), which is, most often, unknown. In such cases, the problem has to be solved based on the training data $(X_1, Y_1), \ldots, (X_n, Y_n)$, which consists of n independent copies of (X, Y). We will be especially interested in the problems in which matrices are large and the optimal prediction rule f_* can be well approximated by a linear function $f_S(\cdot) := \langle S, \cdot \rangle$, where $S \in \mathbb{H}_m$ is a low rank Hermitian matrix, that is, when there exists a low rank matrix S (an oracle) such that the excess risk $\mathcal{E}(f_S)$ is small. Here, and in what follows, $\langle \cdot, \cdot \rangle$ denotes the Hilbert-Schmidt (Frobenius) inner product in $\mathbb{H}_m$. In such problems, we would like to find an estimator $\hat{S}$ based on the training data $(X_1, Y_1), \ldots, (X_n, Y_n)$ such that the excess risk $\mathcal{E}(f_{\hat{S}})$ of the estimator can be bounded from above by the excess risk $\mathcal{E}(f_S)$ of an arbitrary oracle $S \in \mathbb{H}_m$ plus an error term that properly depends on the rank of the oracle. The resulting bounds on the excess risk $\mathcal{E}(f_{\hat{S}})$ of the estimator $\hat{S}$ are supposed to hold with a guaranteed high probability and they are often called "low rank oracle inequalities." We will consider below the rather traditional estimator $\hat{S}$ based on penalized empirical risk minimization with a nuclear norm penalty:

$$\hat{S} := \operatorname{argmin}_{S \in \mathbb{D}} \left[P_n(\ell \bullet f_S) + \varepsilon \|S\|_1 \right], \tag{19.1}$$

where $\mathbb{D} \subset \mathbb{H}_m$ is a closed convex set, $0 \in \mathbb{D}$, P_n is the empirical distribution based on the training data $(X_1, Y_1), \ldots, (X_n, Y_n)$ and

$$P_n(\ell \bullet f_S) = n^{-1} \sum_{j=1}^{n} \ell(Y_j; f_S(X_j))$$

is the corresponding empirical risk with respect to the loss ℓ, $\|S\|_1 := \operatorname{tr}(|S|) = \operatorname{tr}(\sqrt{S^2})$ is the nuclear norm of S and $\varepsilon \geq 0$ is the regularization parameter. Clearly, optimization problem (19.1) is convex. In fact, it is a standard convex relaxation of penalized empirical risk minimization with a penalty proportional to the rank of S, denoted in what follows by $\operatorname{rank}(S)$, which is not a computationally tractable problem. Such convex relaxations have been extensively studied in the recent years (see [3–6, 8, 10–13] and references therein).

To state our main result (a sharp low rank oracle inequality for the estimator $\hat{S}$), we first introduce some assumptions and notations. In what follows, assume that for some constant $a > 0$, $|\langle S, X \rangle| \leq a$ a.s., $S \in \mathbb{D}$. It will be also assumed that ℓ is a convex loss of *quadratic type*. More precisely, suppose that, for all $y \in T$, $\ell(y, \cdot)$ is a twice continuously differentiable convex function in $[-a, a]$ with $Q := \sup_{y \in T} \ell(y; 0) < +\infty$,

$$L(a) := \sup_{y \in T} \sup_{u \in [-a,a]} \left[|\ell'(y; 0)| + \ell''(y; u) a \right] < +\infty$$

and

$$\tau(a) := \inf_{y \in T} \inf_{u \in [-a,a]} \ell''(y; u) > 0.$$

Here ℓ', ℓ'' denote the first and the second derivatives of the loss $\ell(y, u)$ with respect to u. Many important losses in regression and in large margin classification problems are of quadratic type. In particular, if $\ell(y; u) = (y - u)^2$, $y, u \in [-a, a]$ (regression with quadratic loss and with bounded response), then $L(a) = 4a$ and $\tau(a) = 2$. Exponential loss $\ell(y, u) = e^{-yu}$, $y \in \{-1, 1\}$, $u \in [-a, a]$, often used in large margin methods for binary classification, is also of quadratic type.

In what follows, $\|\cdot\|_2$ denotes the Hilbert–Schmidt (Frobenius) norm of Hermitian matrices (generated by the inner product $\langle \cdot, \cdot \rangle$) and $\|\cdot\|$ denotes the operator norm.

We will use certain characteristics of matrices $S \in \mathbb{D}$ that are related to matrix versions of the restricted isometry property (see, e.g., [8], Chap. 9 and references therein). Let $S \in \mathbb{D}$ be a matrix with spectral representation $S = \sum_{j=1}^{r} \lambda_j (\phi_j \otimes \phi_j)$, where $r := \mathrm{rank}(S)$, λ_j are non-zero eigenvalues of S (repeated with their multiplicities) and $\phi_j \in \mathbb{C}^m$ are the corresponding orthonormal eigenvectors. In what follows, we denote

$$\mathrm{sign}(S) := \sum_{j=1}^{r} \mathrm{sign}(\lambda_j)(\phi_j \otimes \phi_j), \quad L := \mathrm{supp}(S) := \mathrm{l.s.}(\phi_1, \ldots, \phi_r).$$

Let $\mathcal{P}_L, \mathcal{P}_L^\perp$ be the following orthogonal projectors in the space $(\mathbb{H}_m, \langle \cdot, \cdot \rangle)$:

$$\mathcal{P}_L(A) := A - P_{L^\perp} A P_{L^\perp}, \quad \mathcal{P}_L^\perp(A) := P_{L^\perp} A P_{L^\perp}, \quad A \in \mathbb{H}_m$$

(here $L^\perp$ is the orthogonal complement of L). Clearly, we have $A = \mathcal{P}_L A + \mathcal{P}_L^\perp A$, $A \in \mathbb{H}_m$, providing a decomposition of a matrix A into a "low rank part" $\mathcal{P}_L A$ and a "high rank part" $\mathcal{P}_L^\perp A$. Given $b > 0$, define the following cone in the space $\mathbb{H}_m$

$$\mathcal{K}(\mathbb{D}; L; b) := \left\{ A \in \mathrm{l.s.}(\mathbb{D}) : \|\mathcal{P}_L^\perp(A)\|_1 \leq b \|\mathcal{P}_L(A)\|_1 \right\}$$

that consists of matrices A with a "dominant" low rank part. Let

$$\beta^{(b)}(\mathbb{D}; L; \Pi) := \inf\{\beta > 0 : \|\mathcal{P}_L(A)\|_2 \leq \beta \|f_A\|_{L_2(\Pi)}, A \in \mathcal{K}(\mathbb{D}; L; b)\}.$$

This quantity is known to be bounded from above by a constant in the case when the matrix form of the "distribution-dependent" restricted isometry condition holds for $r = \text{rank}(S)$ (see [8], Sect. 9.1). In what follows, we will use the following characteristic of oracle S:

$$\beta(S) := \beta^{(5)}(\mathbb{D}; L; \Pi), \quad L := \text{supp}(S).$$

For arbitrary $t > 0$ and $S \in \mathbb{D}$, denote

$$t(S; \varepsilon) := t + 3\log\left(B\log_2\left(\|S\|_1 \vee n \vee \varepsilon \vee Q \vee a^{-1} \vee (L(a))^{-1} \vee 2\right)\right),$$

where $B > 0$ is a constant. Let

$$\Delta := \mathbb{E}\left\|\frac{1}{\sqrt{n}}\sum_{j=1}^{n}\varepsilon_j X_j\right\|,$$

where $\{\varepsilon_j\}$ are i.i.d. Rademacher random variables independent of $\{X_j\}$.

Theorem 19.1. *There exist a numerical constant $B > 0$ in the definition of $t(S; \varepsilon)$ and numerical constants $C, D > 0$ such that for all $t > 0$ and all*

$$\varepsilon \geq \frac{DL(a)\Delta}{\sqrt{n}}, \tag{19.2}$$

with probability at least $1 - e^{-t}$,

$$\mathcal{E}(f_{\hat{S}}) \leq \inf_{S \in \mathbb{D}}\left[\mathcal{E}(f_S) + \left(\frac{3}{\tau(a)}\beta^2(S)\text{rank}(S)\varepsilon^2 \bigwedge 2\varepsilon\|S\|_1\right) + C(a)\frac{t(S; \varepsilon)}{n}\right], \tag{19.3}$$

where

$$C(a) := C\left(\frac{L^2(a)}{\tau(a)} \bigvee L(a)a\right).$$

To control the size of expectation Δ involved in the threshold (19.2) on ε one can use exponential inequalities for sums of independent random matrices that go back to [1]. Namely, the following upper bound easily follows from a noncommutative Bernstein type inequality (see [14]) by integrating its exponential tail bounds:

$$\Delta \leq 4\left(\sigma_X \sqrt{\log(2m)} \bigvee U_X \frac{\log(2m)}{\sqrt{n}}\right),$$

where $\sigma_X^2 := \|\mathbb{E} X^2\|$ and $U_X := \big\|\|X\|\big\|_{L_\infty}$. This bound can be easily applied to various specific sampling models used in low rank matrix recovery, such as sampling from an orthonormal basis that includes, in particular, matrix completion (see, e.g., [8], Chap. 9), leading to more concrete results.

The main feature of oracle inequality (19.3) is that it involves the approximation error term $\mathcal{E}(f_S)$ (the excess risk of the oracle S) with constant equal to 1. In this sense, bound (19.3) is what is usually called a *sharp oracle inequality*. Most low rank oracle inequalities for the nuclear norm penalization method proved in the recent literature are not sharp in the sense that the oracle excess risk $\mathcal{E}(f_S)$ is involved in these bounds with a constant strictly larger than 1. Sharp oracle inequalities are especially important in the cases when for all oracles in $S \in \mathbb{D}$ the approximation error is not particularly small. The first sharp oracle inequalities for the nuclear norm penalization method were proved in [10]. This was done for a "linearized version" of the least squares method with nuclear norm penalty. Under the boundedness assumption $|\langle S, X \rangle| \leq a$ a.s., $S \in \mathbb{D}$ for some $a > 0$ (the same assumption is used in this chapter), the error bounds without an approximation error term for the usual matrix Lasso (that is, the nuclear norm penalized least squares method) were proved in [7]. Earlier, the same problem was studied in [11] under additional assumptions on the so-called "spikiness" of the target matrices. In [9], a sharp oracle inequality for the same method was proved in the case of the noisy matrix completion problem with uniform design (in fact, this result was deduced from more general oracle bounds for estimators of low rank smooth kernels on graphs). In this chapter, we establish sharp oracle inequalities for a version of the problem with more general losses of quadratic type and for general design distributions. Note also that the main part of the random error term of bound (19.3) (that is, the term $\frac{3}{\tau(a)}\beta^2(S)\mathrm{rank}(S)\varepsilon^2 \bigwedge 2\varepsilon\|S\|_1$) depends correctly on the rank of the oracle. This follows from the minimax lower bounds proved in [10] (in fact, the form of the random error term in (19.3) is the same as in that paper).

19.2 Proof of Theorem 19.1

Proof. We start with the following condition that is necessary for $\hat{S}$ to be a solution of convex optimization problem (19.1): for some $\hat{V} \in \partial \|\hat{S}\|_1$,

$$P_n(\ell' \bullet f_{\hat{S}})(f_{\hat{S}} - f_S) + \varepsilon \langle \hat{V}, \hat{S} - S \rangle \leq 0, S \in \mathbb{D}$$

(see, e.g., [2], Chap. 2, Corollary 6; see also [8], pp. 198–199). This implies that, for all $S \in \mathbb{D}$

$$P(\ell' \bullet f_{\hat{S}})(f_{\hat{S}} - f_S) + \varepsilon \langle \hat{V}, \hat{S} - S \rangle \leq (P - P_n)(\ell' \bullet f_{\hat{S}})(f_{\hat{S}} - f_S). \quad (19.4)$$

Since both $\hat{S}, S \in \mathbb{D}$, we have $|f_{\hat{S}}(X)| \leq a, |f_S(X)| \leq a$ a.s., and since ℓ is a loss of quadratic type, it is easy to check that

$$P(\ell' \bullet f_{\hat{S}})(f_{\hat{S}} - f_S) \geq P(\ell \bullet f_{\hat{S}}) - P(\ell \bullet f_S) + \frac{1}{2}\tau(a)\|f_{\hat{S}} - f_S\|_{L_2(\Pi)}^2. \quad (19.5)$$

If $P(\ell \bullet f_{\hat{S}}) \leq P(\ell \bullet f_S)$, the oracle inequality of the theorem holds trivially. So, we assume in what follows that $P(\ell \bullet f_{\hat{S}}) > P(\ell \bullet f_S)$. Inequalities (19.4) and (19.5) imply that

$$P(\ell \bullet f_{\hat{S}}) + \frac{1}{2}\tau(a)\|f_{\hat{S}} - f_S\|_{L_2(\Pi)}^2 + \varepsilon \langle \hat{V}, \hat{S} - S \rangle$$
$$\leq P(\ell \bullet f_S) + (P - P_n)(\ell' \bullet f_{\hat{S}})(f_{\hat{S}} - f_S). \quad (19.6)$$

The following characterization of the subdifferential of the nuclear norm is well known:

$$\partial \|S\|_1 = \{\mathrm{sign}(S) + \mathcal{P}_L^\perp(M) : M \in \mathbb{H}_m, \|M\| \leq 1\},$$

where $L = \mathrm{supp}(S)$ (see, e.g., [8], Appendix A.4). By the duality between the operator and nuclear norms, there exists $M \in \mathbb{H}_m$ with $\|M\| \leq 1$ such that

$$\langle \mathcal{P}_L^\perp(M), \hat{S} - S \rangle = \langle M, \mathcal{P}_L^\perp(\hat{S} - S) \rangle = \|\mathcal{P}_L^\perp(\hat{S} - S)\|_1 = \|\mathcal{P}_L^\perp \hat{S}\|_1.$$

Then, by the monotonicity of subdifferentials of convex functions, we have, for $V = \mathrm{sign}(S) + \mathcal{P}_L^\perp(M) \in \partial \|S\|_1$, that

$$\langle \mathrm{sign}(S), \hat{S} - S \rangle + \|\mathcal{P}_L^\perp \hat{S}\|_1 = \langle V, \hat{S} - S \rangle \leq \langle \hat{V}, \hat{S} - S \rangle.$$

We now substitute the last bound in (19.6) to get

$$P(\ell \bullet f_{\hat{S}}) + \frac{1}{2}\tau(a)\|f_{\hat{S}} - f_S\|_{L_2(\Pi)}^2 + \varepsilon \|\mathcal{P}_L^\perp \hat{S}\|_1$$
$$\leq P(\ell \bullet f_S) + \varepsilon \langle \mathrm{sign}(S), S - \hat{S} \rangle + (P - P_n)(\ell' \bullet f_{\hat{S}})(f_{\hat{S}} - f_S). \quad (19.7)$$

The main part of the proof is a derivation of an upper bound on the empirical process $(P - P_n)(\ell' \bullet f_{\hat{S}})(f_{\hat{S}} - f_S)$. For a given $S \in \mathbb{D}$ and for $\delta_1, \delta_2 \geq 0$, denote

$$\mathcal{A}(\delta_1, \delta_2) := \{A \in \mathbb{D} : A - S \in \mathcal{K}(\mathbb{D}; L; 5),$$
$$\|f_A - f_S\|_{L_2(\Pi)} \leq \delta_1, \|\mathcal{P}_L^\perp A\|_1 \leq \delta_2\},$$

$$\tilde{\mathcal{A}}(\delta_1, \delta_2, \delta_3) := \{A \in \mathbb{D} : \|f_A - f_S\|_{L_2(\Pi)} \leq \delta_1,$$
$$\|\mathcal{P}_L^\perp A\|_1 \leq \delta_2, \|\mathcal{P}_L(A - S)\|_1 \leq \delta_3\},$$
$$\check{\mathcal{A}}(\delta_1, \delta_4) := \{A \in \mathbb{D} : \|f_A - f_S\|_{L_2(\Pi)} \leq \delta_1, \|A - S\|_1 \leq \delta_4\},$$

and

$$\alpha_n(\delta_1, \delta_2) := \sup\{|(P_n - P)(\ell' \bullet f_A)(f_A - f_S)| : A \in \mathcal{A}(\delta_1, \delta_2)\},$$
$$\tilde{\alpha}_n(\delta_1, \delta_2, \delta_2) := \sup\{|(P_n - P)(\ell' \bullet f_A)(f_A - f_S)| : A \in \tilde{\mathcal{A}}(\delta_1, \delta_2, \delta_3)\}.$$
$$\check{\alpha}_n(\delta_1, \delta_4) := \sup\{|(P_n - P)(\ell' \bullet f_A)(f_A - f_S)| : A \in \check{\mathcal{A}}(\delta_1, \delta_4)\}.$$

Lemma 19.1. *Suppose* $0 < \delta_k^- < \delta_k^+$, $k = 1, 2, 3, 4$. *Let* $t > 0$ *and*

$$\bar{t} := t + \sum_{k=1}^{2} \log\left([\log_2(\delta_k^+/\delta_k^-)] + 2\right) + \log 3,$$

$$\tilde{t} := t + \sum_{k=1}^{3} \log\left([\log_2(\delta_k^+/\delta_k^-)] + 2\right) + \log 3,$$

$$\check{t} := t + \sum_{k=1,4} \log\left([\log_2(\delta_k^+/\delta_k^-)] + 2\right) + \log 3.$$

Then, with probability at least $1 - e^{-t}$, *for all* $\delta_k \in [\delta_k^-, \delta_k^+]$, $k = 1, 2, 3$,

$$\alpha_n(\delta_1, \delta_2) \leq 2C_1 L(a) \mathbb{E}\|\Xi\|(\sqrt{\text{rank}(S)}\beta(S)\delta_1 + \delta_2)$$
$$+ 4L(a)\delta_1\sqrt{\frac{\bar{t}}{n}} + 4L(a)a\frac{\bar{t}}{n}, \quad (19.8)$$

$$\tilde{\alpha}_n(\delta_1, \delta_2, \delta_3) \leq 2C_2 L(a) \mathbb{E}\|\Xi\|(\delta_2 + \delta_3) + 4L(a)\delta_1\sqrt{\frac{\tilde{t}}{n}} + 4L(a)a\frac{\tilde{t}}{n},$$

and

$$\check{\alpha}_n(\delta_1, \delta_4) \leq 2C_2 L(a) \mathbb{E}\|\Xi\|\delta_4 + 4L(a)\delta_1\sqrt{\frac{\check{t}}{n}} + 4L(a)a\frac{\check{t}}{n},$$

where $C_1, C_2 > 0$ *are numerical constants.*

Proof. We will prove in detail only the first bound (19.8). Talagrand's concentration inequality (in Bousquet's form; see [8], p. 25) implies that, for all $\delta_1, \delta_2 > 0$, with probability at least $1 - e^{-t}$

$$\alpha_n(\delta_1, \delta_2) \leq 2\mathbb{E}\alpha_n(\delta_1, \delta_2) + 2L(a)\delta_1\sqrt{\frac{t}{n}} + 4L(a)a\frac{t}{n},$$

where we also used the bounds

$$|(\ell' \bullet f_A)(f_A - f_S)| \leq 2L(a)a,$$
$$P(\ell' \bullet f_A)^2(f_A - f_S)^2 \leq L^2(a)\|f_A - f_S\|^2_{L_2(\Pi)} \leq L^2(a)\delta_1^2$$

that hold under the assumptions on the loss. The next step is to use standard Rademacher symmetrization and contraction inequalities (see, e.g., [8], Sects. 2.1 and 2.2) to get

$$\mathbb{E}\alpha_n(\delta_1, \delta_2) \leq 16L(a)\mathbb{E}\sup\{|R_n(f_A - f_S)| : A \in \mathcal{A}(\delta_1, \delta_2)\}, \tag{19.9}$$

where $R_n(f) := \sum_{j=1}^n \varepsilon_j f(X_j)$, $\{\varepsilon_j\}$ being i.i.d. Rademacher random variables independent of $\{(X_j, Y_j)\}$, and where we also used a simple fact that the Lipschitz constant of the function $u \mapsto \ell'(f_S + u)u$ is upper bounded by $4L(a)$. We will bound the expected sup-norm of the Rademacher process in the right-hand side of (19.9). Observe that

$$R_n(f_A - f_S) = \langle \Xi, A - S \rangle, \quad \Xi := n^{-1}\sum_{j=1}^n \varepsilon_j X_j,$$

which implies

$$|R_n(f_A - f_S)| \leq |\langle \mathcal{P}_L\Xi, \mathcal{P}_L(A - S)\rangle| + |\langle \Xi, \mathcal{P}_L^\perp(A - S)\rangle| \tag{19.10}$$
$$\leq \|\mathcal{P}_L\Xi\|_2\|\mathcal{P}_L(A - S)\|_2 + \|\Xi\|\|\mathcal{P}_L^\perp A\|_1$$
$$\leq 2\sqrt{2\mathrm{rank}(S)}\beta(S)\|\Xi\|\|f_A - f_S\|_{L_2(\Pi)} + \|\Xi\|\|\mathcal{P}_L^\perp A\|_1,$$

where we used the facts that $A - S \in \mathcal{K}(\mathbb{D}; L; 5)$ and also that

$$\mathrm{rank}(\mathcal{P}_L\Xi) \leq 2\mathrm{rank}(S), \quad \|\mathcal{P}_L\Xi\|_2 \leq 2\sqrt{\mathrm{rank}(\mathcal{P}_L\Xi)}\|\Xi\|.$$

Therefore,

$$\mathbb{E}\sup\{|R_n(f_A - f_S)| : A \in \mathcal{A}(\delta_1, \delta_2)\}$$
$$\leq \mathbb{E}\|\Xi\|(2\sqrt{2\mathrm{rank}(S)}\beta(S)\delta_1 + \delta_2). \tag{19.11}$$

It follows that with some numerical constant $C_1 > 0$ and with probability at least $1 - e^{-t}$,

$$\alpha_n(\delta_1,\delta_2) \le C_1 L(a)\mathbb{E}\|\Xi\|(\sqrt{\operatorname{rank}(S)}\beta(S)\delta_1 + \delta_2)$$
$$+ 2L(a)\delta_1\sqrt{\frac{t}{n}} + 4L(a)a\frac{t}{n}.$$

We will make this bound uniform in $\delta_k \in [\delta_k^-, \delta_k^+]$. To this end, let $\delta_k^j := \delta_k^+ 2^{-j}$, $j = 0,\ldots,[\log_2(\delta_k^+/\delta_k^-)] + 1$. By the union bound, with probability at least $1 - \frac{1}{3}e^{-t}$, for all $j_k = 0,\ldots,[\log_2(\delta_k^+/\delta_k^-)] + 1$, $k = 1, 2$,

$$\alpha_n(\delta_1^{j_1}, \delta_2^{j_2}) \le C_1 L(a)\mathbb{E}\|\Xi\|(\sqrt{\operatorname{rank}(S)}\beta(S)\delta_1^{j_1} + \delta_2^{j_2})$$
$$+ 2L(a)\delta_1^{j_1}\sqrt{\frac{\bar{t}}{n}} + 4L(a)a\frac{\bar{t}}{n},$$

which implies that, for all $\delta_k \in [\delta_k^-, \delta_k^+]$, $k = 1, 2$,

$$\alpha_n(\delta_1,\delta_2) \le 2C_1 L(a)\mathbb{E}\|\Xi\|(\sqrt{\operatorname{rank}(S)}\beta(S)\delta_1 + \delta_2)$$
$$+ 4L(a)\delta_1\sqrt{\frac{\bar{t}}{n}} + 4L(a)a\frac{\bar{t}}{n}.$$

The proof of the second and the third bounds is similar. For instance, in the case of the second bound, the only difference is that instead of (19.10) we use

$$|R_n(f_A - f_S)| \le \|\Xi\|(\|\mathcal{P}_L(A-S)\|_1 + \|\mathcal{P}_L^\perp(A-S)\|_1),$$

which yields (instead of (19.11))

$$\mathbb{E}\sup\{|R_n(f_A - f_S)| : A \in \tilde{\mathcal{A}}(\delta_1,\delta_2,\delta_3)\} \le \mathbb{E}\|\Xi\|(\delta_2 + \delta_3).$$

This completes the proof of Lemma 19.1.

Note that

$$(P - P_n)(\ell' \bullet f_{\hat{S}})(f_{\hat{S}} - f_S) \le \tilde{\alpha}_n(\|f_{\hat{S}} - f_S\|_{L_2(\Pi)}; \|\mathcal{P}_L^\perp \hat{S}\|_1; \|\mathcal{P}_L(\hat{S} - S)\|_1), \quad (19.12)$$

$$(P - P_n)(\ell' \bullet f_{\hat{S}})(f_{\hat{S}} - f_S) \le \check{\alpha}_n(\|f_{\hat{S}} - f_S\|_{L_2(\Pi)}; \|\hat{S} - S\|_1), \quad (19.13)$$

and also, if $\hat{S} - S \in \mathcal{K}(\mathbb{D}; L; b)$, then

$$(P - P_n)(\ell' \bullet f_{\hat{S}})(f_{\hat{S}} - f_S) \le \alpha_n(\|f_{\hat{S}} - f_S\|_{L_2(\Pi)}; \|\mathcal{P}_L^\perp \hat{S}\|_1). \quad (19.14)$$

Assume for a while that

$$\|f_{\hat{S}} - f_S\|_{L_2(\Pi)} \in [\delta_1^-, \delta_1^+], \|\mathcal{P}_L^\perp \hat{S}\|_1 \in [\delta_2^-, \delta_2^+], \|\mathcal{P}_L(\hat{S} - S)\|_1 \in [\delta_3^-, \delta_3^+]. \tag{19.15}$$

First, we substitute (19.13) in bound (19.6) and use the upper bound on $\check{\alpha}_n$ of Lemma 19.1. Observe also that, since $\hat{V} \in \partial \|\hat{S}\|_1$,

$$\langle \hat{V}, S - \hat{S} \rangle \leq \|S\|_1 - \|\hat{S}\|_1.$$

Therefore, we get

$$P(\ell \bullet f_{\hat{S}}) + \frac{1}{2}\tau(a)\|f_{\hat{S}} - f_S\|^2_{L_2(\Pi)} \tag{19.16}$$

$$\leq P(\ell \bullet f_S) + \varepsilon(\|S\|_1 - \|\hat{S}\|_1) + \check{\alpha}_n(\|f_{\hat{S}} - f_S\|_{L_2(\Pi)}; \|\hat{S} - S\|_1)$$

$$\leq P(\ell \bullet f_S) + \varepsilon(\|S\|_1 - \|\hat{S}\|_1) + 2C_2 L(a)\mathbb{E}\|\Xi\| \|\hat{S} - S\|_1$$

$$+ 4L(a)\|f_{\hat{S}} - f_S\|_{L_2(\Pi)} \sqrt{\frac{\check{t}}{n}} + 4L(a)a\frac{\check{t}}{n}.$$

Assume that the constant D in the condition on ε satisfies $D \geq 8C_2$. Then, we have

$$\varepsilon \geq DL(a)\Delta n^{-1/2} \geq 8C_2 L(a)\mathbb{E}\|\Xi\|. \tag{19.17}$$

Using the bound

$$4L(a)\|f_{\hat{S}} - f_S\|_{L_2(\Pi)} \sqrt{\frac{\check{t}}{n}} \leq \frac{1}{4}\tau(a)\|f_{\hat{S}} - f_S\|^2_{L_2(\Pi)} + \frac{8L^2(a)}{\tau(a)}\frac{\check{t}}{n},$$

we get from (19.16)

$$P(\ell \bullet f_{\hat{S}}) \leq P(\ell \bullet f_S) + \varepsilon(\|S\|_1 - \|\hat{S}\|_1) \tag{19.18}$$

$$+ \varepsilon\|\hat{S} - S\|_1 + \left(\frac{8L^2(a)}{\tau(a)} + 4L(a)a\right)\frac{\check{t}}{n}$$

$$\leq P(\ell \bullet f_S) + 2\varepsilon\|S\|_1 + \left(\frac{8L^2(a)}{\tau(a)} + 4L(a)a\right)\frac{\check{t}}{n}.$$

We will now substitute (19.12) in bound (19.7) and use the upper bound on $\tilde{\alpha}_n$ of Lemma 19.1. We will also bound $\langle \text{sign}(S), S - \hat{S} \rangle$ as follows:

$$|\langle \text{sign}(S), S - \hat{S} \rangle| = |\langle \text{sign}(S), \mathcal{P}_L(S - \hat{S}) \rangle|$$

$$\leq \|\text{sign}(S)\| \|\mathcal{P}_L(\hat{S} - S)\|_1 \leq \|\mathcal{P}_L(\hat{S} - S)\|_1. \tag{19.19}$$

We get

$$P(\ell \bullet f_{\hat{S}}) + \frac{1}{2}\tau(a)\|f_{\hat{S}} - f_S\|_{L_2(\Pi)}^2 + \varepsilon\|\mathcal{P}_L^\perp(\hat{S} - S)\|_1 \quad (19.20)$$

$$\leq P(\ell \bullet f_S) + \varepsilon\|\mathcal{P}_L(\hat{S} - S)\|_1 + \tilde{\alpha}_n(\|f_{\hat{S}} - f_S\|_{L_2(\Pi)}; \|\mathcal{P}_L^\perp \hat{S}\|_1; \|\mathcal{P}_L(\hat{S} - S)\|_1)$$

$$\leq P(\ell \bullet f_S) + \varepsilon\|\mathcal{P}_L(\hat{S} - S)\|_1 + 2C_2 L(a)\mathbb{E}\|\Xi\|(\|\mathcal{P}_L^\perp \hat{S}\|_1 + \|\mathcal{P}_L(\hat{S} - S)\|_1)$$

$$+ 4L(a)\|f_{\hat{S}} - f_S\|_{L_2(\Pi)}\sqrt{\frac{\tilde{t}}{n}} + 4L(a)a\frac{\tilde{t}}{n}.$$

We still assume that $D \geq 8C_2$ and, thus, (19.17) holds. Using the bound

$$4L(a)\|f_{\hat{S}} - f_S\|_{L_2(\Pi)}\sqrt{\frac{\tilde{t}}{n}} \leq \frac{1}{4}\tau(a)\|f_{\hat{S}} - f_S\|_{L_2(\Pi)}^2 + \frac{8L^2(a)}{\tau(a)}\frac{\tilde{t}}{n},$$

we get from (19.20)

$$P(\ell \bullet f_{\hat{S}}) + \frac{1}{4}\tau(a)\|f_{\hat{S}} - f_S\|_{L_2(\Pi)}^2 + \varepsilon\|\mathcal{P}_L^\perp(\hat{S} - S)\|_1 \quad (19.21)$$

$$\leq P(\ell \bullet f_S) + \varepsilon\|\mathcal{P}_L(\hat{S} - S)\|_1 + \frac{\varepsilon}{4}(\|\mathcal{P}_L^\perp \hat{S}\|_1 + \|\mathcal{P}_L(\hat{S} - S)\|_1)$$

$$\left(\frac{8L^2(a)}{\tau(a)} + 4L(a)a\right)\frac{\tilde{t}}{n}.$$

If

$$\left(\frac{8L^2(a)}{\tau(a)} + 4L(a)a\right)\frac{\tilde{t}}{n} \geq \varepsilon\|\mathcal{P}_L(\hat{S} - S)\|_1 + \frac{\varepsilon}{4}(\|\mathcal{P}_L^\perp \hat{S}\|_1 + \|\mathcal{P}_L(\hat{S} - S)\|_1),$$

we conclude that

$$P(\ell \bullet f_{\hat{S}}) \leq P(\ell \bullet f_S) + \left(\frac{16L^2(a)}{\tau(a)} + 8L(a)a\right)\frac{\tilde{t}}{n}, \quad (19.22)$$

which suffices to prove the bound of the theorem. Otherwise, we use the assumption that $P(\ell \bullet f_{\hat{S}}) > P(\ell \bullet f_S)$ to get the following bound from (19.21):

$$\varepsilon\|\mathcal{P}_L^\perp(\hat{S} - S)\|_1 \leq 2\varepsilon\|\mathcal{P}_L(\hat{S} - S)\|_1 + \frac{\varepsilon}{2}(\|\mathcal{P}_L^\perp(\hat{S} - S)\|_1 + \|\mathcal{P}_L(\hat{S} - S)\|_1).$$

This yields

$$\frac{1}{2}\varepsilon\|\mathcal{P}_L^\perp(\hat{S} - S)\|_1 \leq \frac{5}{2}\varepsilon\|\mathcal{P}_L(\hat{S} - S)\|_1,$$

and, hence, $\hat{S} - S \in \mathcal{K}(\mathbb{D}; L; 5)$. This fact allows us to use the bound on α_n of Lemma 19.1. We can modify (19.19) as follows:

$$|\langle \operatorname{sign}(S), S - \hat{S}\rangle| = |\langle \operatorname{sign}(S), \mathcal{P}_L(S - \hat{S})\rangle|$$
$$\leq \|\operatorname{sign}(S)\|_2 \|\mathcal{P}_L(\hat{S} - S)\|_2 \leq \sqrt{\operatorname{rank}(S)} \beta(S) \|f_{\hat{S}} - f_S\|_{L_2(\Pi)};$$

and, instead of (19.20), we get

$$P(\ell \bullet f_{\hat{S}}) + \frac{1}{2}\tau(a)\|f_{\hat{S}} - f_S\|_{L_2(\Pi)}^2 + \varepsilon\|\mathcal{P}_L^\perp \hat{S}\|_1 \qquad (19.23)$$
$$\leq P(\ell \bullet f_S) + \varepsilon\sqrt{\operatorname{rank}(S)}\beta(S)\|f_{\hat{S}} - f_S\|_{L_2(\Pi)} +$$
$$2C_1 L(a)\mathbb{E}\|\Xi\|(\sqrt{\operatorname{rank}(S)}\beta(S)\|f_{\hat{S}} - f_S\|_{L_2(\Pi)} + \|\mathcal{P}_L^\perp \hat{S}\|_1) +$$
$$+4L(a)\|f_{\hat{S}} - f_S\|_{L_2(\Pi)}\sqrt{\frac{\bar{t}}{n}} + 4L(a)a\frac{\bar{t}}{n}.$$

If $D \geq 2C_1$, we have $\varepsilon \geq 2C_1 L(a)\mathbb{E}\|\Xi\|$, and (19.23) implies that

$$P(\ell \bullet f_{\hat{S}}) + \frac{1}{2}\tau(a)\|f_{\hat{S}} - f_S\|_{L_2(\Pi)}^2$$
$$\leq P(\ell \bullet f_S) + \frac{3}{2\tau(a)}\beta^2(S)\operatorname{rank}(S)\varepsilon^2 + \frac{1}{6}\tau(a)\|f_{\hat{S}} - f_S\|_{L_2(\Pi)}^2 +$$
$$\frac{3}{2\tau(a)}\beta^2(S)\operatorname{rank}(S)\varepsilon^2 + \frac{1}{6}\tau(a)\|f_{\hat{S}} - f_S\|_{L_2(\Pi)}^2 +$$
$$+\frac{24L^2(a)}{\tau(a)}\frac{\bar{t}}{n} + \frac{1}{6}\tau(a)\|f_{\hat{S}} - f_S\|_{L_2(\Pi)}^2 + 4L(a)a\frac{\bar{t}}{n}.$$

Therefore, we have

$$P(\ell \bullet f_{\hat{S}}) \leq P(\ell \bullet f_S)$$
$$+ \frac{3}{\tau(a)}\beta^2(S)\operatorname{rank}(S)\varepsilon^2 + \left(\frac{24L^2(a)}{\tau(a)} + 4L(a)a\right)\frac{\bar{t}}{n}. \qquad (19.24)$$

The bound of the theorem will follow from (19.18), (19.22) and (19.24) (provided that conditions (19.15) hold).

We have to choose the numbers $\delta_k^-, \delta_k^+, k = 1, 2, 3, 4$, and establish the bound of the theorem when conditions (19.15) do not hold. First note that, by the definition of $\hat{S}$,

$$P_n(\ell \bullet \hat{S}) + \varepsilon\|\hat{S}\|_1 \leq P_n(\ell \bullet 0) \leq Q,$$

implying that $\|\hat{S}\|_1 \leq \frac{Q}{\varepsilon}$. Next note that

$$\|\mathcal{P}_L^\perp \hat{S}\|_1 = \|P_{L^\perp} \hat{S} P_{L^\perp}\|_1 \leq \|\hat{S}\|_1 \leq \frac{Q}{\varepsilon}$$

and

$$\|\mathcal{P}_L(\hat{S} - S)\|_1 \leq 2\|\hat{S} - S\|_1 \leq \frac{2Q}{\varepsilon} + 2\|S\|_1.$$

Obviously, we also have

$$\|\hat{S} - S\|_1 \leq \frac{Q}{\varepsilon} + \|S\|_1.$$

Finally, we have $\|f_{\hat{S}} - f_S\|_{L_2(\Pi)} \leq 2a$ (since $\hat{S}, S \in \mathbb{D}$ and $\|f_{\hat{S}}\|_{L_\infty} \leq a, \|f_S\|_{L_\infty} \leq a$). Due to these facts, we can take

$$\delta_1^+ := 2a, \ \delta_2^+ := \frac{Q}{\varepsilon}, \ \delta_3^+ := \frac{2Q}{\varepsilon} + 2\|S\|_1, \delta_4^+ := \frac{Q}{\varepsilon} + \|S\|_1,$$

and, with this choice, δ_k^+, $k = 1, 2, 3, 4$, are upper bounds on the corresponding norms in (19.15). We will also choose

$$\delta_1^- := \frac{a}{\sqrt{n}}, \ \delta_2^- := \frac{L(a)a}{n\varepsilon} \wedge (\delta_2^+/2), a$$

$$\delta_3^- := \frac{L(a)a}{n\varepsilon} \wedge (\delta_3^+/2), \ \delta_4^- := \frac{L(a)a}{n\varepsilon} \wedge (\delta_4^+/2).$$

It is not hard to see that

$$\bar{t} \vee \check{t} \vee \tilde{t} \leq t(S; \varepsilon)$$

for a proper choice of numerical constant B in the definition of $t(S; \varepsilon)$. When conditions (19.15) do not hold (which means that at least one of the numbers δ_k^-, $k = 1, 2, 3, 4$, is not a lower bound on the corresponding norm), we still can use the bounds

$$(P - P_n)(\ell' \bullet f_{\hat{S}})(f_{\hat{S}} - f_S)$$
$$\leq \tilde{\alpha}_n(\|f_{\hat{S}} - f_S\|_{L_2(\Pi)} \vee \delta_1^-; \|\mathcal{P}_L^\perp \hat{S}\|_1 \vee \delta_2^-; \|\mathcal{P}_L(\hat{S} - S)\|_1 \vee \delta_3^-)$$

$$(P - P_n)(\ell' \bullet f_{\hat{S}})(f_{\hat{S}} - f_S) \leq \check{\alpha}_n(\|f_{\hat{S}} - f_S\|_{L_2(\Pi)} \vee \delta_1^-; \|\hat{S} - S\|_1 \vee \delta_4^-)$$

instead of (19.12) and (19.13), and, in the case when $\hat{S} - S \in \mathcal{K}(\mathbb{D}; L; 5)$, we can use the bound

$$(P - P_n)(\ell' \bullet f_{\hat{S}})(f_{\hat{S}} - f_S) \leq \alpha_n(\|f_{\hat{S}} - f_S\|_{L^2(\Pi)} \vee \delta_1^-; \|\mathcal{P}_L^\perp \hat{S}\|_1 \vee \delta_2^-)$$

instead of bound (19.14). It is easy now to modify the proof of (19.16)–(19.24) to show that in this case we still have

$$P(\ell \bullet f_{\hat{S}}) \leq P(\ell \bullet f_S) + \left(\frac{3}{\tau(a)}\beta^2(S)\mathrm{rank}(S)\varepsilon^2 \bigwedge 2\varepsilon\|S\|_1\right)$$
$$+ C\left(\frac{L^2(a)}{\tau(a)} \bigvee L(a)a\right)\frac{t(S;\varepsilon)}{n},$$

which holds with probability at least $1 - e^{-t}$ and implies the bound of the theorem.

Acknowledgements This work was partially supported by NSF Grants DMS-1207808, DMS-0906880, and CCF-0808863.

References

1. Ahlswede, R., Winter, A.: Strong converse for identification via quantum channels. IEEE Trans. Inf. Theory **48**, 569–679 (2002)
2. Aubin, J.P., Ekeland, I.: Applied Nonlinear Analysis. Wiley, New York (1984)
3. Candes, E., Plan, Y.: Tight oracle bounds for low-rank matrix recovery from a minimal number of random measurements. IEEE Trans. Inf. Theory **57**(4), 2342–2359 (2011)
4. Candes, E., Recht, B.: Exact matrix completion via convex optimization. Found. Comput. Math. **9**(6), 717–777 (2009)
5. Candes, E., Tao, T.: The power of convex relaxation: near-optimal matrix completion. IEEE Trans. Inf. Theory **56**, 2053–2080 (2010)
6. Gross, D.: Recovering low-rank matrices from few coefficients in any basis. IEEE Trans. Inf. Theory **57**(3), 1548–1566 (2011)
7. Klopp, O.: Noisy low-rank matrix completion with general sampling distribution (2012, preprint). arXiv 1203:0108
8. Koltchinskii, V.: Oracle inequalities in empirical risk minimization and sparse recovery problems. Ecole d'ete de Probabilités de Saint-Flour 2008. Lecture Notes in Mathematics. Springer, Berlin/Heidelberg (2011)
9. Koltchinskii, V., Rangel, P.: Low rank estimation of smooth kernels on graphs. Annu. Stat. **41**(2), 604–640 (2013)
10. Koltchinskii, V., Lounici, K., Tsybakov, A.: Nuclear norm penalization and optimal rates for noisy matrix completion. Ann. Stat. **39**(5), 2302–2329 (2011)
11. Negahban, S., Wainwright, M.: Restricted strong convexity and weighted matrix completion: optimal bounds with noise. J. Mach. Learn. Res. **13**, 1665–1697 (2012)
12. Recht, B., Fazel, M., Parrilo, P.: Guaranteed minimum rank solutions of matrix equations via nuclear norm minimization. SIAM Rev. **52**(3), 471–501 (2010)
13. Rohde, A., Tsybakov, A.: Estimation of high-dimensional low rank matrices. Ann. Stat. **39**, 887–930 (2011)
14. Tropp, J.A.: User-friendly tail bounds for sums of random matrices. Found. Comput. Math. **12**, 389–439 (2012)
15. Vapnik, V.N.: Statistical Learning Theory. Wiley, New York (1998)

Chapter 20
On the Consistency of the Bootstrap Approach for Support Vector Machines and Related Kernel-Based Methods

Andreas Christmann and Robert Hable

Abstract It is shown that bootstrap approximations of support vector machines (SVMs) based on a general convex and smooth loss function and on a general kernel are consistent. This result is useful for approximating the unknown finite sample distribution of SVMs by the bootstrap approach.

20.1 Introduction

Support vector machines and related kernel-based methods can be considered as a hot topic in machine learning because they have good statistical and numerical properties under weak assumptions and have demonstrated their often good generalization properties in many applications; see, e.g., [10,14,15], and [12]. To the best of our knowledge, the original SVM approach in [1] was derived from the generalized portrait algorithm invented earlier in [16]. Throughout the chapter, the term SVM will be used in the broad sense, i.e., for a general convex loss function and a general kernel.

SVMs based on many standard kernels, as such the Gaussian RBF kernel, are nonparametric methods. The finite sample distribution of many nonparametric methods is unfortunately unknown because the distribution P from which the data were generated is usually completely unknown and because there are often only asymptotical results describing the consistency or the rate of convergence of such methods known so far. Furthermore, there is in general *no* uniform rate of convergence for such nonparametric methods due to the famous no-free-lunch theorem; see [5] and [6]. Informally speaking, the no-free-lunch theorem states that, for sufficiently malign distributions, the average risk of any statistical (classification)

A. Christmann (✉) · R. Hable
Department of Mathematics, University of Bayreuth, Lehrstuhl für Stochastik, D-95440, Bayreuth, Germany
e-mail: andreas.christmann@uni-bayreuth.de; robert.hable@uni-bayreuth.de

method may tend arbitrarily slowly to zero. These facts are true for SVMs. SVMs are known to be universally consistent and fast rates of convergence are known for broad *subsets* of all probability distributions. The asymptotic normality of SVMs was shown recently by [8] under certain conditions.

Here, we apply a different approach to SVMs, namely Efron's bootstrap. The goal of this chapter is to show that bootstrap approximations of SVMs which are based on a general convex and smooth loss function and a general smooth kernel are consistent under mild assumptions; more precisely, convergence in outer probability is shown. This result is useful for drawing statistical decisions based on SVMs, e.g., confidence intervals, tolerance intervals and so on.

We mention that both the sequence of SVMs and the sequence of their corresponding risks are qualitatively robust under mild assumptions; see [2]. Hence, Efron's bootstrap approach turns out to be quite successful for SVMs from several points of view.

The rest of the chapter has the following structure. Section 20.2 gives a brief introduction to SVMs. Section 20.3 gives our main result on the consistency of bootstrap SVMs. The last section contains the proof of the main result and the tools we need.

20.2 Support Vector Machines

Current statistical applications are characterized by a wealth of large and high-dimensional data sets. In classification and in regression problems there is a variable of main interest, often called "output values" or "response", and a number of potential explanatory variables, which are often called "input values". These input values are used to model the observed output values or to predict future output values. The observations consist of n pairs $(x_1, y_1), \ldots, (x_n, y_n)$, which will be assumed to be independent realizations of a random pair (X, Y). We are interested in minimizing the risk or obtaining a function $f : \mathcal{X} \to \mathcal{Y}$ such that $f(x)$ is a good predictor for the response y if $X = x$ is observed. The prediction should be made in an automatic way. We refer to this process of determining a prediction method as "statistical machine learning"; see, e.g., [3, 10, 11, 14, 15]. Here, by "good predictor" we mean that f minimizes the expected loss, i.e., the risk,

$$\mathcal{R}_{L,\mathrm{P}}(f) = \mathbb{E}_\mathrm{P}\left[L(X, Y, f(X))\right],$$

where P denotes the unknown joint distribution of the random pair (X, Y) and $L : \mathcal{X} \times \mathcal{Y} \times \mathbb{R} \to [0, +\infty)$ is a fixed loss function. As a simple example, the least squares loss $L(X, Y, f(X)) = (Y - f(X))^2$ yields the optimal predictor $f(x) = \mathbb{E}_\mathrm{P}(Y|X = x)$, $x \in \mathcal{X}$. Because P is unknown, we can neither compute nor minimize the risk $\mathcal{R}_{L,\mathrm{P}}(f)$ directly.

Support vector machines—see [1, 14–16]—provide a highly versatile framework to perform statistical machine learning in a wide variety of setups. The minimization of regularized empirical risks over reproducing kernel Hilbert spaces was already considered, e.g., by Poggio and Girosi [9]. Given a kernel $k : \mathcal{X} \times \mathcal{X} \to \mathbb{R}$ we consider predictors $f \in H$, where H denotes the corresponding reproducing kernel Hilbert space of functions from $\mathcal{X}$ to $\mathbb{R}$. The space H includes, for example, all functions of the form $f(x) = \sum_{j=1}^{m} \alpha_j k(x, x_j)$ where the x_j are arbitrary elements in $\mathcal{X}$ and $\alpha_j \in \mathbb{R}$, $1 \leq j \leq m$. To avoid overfitting, a support vector machine $f_{L,P,\lambda}$ is defined as the solution of a regularized risk minimization problem. More precisely,

$$f_{L,P,\lambda} = \arg \inf_{f \in H} \mathbb{E}_P L(X, Y, f(X)) + \lambda \|f\|_H^2 , \qquad (20.1)$$

where $\lambda \in (0, \infty)$ is the regularization parameter. For a sample $D = ((x_1, y_1), \ldots, (x_n, y_n))$ the corresponding estimated function is given by

$$f_{L,D_n,\lambda} = \arg \inf_{f \in H} \frac{1}{n} \sum_{i=1}^{n} L(x_i, y_i, f(x_i)) + \lambda \|f\|_H^2 , \qquad (20.2)$$

where D_n denotes the empirical distribution based on D (see (20.3) below). Note that the optimization problem (20.2) corresponds to (20.1) when using D_n instead of P.

Efficient algorithms to compute $\hat{f}_n := f_{L,D_n,\lambda}$ exist for a number of different loss functions. However, there are often good reasons to consider other convex loss functions, e.g., the hinge loss $L(X, Y, f(X)) = \max\{1 - Y \cdot f(X), 0\}$ for binary classification purposes or the ϵ-insensitive loss $L(X, Y, f(X)) = \max\{0, |Y - f(X)| - \epsilon\}$ for regression purposes, where $\epsilon > 0$. As these loss functions are not differentiable, the logistic loss functions $L(X, Y, f(X)) = \ln(1 + \exp(-Y \cdot f(X)))$ and $L(X, Y, f(X)) = -\ln(4e^{Y-f(X)}/(1 + e^{Y-f(X)})^2)$ and Huber-type loss functions are also used in practice. These loss functions can be considered as smoothed versions of the previous two loss functions.

An important component of statistical analyses concerns quantifying and incorporating uncertainty (e.g., sampling variability) in the reported estimates. For example, one may want to include confidence bounds along the individual predicted values $\hat{f}_n(x_i)$ obtained from (20.2). Unfortunately, the sampling distribution of the estimated function $\hat{f}_n$ is unknown. Recently, [8] derived the asymptotic distribution of SVMs under some mild conditions. Asymptotic confidence intervals based on those general results are always symmetric.

Here, we are interested in approximating the finite sample distribution of SVMs by Efron's bootstrap approach, because confidence intervals based on the bootstrap approach can be asymmetric. The bootstrap [7] provides an alternative way to estimate the sampling distribution of a wide variety of estimators. To fix ideas, consider a functional $S : \mathcal{M} \to \mathcal{W}$, where $\mathcal{M}$ is a set of probability measures and $\mathcal{W}$ denotes a metric space. Many estimators can be included in this framework. Simple examples include the sample mean (with functional $S(P) = \int Z \, dP$) and M-estimators (with functional defined implicitly as the solution to the equation

$\mathbb{E}_P \Psi(Z, S(P)) = 0$). Let $\mathcal{B}(\mathcal{Z})$ be the Borel σ-algebra on $\mathcal{Z} = \mathcal{X} \times \mathcal{Y}$ and denote the set of all Borel probability measures on $(\mathcal{Z}, \mathcal{B}(\mathcal{Z}))$ by $\mathcal{M}_1(\mathcal{Z}, \mathcal{B}(\mathcal{Z}))$. Then, it follows that (20.1) defines an operator

$$S : \mathcal{M}_1(\mathcal{Z}, \mathcal{B}(\mathcal{Z})) \to H, \qquad S(P) = f_{L,P,\lambda},$$

i.e. the support vector machine. Moreover, the estimator in (20.2) satisfies

$$f_{L,D_n,\lambda} = S(D_n)$$

where

$$D_n = \frac{1}{n} \sum_{i=1}^n \delta_{(x_i, y_i)} \qquad (20.3)$$

is the empirical distribution based on the sample $D = ((x_1, y_1), \ldots, (x_n, y_n))$ and $\delta_{(x_i, y_i)}$ denotes the Dirac measure at the point (x_i, y_i).

More generally, let $Z_i = (X_i, Y_i)$, $i = 1, \ldots, n$, be independent and identically distributed (i.i.d.) random variables with distribution P, and let

$$S_n(Z_1, \ldots, Z_n) = S(\mathbb{P}_n)$$

be the corresponding estimator, where

$$\mathbb{P}_n = \frac{1}{n} \sum_{i=1}^n \delta_{Z_i}.$$

Denote the distribution of $S(\mathbb{P}_n)$ by $\mathcal{L}_n(S; P) = \mathcal{L}(S(\mathbb{P}_n))$. If P was known to us, we could estimate this sampling distribution by drawing a large number of random samples from P and evaluating our estimator on them. The basic idea of Efron's bootstrap approach is to replace the unknown distribution P by an estimate $\hat{P}$. Here we will consider the natural non-parametric estimator given by the sample empirical distribution $\mathbb{P}_n$. In other words, we estimate the distribution of our estimator of interest by its sampling distribution when the data are generated by $\mathbb{P}_n$. In symbols, the bootstrap proposes using

$$\widehat{\mathcal{L}_n(S; P)} = \mathcal{L}_n(S; \mathbb{P}_n).$$

Since this distribution is generally unknown, in practice one uses Monte Carlo simulation to estimate it by repeatedly evaluating the estimator on samples drawn from D_n. Note that drawing a sample from D_n means that n observations are drawn *with replacement* from the original n observations $(x_1, y_1), \ldots, (x_n, y_n)$.

20.3 Consistency of Bootstrap SVMs

In this section it will be shown under appropriate assumptions that the weak consistency of bootstrap estimators carries over to the Hadamard-differentiable SVM functional in the sense that the sequence of "conditional random laws" (given $(X_1, Y_1), (X_2, Y_2), \ldots$) of $\sqrt{n}(f_{L,\hat{\mathbb{P}}_n,\lambda} - f_{L,\mathbb{P}_n,\lambda})$ is asymptotically consistent in probability for estimating the laws of the random elements $\sqrt{n}(f_{L,\mathbb{P}_n,\lambda} - f_{L,\mathbb{P},\lambda})$. In other words, if n is large, the "random distribution"

$$\mathscr{L}(\sqrt{n}(f_{L,\hat{\mathbb{P}}_n,\lambda} - f_{L,\mathbb{P}_n,\lambda}))$$

based on bootstrapping an SVM can be considered as a valid approximation of the unknown finite sample distribution

$$\mathscr{L}(\sqrt{n}(f_{L,\mathbb{P}_n,\lambda} - f_{L,\mathbb{P},\lambda})).$$

Assumption 1. *Let $\mathcal{X} \subset \mathbb{R}^d$ be closed and bounded and let $\mathcal{Y} \subset \mathbb{R}$ be closed. Assume that $k : \mathcal{X} \times \mathcal{X} \to \mathbb{R}$ is the restriction of an m-times continuously differentiable kernel $\tilde{k} : \mathbb{R}^d \times \mathbb{R}^d \to \mathbb{R}$ such that $m > d/2$ and $k \neq 0$. Let H be the RKHS of k and let P be a probability distribution on $(\mathcal{X} \times \mathcal{Y}, \mathcal{B}(\mathcal{X} \times \mathcal{Y}))$. Let $L : \mathcal{X} \times \mathcal{Y} \times \mathbb{R} \to [0, \infty)$ be a convex, P-square-integrable Nemitski loss function of order $p \in [1, \infty)$ such that the partial derivatives*

$$L'(x, y, t) := \frac{\partial L}{\partial t}(x, y, t) \quad \text{and} \quad L''(x, y, t) := \frac{\partial^2 L}{\partial^2 t}(x, y, t)$$

exist for every $(x, y, t) \in \mathcal{X} \times \mathcal{Y} \times \mathbb{R}$. Assume that the maps

$$(x, y, t) \mapsto L'(x, y, t) \quad \text{and} \quad (x, y, t) \mapsto L''(x, y, t)$$

are continuous. Furthermore, assume that for every $a \in (0, \infty)$, there is a $b'_a \in L_2(\mathrm{P})$ and a constant $b''_a \in [0, \infty)$ such that, for every $(x, y) \in \mathcal{X} \times \mathcal{Y}$,

$$\sup_{t \in [-a,a]} |L'(x, y, t)| \leq b'_a(x, y) \quad \text{and} \quad \sup_{t \in [-a,a]} |L''(x, y, t)| \leq b''_a. \quad (20.4)$$

The conditions on the kernel k in Assumption 1 are satisfied for many common kernels, e.g., Gaussian RBF kernel, exponential kernel, polynomial kernel, and linear kernel, but also Wendland kernels $k_{d,\ell}$ based on certain univariate polynomials $p_{d,\ell}$ of degree $\lfloor d/2 \rfloor + 3\ell + 1$ for $\ell \in \mathbb{N}$ such that $\ell > d/4$; see [17].

The conditions on the loss function L in Assumption 1 are satisfied, e.g., for the logistic loss for classification or for regression; however, the popular non-smooth loss functions hinge, ε-insensitive, and pinball are not covered. However, [8, Remark 3.5] described an analytical method to approximate such non-smooth loss functions up to an arbitrarily good precision $\epsilon > 0$ by a convex P-square integrable Nemitski loss function of order $p \in [1, \infty)$.

We can now state our result on the consistency of the bootstrap approach for SVMs.

Theorem 20.1. *Let Assumption 1 be satisfied. Let $\lambda \in (0, \infty)$. Then*

$$\sup_{h \in BL_1(H)} \left| \mathbb{E}_M h\big(\sqrt{n}(f_{L,\hat{\mathbb{P}}_n,\lambda} - f_{L,\mathbb{P}_n,\lambda})\big) - \mathbb{E}h(S'_{\mathbb{P}}(\mathbb{G})) \right| \to 0,$$

$$\mathbb{E}_M h\big(\sqrt{n}(f_{L,\hat{\mathbb{P}}_n,\lambda} - f_{L,\mathbb{P}_n,\lambda})\big)^* - \mathbb{E}_M h\big(\sqrt{n}(f_{L,\hat{\mathbb{P}}_n,\lambda} - f_{L,\mathbb{P}_n,\lambda})\big)_* \to 0,$$

converges in outer probability, where $\mathbb{G}$ is a tight Borel-measurable Gaussian process, $S'_{\mathbb{P}}$ is a continuous linear operator with

$$S'_{\mathbb{P}}(Q) = -K_{\mathbb{P}}^{-1}\big(\mathbb{E}_Q(L'(X, Y, f_{L,\mathbb{P},\lambda}(X))\Phi(X))\big),$$

where $Q \in \mathcal{M}_1(\mathcal{X} \times \mathcal{Y})$ such that $b \in L_2(Q)$ and $b'_a \in L_2(Q)$ for all $a \in (0, \infty)$, and

$$K_{\mathbb{P}} : H \to H, \quad f \mapsto 2\lambda f + \mathbb{E}_{\mathbb{P}}\big(L''(X, Y, f_{L,\mathbb{P},\lambda}(X)) f(X) \Phi(X)\big)$$

is a continuous linear operator which is invertible.

For details on $K_{\mathbb{P}}$, $S'_{\mathbb{P}}$, and $\mathbb{G}$ we refer to Lemma 20.1, Theorem 20.5, and Lemma 20.2.

20.4 Proofs

20.4.1 Tools for the Proof of Theorem 20.1

We will need two general results on bootstrap methods proven in [13] and adopt their notation; see [13, Chaps. 3.6 and 3.9]. Let $\mathbb{P}_n$ be the empirical measure of an i.i.d. sample $Z_1, \ldots Z_n$ from a probability distribution P. The *empirical process* is the signed measure

$$\mathbb{G}_n = \sqrt{n}(\mathbb{P}_n - \mathbb{P}).$$

Given the sample values, let $\hat{Z}_1, \ldots, \hat{Z}_n$ be an i.i.d. sample from $\hat{\mathbb{P}}_n$. The *bootstrap empirical distribution* is the empirical measure $\hat{\mathbb{P}}_n := n^{-1} \sum_{i=1}^n \delta_{\hat{Z}_i}$, and the *bootstrap empirical process* is

$$\hat{\mathbb{G}}_n = \sqrt{n}(\hat{\mathbb{P}}_n - \mathbb{P}_n) = \frac{1}{\sqrt{n}} \sum_{i=1}^n (M_{ni} - 1)\delta_{Z_i},$$

where M_{ni} is the number of times that Z_i is "redrawn" from the original sample $Z_1, \ldots Z_n$; $M := (M_{n1}, \ldots, M_{nn})$ is stochastically independent of $Z_1, \ldots, Z_n$ and multinomially distributed with parameters n and probabilities $\frac{1}{n}, \ldots, \frac{1}{n}$. If outer expectations are computed, stochastic independence is understood in terms of a product probability space. Let $Z_1, Z_2, \ldots$ be the coordinate projections on the first ∞ coordinates of the product space $(\mathcal{Z}^\infty, \mathcal{B}(\mathcal{Z}), P^\infty) \times (\tilde{\mathcal{Z}}, \mathcal{C}, Q)$ and let the multinomial vectors M depend on the last factor only; see [13, p. 345f].

The following theorem shows (conditional) weak convergence for the empirical bootstrap, where the symbol $\rightsquigarrow$ denotes the weak convergence of finite measures. We will need only the equivalence between (*i*) and (*iii*) from this theorem and list part (*ii*) only for the sake of completeness.

Theorem 20.2 ([13, Theorem 3.6.2, p. 347]). *Let $\mathcal{F}$ be a class of measurable functions with finite envelope function. Define* $\mathbb{Y}_n := n^{-1/2} \sum_{i=1}^n (M_{N_n,i} - 1)(\delta_{Z_i} - P)$. *The following statements are equivalent:*

(*i*) *$\mathcal{F}$ is Donsker and $P^* \|f - Pf\|_\mathcal{F}^2 < \infty$;*
(*ii*) *$\sup_{h \in BL_1} |\mathbb{E}_{M,N} h(\hat{\mathbb{Y}}_n) - \mathbb{E}h(\mathbb{G})|$ converges outer almost surely to zero and the sequence $\mathbb{E}_{M,N} h(\hat{\mathbb{Y}}_n)^* - \mathbb{E}_{M,N} h(\hat{\mathbb{Y}}_n)_*$ converges almost surely to zero for every $h \in BL_1$.*
(*iii*) *$\sup_{h \in BL_1} |\mathbb{E}_M h(\hat{\mathbb{G}}_n) - \mathbb{E}h(\mathbb{G})|$ converges outer almost surely to zero and the sequence $\mathbb{E}_M h(\hat{\mathbb{G}}_n)^* - \mathbb{E}_M h(\hat{\mathbb{G}}_n)_*$ converges almost surely to zero for every $h \in BL_1$.*

Here the asterisks denote the measurable cover functions with respect to M, N, and $Z_1, Z_2, \ldots$ jointly.

Consider sequences of random elements $\mathbb{P}_n = \mathbb{P}_n(Z_n)$ and $\hat{\mathbb{P}}_n = \hat{\mathbb{P}}_n(Z_n, M_n)$ in a normed space $\mathbb{D}$ such that the sequence $\sqrt{n}(\mathbb{P}_n - P)$ converges unconditionally and the sequence $\sqrt{n}(\hat{\mathbb{P}}_n - \mathbb{P}_n)$ converges conditionally on Z_n in distribution to a tight random element $\mathbb{G}$. A precise formulation of the second assumption is

$$\sup_{h \in BL_1(\mathbb{D})} \left|\mathbb{E}_M h(\sqrt{n}(\hat{\mathbb{P}}_n - \mathbb{P}_n)) - \mathbb{E}h(\mathbb{G})\right| \to 0, \tag{20.5}$$

$$\mathbb{E}_M h(\sqrt{n}(\hat{\mathbb{P}}_n - \mathbb{P}_n))^* - \mathbb{E}_M h(\sqrt{n}(\hat{\mathbb{P}}_n - \mathbb{P}_n))_* \to 0, \tag{20.6}$$

in outer probability, with h ranging over the bounded Lipschitz functions; see [13, p. 378, Formula (3.9.9)]. The next theorem shows that under appropriate assumptions, weak consistency of the bootstrap estimators carries over to any Hadamard-differentiable functional in the sense that the sequence of "conditional random laws" (given $Z_1, Z_2, \ldots$) of $\sqrt{n}(\phi(\hat{\mathbb{P}}_n) - \phi(\mathbb{P}_n))$ is asymptotically consistent in probability for estimating the laws of the random elements $\sqrt{n}(\phi(\mathbb{P}_n) - \phi(P))$; see [13, p. 378].

Theorem 20.3 ([13, Theorem 3.9.11, p. 378]). *(Delta method for bootstrap in probability) Let $\mathbb{D}$ and $\mathbb{E}$ be normed spaces. Let $\phi : \mathbb{D}_\phi \subset \mathbb{D} \to \mathbb{E}$ be Hadamard-differentiable at P tangentially to a subspace $\mathbb{D}_0$. Let $\mathbb{P}_n$ and $\hat{\mathbb{P}}_n$ be maps as indicated* *previously with values in $\mathbb{D}_\phi$ such that $\mathbb{G}_n := \sqrt{n}(\mathbb{P}_n - P) \rightsquigarrow \mathbb{G}$ and that (20.5)–(20.6) hold in outer probability, where $\mathbb{G}$ is separable and takes its values from $\mathbb{D}_0$. Then*

$$\sup_{h \in \mathrm{BL}_1(\mathbb{E})} \left| \mathbb{E}_M h\left(\sqrt{n}(\phi(\hat{\mathbb{P}}_n) - \phi(\mathbb{P}_n))\right) - \mathbb{E}h(\phi'_\mathrm{P}(\mathbb{G})) \right| \to 0,$$

$$\mathbb{E}_M h\left(\sqrt{n}(\phi(\hat{\mathbb{P}}_n) - \phi(\mathbb{P}_n))\right)^* - \mathbb{E}_M h\left(\sqrt{n}(\phi(\hat{\mathbb{P}}_n) - \phi(\mathbb{P}_n))\right)_* \to 0,$$

holds in outer probability.

As was pointed out by van der Vaart and Wellner [13, p. 378], consistency in probability appears to be sufficient for (many) statistical purposes and the theorem above shows this is retained under Hadamard-differentiability at the single distribution P.

We now list some results from [8], which will also be essential for the proof of Theorem 20.1.

Theorem 20.4 ([8, Theorem 3.1]). *Let Assumption 1 be satisfied. Then, for every regularizing parameter $\lambda_0 \in (0, \infty)$, there is a tight, Borel-measurable Gaussian process $\mathbb{H} : \Omega \to H$, $\omega \to \mathbb{H}(\omega)$, such that*

$$\sqrt{n}\left(f_{L,\mathbf{D}_n,\lambda_{\mathbf{D}_n}} - f_{L,\mathrm{P},\lambda_0}\right) \rightsquigarrow \mathbb{H} \quad \text{in } H$$

for every Borel-measurable sequence of random regularization parameters $\lambda_{\mathbf{D}_n}$ with $\sqrt{n}(\lambda_{\mathbf{D}_n} - \lambda_0) \to 0$ in probability. The Gaussian process $\mathbb{H}$ is zero-mean; i.e., $\mathbb{E}\langle f, \mathbb{H}\rangle_H = 0$ for every $f \in H$.

Lemma 20.1 ([8, Lemma A.5]). *For every $F \in B_S$ defined later in (20.8),*

$$K_F : H \to H, \quad f \mapsto 2\lambda_0 f + \int L''(x, y, f_{L,\iota(F),\lambda_0}(x)) f(x)\Phi(x) d\iota(F)(x, y)$$

is a continuous linear operator which is invertible.

Theorem 20.5 ([8, Theorem A.8]). *For every $F_0 \in B_S$ which fulfils $F_0(b) < \mathbb{E}_\mathrm{P}(b) + \lambda_0$, the map $S : B_S \to H$, $F \mapsto f_{\iota(F)}$, is Hadamard-differentiable in F_0 tangentially to the closed linear span $B_0 = \mathrm{cl}(\mathrm{lin}(B_S))$. The derivative in F_0 is a continuous linear operator $S'_{F_0} : B_0 \to H$ such that*

$$S'_{F_0}(G) = -K_{F_0}^{-1}\left(\mathbb{E}_{\iota(G)}(L'(X, Y, f_{L,\iota(F_0),\lambda_0}(X))\Phi(X))\right), \quad \forall\, G \in \mathrm{lin}(B_S).$$

Lemma 20.2 ([8, Lemma A.9]). *For every data set $D_n = ((x_1, y_1), \ldots, (x_n, y_n)) \in (\mathcal{X} \times \mathcal{Y})^n$, let $\mathbb{F}_{D_n}$ denote the element of $\ell_\infty(\mathcal{G})$ which corresponds to the empirical measure $\mathbb{P}_n := \mathbb{P}_{D_n}$. That is, $\mathbb{F}_{D_n}(g) = \int g\, d\mathbb{P}_n = n^{-1}\sum_{i=1}^n g(x_i, y_i)$ for every $g \in \mathcal{G}$. Then*

$$\sqrt{n}\left(\mathbb{F}_{D_n} - \iota^{-1}(\mathrm{P})\right) \rightsquigarrow \mathbb{G} \quad \text{in } \ell_\infty(\mathcal{G}),$$

where $\mathbb{G} : \Omega \to \ell_\infty(\mathscr{G})$ is a tight Borel-measurable Gaussian process such that $\mathbb{G}(\omega) \in B_0$ for every $\omega \in \Omega$.

20.4.2 Proof of Theorem 20.1

The proof relies on the application of Theorem 20.3. Hence, we have to show the following steps:

1. The empirical process $\mathbb{G}_n = \sqrt{n}(\mathbb{P}_n - P)$ weakly converges to a separable Gaussian process $\mathbb{G}$.
2. SVMs are based on a map ϕ which is Hadamard-differentiable at P tangentially to some appropriate subspace.
3. The assumptions (20.5)–(20.6) of Theorem 20.3 are satisfied. For this purpose we will use Theorem 20.2. Actually, we will show that part *(i)* of Theorem 20.2 is satisfied, which gives the equivalence to part *(iii)*, from which we conclude that (20.5)–(20.6) hold true. For the proof that part *(i)* of Theorem 20.2 is satisfied, i.e., that a suitable set $\mathscr{F}$ is a P-Donsker class and that $P^* \| f - Pf \|_{\mathscr{F}}^2 < \infty$, we use several facts recently shown by Hable [8].
4. We put all parts together and apply Theorem 20.3.

Step 1. To apply Theorem 20.3, we first have to specify the considered spaces $\mathbb{D}$, $\mathbb{E}$, $\mathbb{D}_\phi$, $\mathbb{D}_0$ and the map ϕ. As in [8] we use the following notations. Because L is a P-square-integrable Nemitski loss function of order $p \in [1, \infty)$, there is a function $b \in L_2(P)$ such that

$$|L(x, y, t)| \leq b(x, y) + |t|^p, \qquad (x, y, t) \in \mathcal{X} \times \mathcal{Y} \times \mathbb{R}.$$

Let

$$c_0 := \sqrt{\lambda_0^{-1} \mathbb{E}_P(b)} + 1, \qquad (20.7)$$

Define

$$\mathscr{G} := \mathscr{G}_1 \cup \mathscr{G}_2 \cup \mathscr{G}_3,$$

where

$$\mathscr{G}_1 := \{g : \mathcal{X} \times \mathcal{Y} \to \mathbb{R} : \exists z \in \mathbb{R}^{d+1} \text{ such that } g = I_{(-\infty, z]}\}$$

is the set of all indicator functions $I_{(-\infty, z]}$,

$$\mathscr{G}_2 := \left\{ g : \mathcal{X} \times \mathcal{Y} \to \mathbb{R} \ \middle| \ \begin{array}{l} \exists f_0 \in H, \exists f \in H \text{ such that } \|f_0\|_H \leq c_0, \\ \|f\|_H \leq 1, g(x, y) = L'(x, y, f_0(x)) f(x) \ \forall \ (x, y) \end{array} \right\},$$

and
$$\mathscr{G}_3 := \{b\}.$$

Now let $\ell_\infty(\mathscr{G})$ be the set of all bounded functions $F : \mathscr{G} \to \mathbb{R}$ with norm $\|F\|_\infty = \sup_{g \in \mathscr{G}} |F(g)|$. Define

$$B_S := \left\{ F : \mathscr{G} \to \mathbb{R} \, \middle| \, \begin{array}{l} \exists \mu \neq 0 \text{ a finite measure on } \mathcal{X} \times \mathcal{Y} \text{ such that} \\ F(g) = \int g \, d\mu \; \forall \, g \in \mathscr{G}, \\ b \in L_2(\mu), b'_a \in L_2(\mu) \; \forall \, a \in (0, \infty) \end{array} \right\} \quad (20.8)$$

and
$$B_0 := \mathrm{cl}(\mathrm{lin}(B_S))$$

as the closed linear span of B_S in $\ell_\infty(\mathscr{G})$. That is, B_S is a subset of $\ell_\infty(\mathscr{G})$ whose elements correspond to finite measures. Hence probability measures are covered as special cases. The elements of B_S can be interpreted as some kind of generalized distribution functions, because $\mathscr{G}_1 \subset \mathscr{G}$. The assumptions on L and P imply that $\mathscr{G} \to \mathbb{R}$, $g \mapsto \int g \, dP$ is a well-defined element of B_S. For every $F \in B_S$, let $\iota(F)$ denote the corresponding finite measure on $(\mathcal{X} \times \mathcal{Y}, \mathcal{B}(\mathcal{X} \times \mathcal{Y}))$ such that $F(g) = \int g \, d\mu$ for all $g \in \mathscr{G}$. Note that the map ι is well defined, because by definition of B_S, $\iota(F)$ uniquely exists for every $F \in B_S$.

With these notations, we will apply Theorem 20.3 for

$$\mathbb{D} := \ell_\infty(\mathscr{G}), \; \mathbb{E} := H \; (= \text{RKHS of the kernel } k),$$
$$\mathbb{D}_\phi := B_S, \quad \mathbb{D}_0 := B_0 := \mathrm{cl}(\mathrm{lin}(B_S)),$$
$$\lambda_0 \in (0, \infty),$$
$$\phi := S, \quad S : B_S \to H, \; F \mapsto f_{\iota(F)} := f_{L,\iota(F),\lambda_0} :=$$
$$\arg\inf_{f \in H} \int L(x, y, f(x)) \, d\iota(F)(x, y) + \lambda_0 \|f\|_H^2.$$

At first glance this definition of S seems to be somewhat technical. However, this will allow us to use a functional delta method for bootstrap estimators of SVMs with regularization parameter $\lambda = \lambda_0 \in (0, \infty)$.

Lemma 20.2 guarantees that the empirical process $\mathbb{G}_n := \sqrt{n}(\mathbb{P}_n - P)$ weakly converges to a tight Borel-measurable Gaussian process.

Since a σ-compact set in a metric space is separable, separability of a random variable is slightly weaker than tightness; see [13, p. 17]. Therefore, $\mathbb{G}$ in our Theorem 20.1 is indeed separable.

Step 2. Theorem 20.5 showed that the map S indeed satisfies the necessary Hadamard-differentiability at the point $P := \iota^{-1}(F)$.

Step 3. We know that $\mathscr{G}$ is a P-Donsker class; see Lemma 20.2. Hence, an immediate consequence from [13, Theorem 3.6.1, p. 347] is that

$$\sup_{h \in BL_1} |\mathbb{E}_M h(\hat{\mathbb{G}}_n) - \mathbb{E}h(\mathbb{G})| \qquad (20.9)$$

converges in outer probability to 0 and $\hat{\mathbb{G}}_n$ is asymptotically measurable.

However, we will prove a somewhat stronger result, namely that $\mathscr{G}$ is a P-Donsker class and $P^* \|g - Pg\|_{\mathscr{G}}^2 < \infty$, which is part *(i)* of Theorem 20.2, and then part *(iii)* of Theorem 20.2 yields that the term in (20.9) converges even outer almost surely to 0 and the sequence

$$\mathbb{E}_M h(\hat{\mathbb{G}}_n)^* - \mathbb{E}_M h(\hat{\mathbb{G}}_n)_*$$

converges almost surely to 0 for every $h \in BL_1$.

Because $\mathscr{G}$ is a P-Donsker class, it remains to show that $P^* \|g - Pg\|_{\mathscr{G}}^2 < \infty$. Due to

$$P^* \|g - Pg\|_{\mathscr{G}}^2 := \int (\sup_{g \in \mathscr{G}} |g - \mathbb{E}_P(g)|)^2 \, dP^*$$

and $\mathscr{G} = \mathscr{G}_1 \cup \mathscr{G}_2 \cup \mathscr{G}_3$, we obtain the inequality

$$P^* \|g - Pg\|_{\mathscr{G}}^2 \leq P^* \sup_{g \in \mathscr{G}} (g^2 + 2|g| \cdot P|g| + (P|g|)^2)$$

$$\leq P^* \sup_{g \in \mathscr{G}} g^2 + 2 P^* \sup_{g \in \mathscr{G}} (|g| \cdot P|g|) + \sup_{g \in \mathscr{G}} (P|g|)^2$$

$$\leq \sum_{j=1}^{3} \left(P^* \sup_{g \in \mathscr{G}_j} g^2 + 2 P^* \sup_{g \in \mathscr{G}_j} (|g| \cdot P|g|) + \sup_{g \in \mathscr{G}_j} (P|g|)^2 \right).$$

$$(20.10)$$

We will show that each of the three summands on the right-hand side of the last inequality is finite. If $g \in \mathscr{G}_1$, then g equals the indicator function $I_{(-\infty, z]}$ for some $z \in \mathbb{R}^{d+1}$. Hence, $\|g\|_\infty = 1$ and the summand for $j = 1$ is finite. If $g \in \mathscr{G}_3$, then $g = b \in L_2(P)$ because L is by assumption a P-square-integrable Nemitski loss function of order $p \in [1, \infty)$. Hence the summand for $j = 3$ is finite, too. Let us now consider the case where $g \in \mathscr{G}_2$. By definition of $\mathscr{G}_2$, for every $g \in \mathscr{G}_2$ there exist $f, f_0 \in H$ such that $\|f_0\|_H \leq c_0$, $\|f\|_H \leq 1$, and $g = L'_{f_0} f$, where we use the notation $(L'_{f_0} f)(x, y) := L'(x, y, f_0(x)) f(x)$ for all $(x, y) \in \mathcal{X} \times \mathcal{Y}$. Using $\|f\|_\infty \leq \|k\|_\infty \|f\|_H$ for every $f \in H$, we obtain

$$\|f_0\|_H \leq c_0 \Rightarrow \|f_0\|_\infty \leq c_0 \|k\|_\infty \quad \text{and} \quad \|f\|_H \leq 1 \Rightarrow \|f\|_\infty \leq \|k\|_\infty.$$
$$(20.11)$$

Define the constant $a := c_0 \|k\|_\infty$ with c_0 given by (20.7). Hence, for all $(x, y) \in \mathcal{X} \times \mathcal{Y}$,

$$\sup_{f_0 \in H; \|f_0\|_H \le c_0} |L'(x,y,f_0(x))|^2 \le \sup_{f_0 \in H; \|f_0\|_\infty \le a} \sup_{t \in [-a,+a]} |L'(x,y,t)|^2$$

$$\overset{(20.4)}{\le} \sup_{f_0 \in H; \|f_0\|_\infty \le a} (b'_a(x,y))^2. \quad (20.12)$$

Hence we get

$$P^* \sup_{g \in \mathscr{G}_2} g^2$$

$$= \int \sup_{g \in \mathscr{G}_2; \|f_0\|_H \le c_0, \|f\|_H \le 1, g = L'_{f_0} f} |L'(x,y,f_0(x)) f(x)|^2 \, dP^*(x,y)$$

$$\le \int \sup_{f_0 \in H; \|f_0\|_H \le c_0} |L'(x,y,f_0(x))|^2 \sup_{f \in H; \|f\|_H \le 1} |f(x)|^2 \, dP^*(x,y)$$

$$\overset{(20.12),(20.11)}{\le} \|k\|_\infty^2 \int (b'_a)^2 \, dP^* = \|k\|_\infty^2 \int (b'_a)^2 \, dP < \infty,$$

because $b'_a \in L_2(P)$ and $\|k\|_\infty < \infty$ by Assumption 1. With the same arguments we obtain, for every $g \in \mathscr{G}_2$,

$$P|g| \le \int \sup_{g \in \mathscr{G}_2} |g| \, dP^*$$

$$\le \int \sup_{f_0 \in H; \|f_0\|_H \le c_0} |L'(x,y,f_0(x))| \sup_{f \in H; \|f\|_H \le 1} |f(x)| \, dP^*(x,y)$$

$$\overset{(20.12),(20.11)}{\le} \int b'_a(x,y) \|k\|_\infty \, dP^*(x,y)$$

$$\le \|k\|_\infty \int b'_a \, dP < \infty,$$

because $b'_a \in L_2(P)$ and $\|k\|_\infty < \infty$ by Assumption 1. Hence,

$$P^* \sup_{g \in \mathscr{G}_2} (|g| P|g|) \le \|k\|_\infty \int b'_a \, dP \int \sup_{g \in \mathscr{G}_2} |g| \, dP^* \le \|k\|_\infty^2 \left(\int b'_a \, dP \right)^2 < \infty.$$

Therefore, the sum on the right-hand side in (20.10) is finite and thus the assumption $P^* \|g - Pg\|_{\mathscr{G}}^2 < \infty$ is satisfied. This yields by part *(iii)* of Theorem 20.2 that $\sup_{h \in BL_1} |\mathbb{E}_M h(\hat{\mathbb{G}}_n) - \mathbb{E} h(\mathbb{G})|$ converges outer almost surely to 0 and the sequence

$$\mathbb{E}_M h(\hat{\mathbb{G}}_n)^* - \mathbb{E}_M h(\hat{\mathbb{G}}_n)_* \quad (20.13)$$

converges almost surely to 0 for every $h \in BL_1$, where the asterisks denote the measurable cover functions with respect to M and $Z_1, Z_2, \ldots$ jointly.

Step 4. Due to Step 3, the assumption (20.5) of Theorem 20.3 is satisfied. We now show that additionally (20.6) is satisfied, i.e., that the term in (20.13) converges to 0 in outer probability. In general, one cannot conclude that almost sure convergence implies convergence in outer probability; see [13, p. 52]. We know that the term in (20.13) converges almost surely to 0 for every $h \in BL_1$, where the asterisks denote the *measurable* cover functions with respect to M and $(X_1, Y_1), (X_2, Y_2), \ldots$ *jointly*. Hence, for every $h \in BL_1$, the cover functions to be considered in (20.13) are measurable. Additionally, the multinomially distributed random variable M is stochastically independent of $(X_1, Y_1), \ldots, (X_n, Y_n)$ in the bootstrap, where independence is understood in terms of a product probability space; see [13, p. 346] for details. Therefore, an application of the Fubini–Tonelli theorem, see—e.g., [4, p. 174—Theorem 2.4.10], yields that the inner integral $\mathbb{E}_M h(\sqrt{n}(\hat{\mathbb{P}}_n - \mathbb{P}_n))^* - \mathbb{E}_M h(\sqrt{n}(\hat{\mathbb{P}}_n - \mathbb{P}_n))_*$ considered by Fubini–Tonelli is *measurable* for every $n \in \mathbb{N}$ and every $h \in BL_1$. Recall that almost sure convergence of measurable functions implies convergence in probability that is equivalent to convergence in outer probability for measurable functions. Hence we have convergence in outer probability in (20.13). Therefore, all assumptions of Theorem 20.3 are satisfied and the assertion of our theorem follows. ∎

References

1. Boser, B., Guyon, I., Vapnik, V.: A training algorithm for optimal margin classifiers. In: Proceedings of the Fifth Annual Workshop on Computational Learning Theory, COLT'92, Pittsburgh, pp. 144–152. ACM, New York (1992)
2. Christmann, A., Salibían-Barrera, M., Aelst, S.V.: Qualitative robustness of bootstrap approximations for kernel based methods. In: Becker, C., Fried, R., Kuhnt, S. (eds.) Robustness and Complex Data Structures, chapter 16, pp. 263–278. Springer, Heidelberg (2013). (Preprint available on http://arxiv.org/abs/1111.1876)
3. Cucker, F., Zhou, D.: Learning Theory: An Approximation Theory Viewpoint. Cambridge University Press, Cambridge (2007)
4. Denkowski, Z., Migórski, S., Papageorgiou, N.: An Introduction to Nonlinear Analysis: Theory. Kluwer, Boston (2003)
5. Devroye, L.: Any discrimination rule can have an arbitrarily bad probability of error for finite sample size. IEEE Trans. Pattern Anal. Mach. Intell. **4**, 154–157 (1982)
6. Devroye, L., Györfi, L., Lugosi, G.: A Probabilistic Theory of Pattern Recognition. Springer, New York (1996)
7. Efron, B.: Bootstrap methods: another look at the jackknife. Ann. Stat. **7**, 1–26 (1979)
8. Hable, R.: Asymptotic normality of support vector machine variants and other regularized kernel methods. J. Multivar. Anal. **106**, 92–117 (2012)
9. Poggio, T., Girosi, F.: Networks for approximation and learning. Proc. IEEE **78**(9), 1481–1497 (1990)
10. Schölkopf, B., Smola, A.: Learning with Kernels. Support Vector Machines, Regularization, Optimization, and Beyond. MIT, Cambridge (2002)

11. Smale, S., Zhou, D.: Learning theory estimates via integral operators and their approximations. Constr. Approx. **26**, 153–172 (2007)
12. Steinwart, I., Christmann, A.: Support Vector Machines. Springer, New York (2008)
13. van der Vaart, A., Wellner, J.: Weak Convergence and Empirical Processes. Springer, New York (1996)
14. Vapnik, V.: The Nature of Statistical Learning Theory. Springer, New York (1995)
15. Vapnik, V.: Statistical Learning Theory. Wiley, New York (1998)
16. Vapnik, V., Lerner, A.: Pattern recognition using generalized portrait method. Autom. Remote Control **24**, 774–780 (1963)
17. Wendland, H.: Scattered Data Approximation. Cambridge University Press, Cambridge (2005)

Chapter 21
Kernels, Pre-images and Optimization

John C. Snyder, Sebastian Mika, Kieron Burke, and Klaus-Robert Müller

Abstract In the last decade, kernel-based learning has become a state-of-the-art technology in Machine Learning. We briefly review kernel PCA (kPCA) and the pre-image problem that occurs in kPCA. Subsequently, we discuss a novel direction where kernel-based models are used for property optimization. For this purpose, a stable estimation of the model's gradient is essential and non-trivial to achieve. The appropriate use of pre-image projections is key to successful gradient-based optimization—as will be shown for toy and real-world problems from quantum chemistry and physics.

21.1 Introduction

Since the seminal work of Vapnik and collaborators (see [4, 7, 10, 29, 43]), kernel methods have become ubiquitous in the sciences and industry.

Kernel methods have enriched the spectrum of machine learning and statistical methods with a vast new set of non-linear algorithms. Kernel PCA (kPCA) has been established as a blueprint for "kernelizing" linear scalar product-based

---

J.C. Snyder (✉) · K. Burke
Department of Chemistry, Department of Physics, University of California, Irvine, CA 92697, USA
e-mail: jcsnyder@uci.edu; kieron@uci.edu

S. Mika
idalab GmbH, Adalbertstr. 20, 10997 Berlin, Germany
e-mail: mika@idalab.de

K.-R. Müller (✉)
Machine Learning Group, Technical University of Berlin, 10587 Berlin, Germany

Department of Brain and Cognitive Engineering, Korea University, Anam-dong, Seongbuk-gu, Seoul 136-713, Korea
e-mail: klaus-robert.mueller@tu-berlin.de

algorithms, given that a conditionally positive definite kernel is used [34]. The so-called empirical kernel map [35] allows preprocessing of data by projecting it onto the leading kPCA components; thus non-linear variants of algorithms can be constructed via a non-linear transformation.

This chapter begins with a brief review of some concepts in kPCA and the analysis of pre-images. A novel aspect we will discuss is the computation of gradients of a kernel-based model that can be used for the purpose of optimization. The computation of such gradients turns out to be rather tricky; as we will see, the gradients can easily be dominated by noise in irrelevant directions, and thus need to be stabilized. One way of doing so is to apply pre-image methods, which will then allow us to present some interesting applications from the domain of quantum chemistry—an area that was only very recently explored with kernel methods [1, 31, 33, 38, 39].

In the following, we briefly review kernel methods (Sect. 21.2), kPCA (Sect. 21.3), and the pre-image problem (Sect. 21.4). Then, in Sect. 21.5, we show how to use gradient information that is derived from a kernel-based model. A particular difficulty here is that gradient estimates, in many circumstances, are prone to large amounts of noise. Pre-images hold the key to solving this issue and achieving stable gradients, which enable optimization over the data manifold given the kernel-based learning model. This section will also demonstrate optimization with respect to model properties for (a) a toy example and (b) real-world problems from quantum chemistry and physics. Finally we give a brief concluding discussion in Sect. 21.6.

21.2 The Kernel Trick

With regard to the kernel idea behind support vector machines (SVMs) [4, 7, 10, 29, 43] of non-linearizing the linear classifier formulation, Schölkopf, Smola and Müller [34] were the first to realize that this trick can be applied to almost any linear algorithm. The only prerequisite was that one be able to formulate the algorithm in terms of the dot product between data points. The key was the re-discovery of a long known mathematical fact: under certain conditions, $k(\mathbf{x}, \mathbf{x}') : \mathbb{R}^m \times \mathbb{R}^m \to \mathbb{R}$ is equivalent to the dot product in another space F (the feature space).[1] Thus, the kernel function $k(\mathbf{x}, \mathbf{x}')$ can be interpreted as $\Phi(\mathbf{x}) \cdot \Phi(\mathbf{x}')$, where $\Phi : \mathbb{R}^m \to F$ is the map to the feature space.

The consequences were dramatic: it became possible to extend well-understood linear models with a sound theoretical foundation to a much larger class of non-linear models—seemingly for free. However, there are two prominent drawbacks:

---

[1] In general, $\mathbf{x}$ is not restricted to being in $\mathbb{R}^m$ and could be any object.

- While most linear methods scale computationally with the number of input dimensions m (i.e., $\mathcal{O}(m^3)$), most kernel methods scale with the number of samples n (i.e., $\mathcal{O}(n^3)$)—which for many applications is tremendously larger than m. In particular, most kernel methods handle dense $n \times n$ matrices. In the present era of "big data," this rapidly becomes intractable. However, it is often possible to devise clever algorithms or approximations to circumvent this issue (e.g., for SVM, see [19, 21, 30]).
- The solution of such non-linear algorithms is usually expressed as a linear combination of the kernel function $f(\mathbf{x}) = \sum_{j=1}^{n} \alpha_j k(\mathbf{x}_j, \mathbf{x})$, where n is the number of training samples, α_j are weights, and $\mathbf{x}_j$ are the training samples. This is equivalent to the dot product $\Phi(\mathbf{x}) \cdot \Psi$ in the feature space, where $\Psi = \sum_{j=1}^{n} \alpha_j \Phi(\mathbf{x}_j)$. This is not a problem if the application requires only $f(\mathbf{x})$. However, if one would like to interpret Ψ or map back to input space $\mathbb{R}^m$, one needs the idea of pre-images (see Sect. 21.4).

As noted above, Schölkopf et al. [34] exemplified the "kernelization" procedure for the popular PCA algorithm. Meanwhile, a plethora of other algorithms were kernelized, ranging from Linear Discriminants [2, 22, 24], over nonlinear variants of ICA [16] and One Class SVM [36], to Canonical Correlation Analysis [15], Principal Manifolds [37], Relevance Vector Machines [41], and many more. In addition, kernel methods have been devised to analyse other learning machines [26] or trained kernel machines [6, 25]. The new formulation of these algorithms as a linear technique in some kernel feature space provided extremely valuable insights, both from a theoretical point of view as well as from an algorithmic point of view (e.g., the strong connection between mathematical optimization and learning [5]).

21.3 Kernel PCA

Principal Component Analysis (PCA) [11] is an orthogonal basis transformation which is found by diagonalizing the centred covariance matrix of a data set, $\{\mathbf{x}_j \in \mathbb{R}^m, j = 1, \ldots, n\}$, defined by $C = X^\top X / n$, where $X = (\mathbf{x}_1, \ldots, \mathbf{x}_n)^\top$ and the samples are assumed to be centred, i.e., $\sum_{j=1}^{n} \mathbf{x}_j = 0$. The eigenvectors $\mathbf{v}_i$ of C are called the principal components (PCs), and the sample variance along $\mathbf{v}_i$ is given by the corresponding eigenvalue λ_i. Projecting onto the eigenvectors with the largest eigenvalues (i.e., the first q PCs) is optimal in the sense that minimal information is lost. In many applications these directions contain the most interesting information. For example, in data compression, one projects onto the PCs in order to retain as much information as possible, and in de-noising one discards directions with small variance (assuming that low variance is equivalent to noise).

As mentioned in Sect. 21.2, kernel PCA (kPCA) is a non-linear generalization of PCA using kernel functions [34]. To state the result, the principal components are given by $\mathbf{v}_i = \sum_{j=1}^{n} \mathbf{a}_{i,j} \Phi(\mathbf{x}_j)$, where the $\mathbf{a}_i$ are the eigenvectors of the kernel matrix K, given by $K_{ij} = k(\mathbf{x}_i, \mathbf{x}_j)$, sorted in order of decreasing corresponding eigenvalue. Hence kPCA amounts to computing the eigenvectors of the kernel matrix K instead of the covariance matrix C. To project onto the PCs as in linear PCA, we define a projection P_q by

$$P_q \Phi(\mathbf{x}) = \sum_{i=1}^{q} \beta_i \mathbf{v}_i, \qquad (21.1)$$

where the $\beta_i = \mathbf{v}_i \cdot \Phi(\mathbf{x})$ are the projections of $\Phi(\mathbf{x})$ onto the PCs. If q is chosen such that all PCs with non-zero eigenvalues are kept, we can perfectly reconstruct the data (i.e., $P_q \Phi(\mathbf{x}_j) = \Phi(\mathbf{x}_j)$). While the equivalent for linear PCA would often amount to a lower-dimensional representation of the data (i.e., $q < m$), this is less likely for kPCA as the representation in the feature space is much higher dimensional (i.e., $q \leq n$ but often $q \geq m$). If some PCs with non-zero variance are thrown away, kPCA fulfils the PCA property that $P_q \Phi(\mathbf{x})$ will be the optimal least squares approximation to $\Phi(\mathbf{x})$ when restricted to orthogonal projection—but this holds true only in the feature space.

21.4 Pre-images

As already mentioned above, there are many applications for which one needs an optimal reconstruction of $P_q \Phi(\mathbf{x})$ in input space $\mathbb{R}^m$. Examples would be (lossy) compression (e.g., of images) or de-noising. One straightforward approach to this issue was proposed in Ref. [23]. The idea is to find an approximate pre-image $\tilde{\mathbf{x}}$ in input space that will map closest to the projection $P_q \Phi(\mathbf{x})$ in the feature space:

$$\tilde{\mathbf{x}} = \underset{\mathbf{x}' \in \mathbb{R}^n}{\mathrm{argmin}} \, \|\Phi(\mathbf{x}') - P_q \Phi(\mathbf{x})\|^2. \qquad (21.2)$$

It can be shown (see [23, 35] for details) that this equation can be formulated entirely in terms of the kernel $k(\mathbf{x}, \mathbf{x}') = \Phi(\mathbf{x}) \cdot \Phi(\mathbf{x}')$. The pre-image $\tilde{\mathbf{x}}$ can then be optimized using standard gradient descent methods. For kernels of the form $k(\mathbf{x}, \mathbf{x}') = f(\|\mathbf{x} - \mathbf{x}'\|)$ (e.g., Gaussian kernels), [23, 25] devise an iteration scheme to find $\tilde{\mathbf{x}}$.

21.5 Pre-images for Gradient-Based Optimization

In many applications of machine learning, one would like to use the estimator to optimize some property with respect to the data representation. For example, in image compression, one wants to optimize the representation to reduce the size without losing useful information. In neuroscience, one can optimize a stimulus to increase response. For these types of optimization, the quality of the gradient of the estimator is crucial. In certain circumstances, however, the gradient exhibits a high amount of "noise". In the following, we explore a simple example that clearly illustrates the origin of this noise and the problems that it creates in optimization. We then describe how properties of kernel PCA and pre-images offer a solution.

21.5.1 Example: Shape Optimization

The perimeter of simple two-dimensional shapes, represented by single-valued radius r as a function of angle θ, is given exactly by the integral

$$P[r] = \int_0^{2\pi} d\theta \, \sqrt{r(\theta)^2 + r'(\theta)^2},$$

where $r'(\theta) = dr/d\theta$. $P[r]$ is called a *functional* of $r(\theta)$. Now suppose we are not given this formula, but only a set of examples $\{\mathbf{r}_j, P_j\}_{j=1,\ldots,n}$ to learn from, where $\mathbf{r}_j \in \mathbb{R}^m$ are sufficiently dense histogram representations (e.g., 100 bins) of $r(\theta)$ on $\theta \in [0, 2\pi]$. In particular, we are given noise-free examples of ellipses with axes a and b:

$$r(\theta) = ab/\sqrt{b^2 \cos^2 \theta + a^2 \sin^2 \theta}.$$

Given this data, we use kernel ridge regression (KRR) [17] to predict the perimeter of new ellipses:

$$P^{\mathrm{ML}}(\mathbf{r}) = \sum_{j=1}^{n} \alpha_j k(\mathbf{r}, \mathbf{r}_j),$$

where α_j are the weights and k is the kernel. We choose the Gaussian kernel $k(\mathbf{r}, \mathbf{r}') = \exp(-\|\mathbf{r} - \mathbf{r}'\|^2/(2\sigma^2))$, where σ is the length scale. Minimizing the quadratic cost plus regularization $\sum_{j=1}^{n}(P^{\mathrm{ML}}(\mathbf{r}_j) - P_j)^2 + \lambda \boldsymbol{\alpha}^T K \boldsymbol{\alpha}$ yields

$$\boldsymbol{\alpha} = (K + \lambda I)^{-1} \mathbf{P},$$

where $\boldsymbol{\alpha} = (\alpha_1, \ldots, \alpha_n)^T$, $\mathbf{P} = (P_1, \ldots, P_n)^T$, K is the kernel matrix, and λ is a constant known as the noise level [17].

Figure 21.1 shows a sample dataset of 16 ellipses with $(a, b) \in \{1, \frac{4}{3}, \frac{5}{3}, 2\} \times \{1, \frac{4}{3}, \frac{5}{3}, 2\}$ (the model does not account for rotational symmetry, so we distinguish between, e.g., $(1, 2)$ and $(2, 1)$). After cross-validation of hyperparameters, we choose $\sigma = 13$ and $\lambda = 10^{-6}$. Contours of the perimeter values and the percentage error of the model are given in Fig. 21.2 as a function of a and b. The model has less than 0.1 % error within the interpolation region $1 < a, b < 2$.

Now suppose we use our model P^{ML} to find the shape with area $A_0 = 9\pi/4$ and minimum perimeter. Of course, the solution is a circle with radius $r = 3/2$

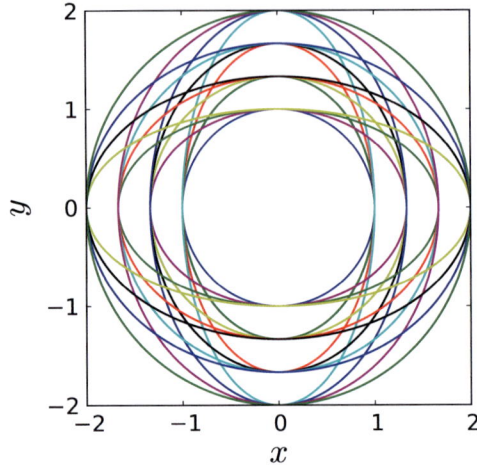

Fig. 21.1 The dataset of 16 ellipses represented in Cartesian coordinates

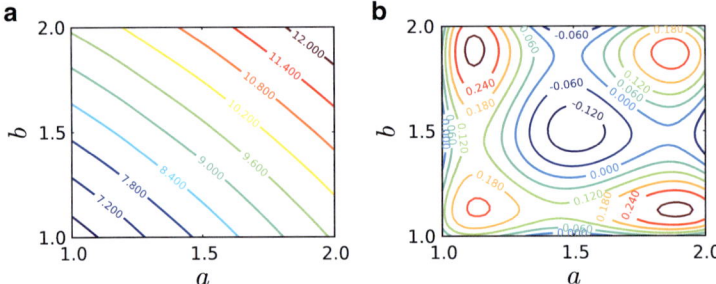

Fig. 21.2 (a) Contour plot of the perimeter of the ellipse as a function of axes lengths a and b. (b) Contour plot of the percentage error $(P^{\mathrm{ML}}(\mathbf{r}) - P[r])/P[r] \times 100\,\%$ of the model

with perimeter 3π, which is well within the interpolation region of the model (but not in the training set). This constrained optimization can be formulated in a variety of ways (see, e.g., [40]). For example, the penalty method enforces the constraint by regularizing deviations of the area from A_0, and solves a series of unconstrained minimization problems, slowly increasing the penalty strength until convergence. Define the penalty function

$$F_p(\mathbf{r}) = P^{\mathrm{ML}}(\mathbf{r}) + p(A(\mathbf{r}) - A_0)^2, \tag{21.3}$$

where the area functional

$$A[r] = \frac{1}{2}\int_0^{2\pi} d\theta\, r(\theta)^2$$

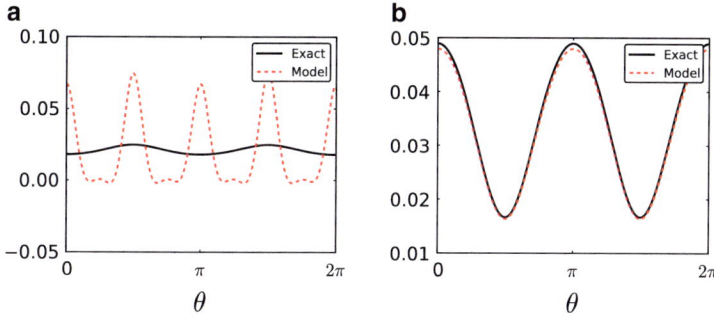

Fig. 21.3 (a) The gradient of the model and exact functional, for $a = 1.2$ and $b = 1.3$. (b) These gradients projected onto the tangent space of the data manifold

can be approximated by a Riemann sum $A(\mathbf{r})$ from our histogram representation of $r(\theta)$. Let $\mathbf{r}^*(p)$ minimize $F_p(\mathbf{r})$. Then the solution to the optimization is given by

$$\mathbf{r}^* = \lim_{p \to \infty} \mathbf{r}^*(p).$$

Standard unconstrained minimization methods can be applied to find a solution for each p [40]. This requires the gradient of the model

$$\nabla P^{\mathrm{ML}}(\mathbf{r}) = \sum_{j=1}^{n} \alpha_j (\mathbf{r}_j - \mathbf{r}) k(\mathbf{r}, \mathbf{r}_j) / \sigma^2, \tag{21.4}$$

while the exact functional derivative of $P[r]$ is given by

$$\frac{\delta P[r]}{\delta r(\theta)} = \frac{r(\theta)^3 + 2r(\theta)r'(\theta)^2 - r(\theta)^2}{(r(\theta)^2 + r'(\theta)^2)^{3/2}}. \tag{21.5}$$

Also, $\nabla A(\mathbf{r}) = \mathbf{r}\Delta\theta$, where $\Delta\theta = 2\pi/m$ is the spacing between bins. Figure 21.3a compares the gradient of the model with the exact functional derivative. The large error in ∇P^{ML} is typical for all shapes within the interpolation region. In addition, these errors are not a result of overfitting—no combination of hyperparameters yields accurate gradients in this case. Increasing the number of training samples does not improve the gradient either.

Figure 21.4a shows a sample optimization in which gradient descent is used to minimize $F_p(\mathbf{r})$, for $p = 5$, starting from $\mathbf{r}$ with $a = 2$, $b = 4/3$. The shape quickly deforms due to the noise in the gradient, leaving the region spanned by the data. Each step in the gradient descent introduces more noise into the shape. One can attempt to remedy this by applying the de-noising procedure described in Sect. 21.4 *during* the optimization:

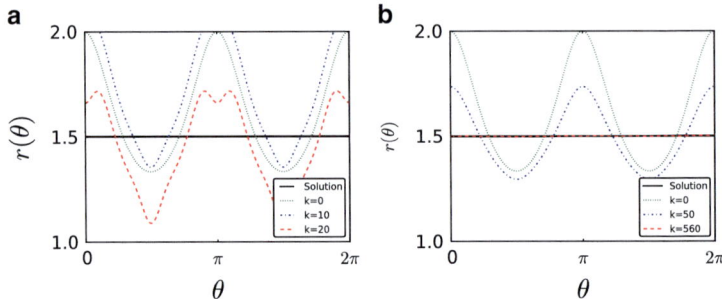

Fig. 21.4 (a) Minimization of Eq. 21.3 for $p = 1$ using the bare gradient of the model. The shape quickly develops spurious wiggles due to the noise in the gradient. The initial guess for the shape was $a = 2$, $b = 4/3$. (b) The same minimization in (a) using denoising with a Gaussian kernel with length scale $\sigma' = 18.1$, keeping $q = 5$ PCs in kPCA, and applying the de-noising $\ell = 5$ times each iteration

---

Modified Gradient Descent Algorithm

* Start from initial guess $\mathbf{r}_0$.
1. Take a step $\mathbf{r}_{k+1} = \mathbf{r}_k - \frac{\epsilon}{k} \frac{\nabla P^{\mathrm{ML}}(\mathbf{r}_k)}{\|\nabla P^{\mathrm{ML}}(\mathbf{r}_k)\|}$, where ϵ is a constant.
2. De-noise $\mathbf{r}_{k+1}$ by replacing it with $\tilde{\mathbf{r}}_{k+1}$. Repeat this ℓ times (depending on how much noise we introduced in the last step).
3. Repeat until $\|\mathbf{r}_{k+1} - \mathbf{r}_k\| < \delta$, where δ is the desired accuracy.

The result is shown in Fig. 21.4b, where the de-noising is performed with a Gaussian kernel with length scale $\sigma' = 18.1$ and we keep $q = 5$ principal components. The minimization gives a decent approximate solution, based solely on our learned model. This method gets us quickly close to the solution, but convergence near the solution is sensitive to the choice of the parameters σ', q, and ℓ. In addition, we find that the optimal parameters depend on the initial guess of the shape as well as on where the solution lies in input space. In the next section, we discuss where this "noise" in the gradient comes from and how to remove it, leading to a much better method for performing the optimization.

21.5.2 Origin of the "Noise"

The noise in the gradient of the model occurs generally when the data set is intrinsically low dimensional, embedded in a high-dimensional input space. Assuming the data is generated by a smooth mapping ψ from an underlying parameter space

$\Theta \subset \mathbb{R}^d$ to input space $\mathbb{R}^m$, we define the *data manifold* as the image $M = \psi(\Theta)$. The noise in the gradient occurs when $d \ll m$. The reasoning is as follows:

- The gradient measures change in the target value in *all* directions in input space, but all the given data lies on M.
- Regression is a method of interpolation, particularly with the Gaussian kernel.
- If we consider a point $\mathbf{x} \in M$ and move in input space while confined to M, the model is given information about how the target value changes (e.g., interpolation).
- If we move orthogonally to the tangent space of M at $\mathbf{x}$, the model has no information about the change in the target value. The "noise" comes from this extrapolation.
- Thus, we should be able to remove the noise if we project the gradient of the model onto the tangent space of M.

For our example, Fig. 21.3b compares the model gradient (Eq. 21.4) with the exact functional derivative (Eq. 21.5) when both are projected onto the tangent space of M at $a = 1.2$, $b = 1.3$. Clearly, the discrepancy is restricted to the orthogonal complement of the tangent space.

Based on this analysis, we can understand how the de-noising optimization worked in the previous section. At each step, the noise in gradient takes $\mathbf{r}$ far off the data manifold, away from the data. The de-noising step effectively returns $\mathbf{r}$ back onto M. However, a smarter way to perform the optimization would be to de-noise the *gradient* of the model, by projecting it onto the tangent space of the manifold at each step of the gradient descent. This would constrain the optimization to lie within the data manifold, never leaving the interpolation region. In general, however, one does not know the structure of the data manifold a-priori! One needs an accurate method to approximate M, at least locally near a given point.

21.5.3 Optimization Constrained to the Data Manifold

Such methods of locally approximating or globally reconstructing the data manifold fall under the general technique of nonlinear dimensionality reduction. This includes kernel PCA (kPCA) [34], Laplacian eigenmaps [3], diffusion maps [9], local linear embedding [32], Hessian local linear embedding [12], and local tangent space alignment [44, 45]. These methods provide a coarse reconstruction of M, but the local linear approximation breaks down when data sampling is too sparse, or M has a high curvature.

In the denoising procedure (see Sects. 21.3 and 21.4) a sample $\mathbf{x} \in X$ is mapped into feature space $\Phi(\mathbf{x})$ and projected onto the first q principal components, $P_q \Phi(\mathbf{x})$ (see Eq. 21.1). Then, the approximate pre-image $\tilde{\mathbf{x}}$ is found (Eq. 21.2). If $\mathbf{x}$ is far from the data manifold, then its representation in the feature space will be poor. The kernel PCA projection error

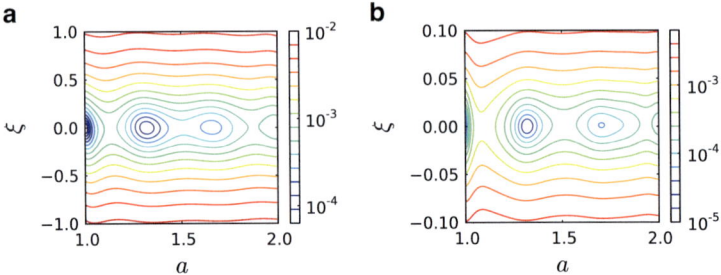

Fig. 21.5 Log contour plots of (**a**) $p_q(\mathbf{r}_{a,1.5} + \xi \mathbf{z})^2$ and (**b**) $d_q(\mathbf{r}_{a,1.5} + \xi \mathbf{z})^2$, where $\mathbf{z}$ is a randomly chosen direction of length 1. The qualitative features are the same for both (**a**) and (**b**), and are independent of the choice of $\mathbf{z}$. The length scale σ' in kPCA was 6.0, chosen as twice the median over all nearest neighbour distances between training samples, and all principal components with nonzero eigenvalues were used ($q = 15$)

$$p_q(\mathbf{x}) = \|\Phi(\mathbf{x}) - P_q \Phi(\mathbf{x})\|$$

and the denoising magnitude

$$d_q(\mathbf{x}) = \|\mathbf{x} - \tilde{\mathbf{x}}\|$$

provide useful information that can be used to characterized the data manifold. For our toy example, these quantities are plotted in Fig. 21.5. The line $\xi = 0$ corresponds to the data manifold. Qualitatively, both quantities are small on M, and increase quickly as one moves away from M. In particular, p_q^2 is flat along the direction of M, and highly convex in directions moving away from M. This information can be used to find the tangent space of M at a point $\mathbf{r}$ as follows:

Nonlinear gradient denoising (NLGD)

1. Compute the Hessian of p_q^2, H, evaluated at a point $\mathbf{r}$ which is known to be on the data manifold M.
2. Compute the eigenvalues $\lambda_1, \ldots, \lambda_m$ and eigenvectors $\mathbf{u}_1, \ldots, \mathbf{u}_m$ of $H(\mathbf{r})$ and order them in order of increasing eigenvalue magnitude.
3. The first d eigenvalues correspond to directions with small curvature. The corresponding d eigenvectors form a basis for the tangent space $T_M(\mathbf{r})$. The remaining eigenvalues will be large and positive.
4. Finally, the projection onto the tangent is given by

$$P_T(\mathbf{r}) = \sum_{j=1}^{d} \mathbf{u}_j \mathbf{u}_j^T.$$

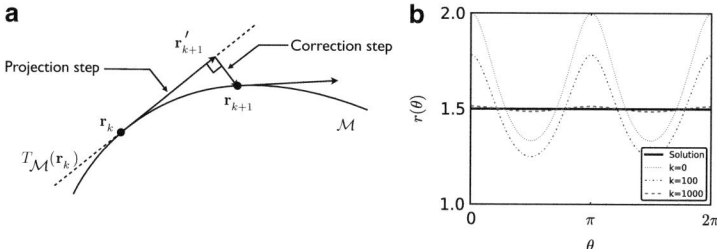

Fig. 21.6 (a) The projection algorithm [40], where the gradient of the model is projected onto the tangent space of the data manifold at each step. Because we move slightly off M, we require a correction step. (b) The same minimization as in Fig. 21.4, using the NLGD projected gradient descent algorithm, keeping all PCs in kPCA with $\sigma' = 6.0$

This procedure can be used to approximate the tangent space of M based solely on the data given. The denoising magnitude can be used likewise in place of the kernel PCA projection error. In all cases we have observed so far, the two give similar results, but p_q^2 is easier to compute.

Applying this to our optimization leads to a new algorithm (see Fig. 21.6a):

NLGD projected gradient descent algorithm

* Start from initial guess $\mathbf{r}_0$.
1. Compute the tangent space $T_M(\mathbf{r}_k)$ of the data manifold M and the NLGD projection $P_T(\mathbf{r}_k)$.
2. **Projection step.** Take a step $\mathbf{r}'_{k+1} = \mathbf{r}_k - \frac{\epsilon}{k} P_T(\mathbf{r}_k) \frac{\nabla P^{\mathrm{ML}}(\mathbf{r}_k)}{\|\nabla P^{\mathrm{ML}}(\mathbf{r}_k)\|}$, where ϵ is a constant.
3. **Correction step.** Minimize $p_q^2(\mathbf{r})$ starting from $\mathbf{r}'_{k+1}$ within the orthogonal complement of $T_M(\mathbf{r}_k)$. Let the solution be $\mathbf{r}_{k+1}$.
4. Repeat until $\|\mathbf{r}_{k+1} - \mathbf{r}_k\| < \delta$, where δ is the desired accuracy.

Applying this to our toy example yields the result in Fig. 21.6b. The sensitivity of the solution on the initial condition $\mathbf{r}_0$ and the parameters σ' and q is removed, and convergence is well conditioned. In the next section, we describe some real applications of this method in recent literature.

21.5.4 Applications in Density Functional Theory

Density functional theory (DFT) is now the most commonly used method for electronic structure calculations in quantum chemistry and solid state physics [8]. DFT attempts to circumvent directly solving the Schrödinger equation by

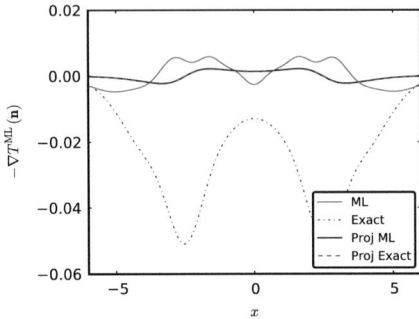

Fig. 21.7 The functional derivative of the kinetic energy functional of the ML model (MLA) compared with the exact. This derivative suffers from the same "noise" we described in Sect. 21.5.2 (i.e., the large deviation between ML and the exact derivative). Using the NLGD technique, the noise was removed by projecting the derivative onto the tangent space of the data manifold

approximating the energy as a functional of the electron density (instead of using the traditional wave function) [18, 20]. Recently, machine learning was used for the first time to directly approximate the kinetic energy density functional of one-dimensional electrons confined to a box (a toy model commonly used to test new approximations) [38]. The authors used kernel ridge regression with a Gaussian kernel to predict the kinetic energy of new densities based on examples of electron densities and their exact kinetic energies. The generalization error of the model was extremely low, but in density functional theory, traditionally an energy functional is useless unless its functional derivative is accurate as well (since ground-state densities are found through a self-consistent minimization of the total energy) [13].

This situation is exactly as described in the toy problem. The inputs (electron densities) are represented as high-dimensional (e.g., 500) vectors while the data is generated from a parameter space of only a few dimensions. The noise in the gradient the authors observed was due to this general phenomenon. To remedy the noise, the authors' solution was to project the gradient of the model on a local linear PCA subspace using only a few principal components. Using this method, they were able to perform accurate optimizations with the ML model, although the final density was slightly sensitive to the initial guess.

In later work [39], the same authors moved on to a more difficult system: a one-dimensional model of chemically bonded diatomics. Again, the gradient was found to be noisy (see Fig. 21.7), and the local linear PCA method of [38] was inaccurate due to the high curvature of the data manifold. Instead, the authors applied the NLGD projected gradient descent algorithm, achieving high accuracy, and were able to compute highly accurate binding curves and molecular forces from a model trained from a sparse sampling of data.

21.6 Conclusion

This chapter has briefly reviewed kernel methods and discussed in particular kernel PCA and the pre-image problem. When a kernel-based model has learned to predict a certain property, say, the atomization energy of a certain compound, then an interesting question is whether we can use the gradient of the model for optimization of some related (e.g., chemical) property. We have shown that the naive use of gradient information fails, due to noise that in many cases contaminates the gradient. Adapting techniques from pre-image computation, we can define projections that make the gradient of the ML model more meaningful, so that it can be used for optimization. A simple toy example illustrates this nonlinear gradient denoising (NLGD) procedure and shows its use for property optimization. We briefly reviewed two real-world applications of NLGD stemming from the domains of quantum chemistry and physics. Other future work will continue along the successful path of applying kernel-based methods in quantum chemistry and physics [1,31,33,38,39] with the aim of contributing to the quest for novel materials and chemical compounds.

Many open challenges need to be resolved in kernel-based learning: all kernel algorithms scale in the number of data points (not in the dimensionality of the data); thus the application of kernel methods for large problems remains an important challenge (see, e.g., [19, 21, 30, 42]). There may be large "big data" problems that are practically only amenable to neural networks (see [27, 28]) or other learning machines that allow for high-throughput streaming (see, e.g., [14]). However, a large number of mid-scale applications in the sciences and technology will remain where kernel methods will be able to contribute with highly accurate and robust predictive models.

Acknowledgements KRM thanks Vladimir N. Vapnik for continuous mentorship and collaboration since their first discussion in April 1995. This wonderful and serendipitous moment has profoundly changed the scientific agenda of KRM. From then on, KRM's IDA group—then at GMD FIRST in Berlin—and later the offspring of this group have contributed actively to the exciting research on kernel methods. KRM acknowledges funding by the DFG, the BMBF, the EU and other sources that have helped in this endeavour. This work is supported by the World Class University Program through the National Research Foundation of Korea, funded by the Ministry of Education, Science, and Technology (grant R31-10008). JS and KB thank the NSF (Grant No. CHE-1240252) for funding.

References

1. Bartók, A.P., Payne, M.C., Kondor, R., Csányi, G.: Gaussian approximation potentials: the accuracy of quantum mechanics, without the electrons. Phys. Rev. Lett. **104**, 136403 (2010)
2. Baudat, G., Anouar, F.: Generalized discriminant analysis using a kernel approach. Neural Comput. **12**(10), 2385–2404 (2000)
3. Belkin, M., Niyogi, P.: Laplacian eigenmaps for dimensionality reduction and data representation. Neural Comput. **15**(6), 1373–1396 (2003)

4. Boser, B., Guyon, I., Vapnik, V.: A training algorithm for optimal margin classifiers. In: Haussler, D. (ed.) Proceedings of the 5th Annual ACM Workshop on Computational Learning Theory, Pittsburgh, pp. 144–152 (1992)
5. Bradley, P., Fayyad, U., Mangasarian, O.: Mathematical programming for data mining: formulations and challenges. J. Comput. **11**(3), 217–238 (1999)
6. Braun, M., Buhmann, J., Müller, K.R.: On relevant dimensions in kernel feature spaces. J. Mach. Learn. Res. **9**, 1875–1908 (2008)
7. Burges, C.: A tutorial on support vector machines for pattern recognition. Knowl. Discov. Data Min. **2**(2), 121–167 (1998)
8. Burke, K.: Perspective on density functional theory. J. Chem. Phys. **136**(15), 150,901 (2012)
9. Coifman, R.R., Lafon, S.: Diffusion maps. Appl. Comput. Harmon. Anal. **21**(1), 5–30 (2006)
10. Cortes, C., Vapnik, V.: Support vector networks. Mach. Learn. **20**, 273–297 (1995)
11. Diamantaras, K., Kung, S.: Principal Component Neural Networks. Wiley, New York (1996)
12. Donoho, D.L., Grimes, C.: Hessian eigenmaps: locally linear embedding techniques for high-dimensional data. Proc. Natl. Acad. Sci. **100**(10), 5591–5596 (2003)
13. Dreizler, R.M., Gross, E.K.U.: Density Functional Theory: An Approach to the Quantum Many-Body Problem. Springer, New York (1990)
14. Farabet, C., Couprie, C., Najman, L., LeCun, Y.: Learning hierarchical features for scene labeling. IEEE Trans. Pattern Anal. Mach. Intell. (2013, in press)
15. Gestel, T.V., Suykens, J.A.K., Brabanter, J.D., Moor, B.D., Vandewalle, J.: Kernel canonical correlation analysis and least squares support vector machines. In: Proceedings of the International Conference on Artificial Neural Networks (ICANN 2001), Vienna, pp. 381–386 (2001)
16. Harmeling, S., Ziehe, A., Kawanabe, M., Müller, K.R.: Kernel-based nonlinear blind source separation. Neural Comput. **15**, 1089–1124 (2003)
17. Hastie, T., Tibshirani, R., Friedman, J.: The Elements of Statistical Learning. Data Mining, Inference, and Prediction, 2nd edn. Springer, New York (2009)
18. Hohenberg, P., Kohn, W.: Inhomogeneous electron gas. Phys. Rev. B **136**(3B), 864–871 (1964)
19. Joachims, T.: Making large-scale SVM learning practical. In: Schölkopf, B., Burges, C., Smola, A. (eds.) Advances in Kernel Methods—Support Vector Learning, pp. 169–184. MIT, Cambridge (1999)
20. Kohn, W., Sham, L.J.: Self-consistent equations including exchange and correlation effects. Phys. Rev. A **140**(4A), 1133–1138 (1965)
21. Laskov, P., Gehl, C., Krüger, S., Müller, K.R.: Incremental support vector learning: analysis, implementation and applications. J. Mach. Learn. Res. **7**, 1909–1936 (2006)
22. Mika, S., Rätsch, G., Weston, J., Schölkopf, B., Müller, K.R.: Fisher discriminant analysis with kernels. In: Hu, Y.H., Larsen, J., Wilson, E., Douglas, S. (eds.) Neural Networks for Signal Processing IX, pp. 41–48. IEEE, New York (1999)
23. Mika, S., Schölkopf, B., Smola, A., Müller, K.R., Scholz, M., Rätsch, G.: Kernel PCA and de-noising in feature spaces. In: Kearns, M., Solla, S., Cohn, D. (eds.) Advances in Neural Information Processing Systems, vol. 11, pp. 536–542. MIT, Cambridge (1999)
24. Mika, S., Rätsch, G., Weston, J., Schölkopf, B., Smola, A., Müller, K.R.: Constructing descriptive and discriminative nonlinear features: Rayleigh coefficients in kernel feature spaces. IEEE Trans. Patterns Anal. Mach. Intell. **25**(5), 623–627 (2003)
25. Montavon, G., Braun, M., Krüger, T., Müller, K.R.: Analyzing local structure in kernel-based learning: explanation, complexity and reliability assessment. IEEE Signal Process. Mag. **30**(4), 62–74 (2013)
26. Montavon, G., Braun, M., Müller, K.R.: A kernel analysis of deep networks. J. Mach. Learn. Res. **12**, 2579–2597 (2011)
27. Montavon, G., Müller, K.R.: Big learning and deep neural networks. In: Montavon, G., Orr, G.B., Müller, K.R. (eds.) Neural Networks: Tricks of the Trade, Lecture Notes in Computer Science, vol. 7700, pp. 419–420. Springer, Berlin/Heidelberg (2012)
28. Montavon, G., Orr, G., Müller, K.R. (eds.): Neural Networks: Tricks of the Trade, vol. 7700. In: LNCS. Springer, New York (2012)

29. Müller, K.R., Mika, S., Rätsch, G., Tsuda, K., Schölkopf, B.: An introduction to kernel-based learning algorithms. IEEE Trans. Neural Netw. **12**(2), 181–201 (2001)
30. Platt, J.: Fast training of support vector machines using sequential minimal optimization. In: Schölkopf, B., Burges, C., Smola, A. (eds.) Advances in Kernel Methods — Support Vector Learning, pp. 185–208. MIT, Cambridge (1999)
31. Pozun, Z.D., Hansen, K., Sheppard, D., Rupp, M., Müller, K.R., Henkelman, G.: Optimizing transition states via kernel-based machine learning. J. Chem. Phys. **136**(17), 174101 (2012)
32. Roweis, S.T., Saul, L.K.: Nonlinear dimensionality reduction by locally linear embedding. Science **290**(5500), 2323–2326 (2000)
33. Rupp, M., Tkatchenko, A., Müller, K.R., von Lilienfeld, O.A.: Fast and accurate modeling of molecular atomization energies with machine learning. Phys. Rev. Lett. **108**(5), 058301 (2012)
34. Schölkopf, B., Smola, A., Müller, K.: Nonlinear component analysis as a kernel eigenvalue problem. Neural comput. **10**(5), 1299–1319 (1998)
35. Scholkopf, B., Mika, S., Burges, C., Knirsch, P., Muller, K.R., Ratsch, G., Smola, A.: Input space versus feature space in kernel-based methods. IEEE Trans. Neural Netw. **10**, 1000–1017 (1999)
36. Schölkopf, B., Platt, J., Shawe-Taylor, J., Smola, A., Williamson, R.: Estimating the support of a high-dimensional distribution. Neural Comput. **13**(7), 1443–1471 (2001)
37. Smola, A., Mika, S., Schölkopf, B., Williamson, R.: Regularized principal manifolds. J. Mach. Learn. Res. **1**, 179–209 (2001)
38. Snyder, J.C., Rupp, M., Hansen, K., Müller, K.R., Burke, K.: Finding density functionals with machine learning. Phys. Rev. Lett. **108**, 253002 (2012)
39. Snyder, J.C., Rupp, M., Hansen, K., Bloooston, L., Müller, K.R., Burke, K.: Orbital-free bond breaking via machine learning. Submitted to J. Chem. Phys. (2013)
40. Snyman, J.A.: Practical Mathematical Optimization. Springer, New York (2005)
41. Tipping, M.: The relevance vector machine. In: Solla, S., Leen, T., Müller, K.R. (eds.) Advances in Neural Information Processing Systems, vol. 12, pp. 652–658. MIT, Cambridge (2000)
42. Tresp, V.: Scaling kernel-based systems to large data sets. Data Min. Knowl. Discov. **5**, 197–211 (2001)
43. Vapnik, V.: The Nature of Statistical Learning Theory. Springer, New York (1995)
44. Wang, J.: Improve local tangent space alignment using various dimensional local coordinates. Neurocomputing **71**(16), 3575–3581 (2008)
45. Zhang, Z.Y., Zha, H.Y.: Principal manifolds and nonlinear dimensionality reduction via tangent space alignment. J. Shanghai University (English Edition) **8**(4), 406–424 (2004)

Chapter 22
Efficient Learning of Sparse Ranking Functions

Mark Stevens, Samy Bengio, and Yoram Singer

Abstract Algorithms for learning to rank can be inefficient when they employ risk functions that use structural information. We describe and analyze a learning algorithm that efficiently learns a ranking function using a domination loss. This loss is designed for problems in which we need to rank a small number of positive examples over a vast number of negative examples. In that context, we propose an efficient coordinate descent approach that scales *linearly* with the number of examples. We then present an extension that incorporates regularization, thus extending Vapnik's notion of regularized empirical risk minimization to ranking learning. We also discuss an extension to the case of multi-value feedback. Experiments performed on several benchmark datasets and large-scale Google internal datasets demonstrate the effectiveness of the learning algorithm in constructing compact models while retaining the empirical performance accuracy.

22.1 Introduction

The past decade's proliferation of search engines and online advertisements has underscored the need for accurate yet efficiently computable ranking functions. Moreover, the emergence of personalized search and targeted advertisement further emphasizes the need for efficient algorithms that can generate a plethora of ranking functions which are used in tandem at serving time. The focus of this chapter is the derivation of such an efficient learning algorithm that yields compact ranking functions while achieving competitive accuracy. Before embarking on a description of our approach we would like to make connections to existing methods that influenced our line of research as well as briefly describe alternative methods for learning to rank. Due to the space constraints and the voluminous amount of work

---

M. Stevens (✉) · S. Bengio · Y. Singer
Google Research, 1600 Amphitheatre Parkway, 94043 Mountain View, CA, USA
e-mail: singer@google.com; stevensm@google.com; bengio@google.com

on this subject, we clearly cannot give a comprehensive overview. The home pages of two recent workshops at NIPS'09, "Learning to Rank" and "Learning with Orderings", are good sources of information on different learning to rank methods and analyses.

The roots of learning to rank go back to the early days of information retrieval (IR), such as to those of the classical IR described by Gerard Salton [10]. One of the early papers to cast the ranking task as a learning problem is "Learning to Order Things" [2]. While the learning algorithm presented in this paper was rather naive as it encompassed the notion of near-perfect ranking "experts", it laid some of the foundations later used by more effective algorithms such as RankSVM [6], RankBoost [4], and PAMIR [5]. These three algorithms, and many other algorithms, reduced the ranking problem to preference learning over pairs. This reduction enabled the usage of existing learning tools with matching generalization analysis that stem from Vladimir Vapnik's work [13–15]. However, the reduction to pairs of instances may result in poor ranking accuracy when the ranking objective is not closely related to the pairs' preference objective. Moreover, the usage of pairs of instances can impose computational burdens in large ranking tasks. The deficiency of preference-based approaches sparked research that tackles non-linear and often non-convex ranking loss functions; see, for instance, [1, 7, 16]. These more recent approaches resulted in improved results. However, they are typically computationally expensive, may converge to a local optimum [1], or are tailored for a specific setting [7]. Moreover, most learning to rank algorithms do not include a natural mechanism for controlling the compactness of the ranking function.

In this chapter we use a loss function called the domination loss [3]. To make this loss applicable to different settings, we extend and generalize the loss by incorporating the notion of margin [14, 15] over pairs of instances and enable the usage of multi-valued feedback. We devise a simple yet effective coordinate descent algorithm that is guaranteed to converge to the unique optimal solution; see, for instance, [8, 12] for related convergence proofs. Although the domination loss is expressed in terms of ordering relations over pairs, by using a bound optimization technique we are able to decompose each coordinate descent step so that the resulting update scales *linearly* with the number of instances that we rank. Furthermore, we show how to incorporate an ℓ_1 regularization term into the objective and the descent process. This term promotes sparse ranking functions. We present empirical results which demonstrate the effectiveness of our approach in building compact ranking functions whose performance matches the state-of-the-art results.

22.2 Problem Setting

We start by establishing the notation used throughout the chapter. Vectors are denoted in bold face, e.g., $\mathbf{x}$, and are assumed to be in column orientation. The transpose of a vector is denoted by $\mathbf{x}^\dagger$. The (vector) expectation of a set of vectors $\{\mathbf{x}_i\}$ with respect to a discrete distribution $\mathbf{p}$ is denoted by $\mathbb{E}_\mathbf{p}[\mathbf{x}] = \sum_j p_j \mathbf{x}_j$.

22 Efficient Learning of Sparse Ranking Functions

We observe a set of instances (e.g., images) where each instance is represented as a vector in $\mathbb{R}^n$. The ith instance, denoted by $\mathbf{x}_i$, is associated with a quality feedback, denoted $\tau_i \in \mathbb{R}$. We describe our derivation for a single query as the extension to multiple queries is straightforward.

Given the feedback set for instances observed for a query, we wish to learn a *ranking* function for that query, $f : \mathbb{R}^n \to \mathbb{R}$, that is consistent with the feedback as often as possible. Concretely, we would like the function f to induce an ordering such that $\tau_i > \tau_j \Rightarrow f(\mathbf{x}_i) > f(\mathbf{x}_j)$. We use the domination loss as a surrogate convex loss function to promote the ordering requirements. For an instance $\mathbf{x}_i$ we denote by $\mathcal{D}(i)$ the set of instances that should be ranked below it according to the feedback, $\mathcal{D}(i) = \{j \mid \tau_i > \tau_j\}$. The *combinatorial* domination loss is 1 if there exists $j \in \mathcal{D}(i)$ such that $f(\mathbf{x}_j) > f(\mathbf{x}_i)$. That is, the requirement that the ith instance dominates all the instances in $\mathcal{D}(i)$ is violated. To alleviate the intractability problems that arise when using a combinatorial loss, we use the following convex relaxation for a single instance,

$$\ell_{\mathcal{D}}(\mathbf{x}_i; f) = \log \left(1 + \sum_{j \in \mathcal{D}(i)} e^{f(\mathbf{x}_j) - f(\mathbf{x}_i) + \Delta(i,j)} \right).$$

Here, $\Delta(i, j) \in \mathbb{R}_+$ denotes a margin requirement between the ith and the jth instances. This function enables us to express richer relaxation requirements, which are often necessary in retrieval applications. For example, for the query dog we would like an image with a single nicely captured dog to attain a large margin over irrelevant images. In contrast, an image with multiple animals, including dogs, should attain only a modest margin over the irrelevant images.

22.3 An Efficient Coordinate Descent Algorithm

In this section we focus on a special case in which $\tau_i \in \{-1, +1\}$. That is, each instance is either positive (good, relevant) or negative (bad, irrelevant). In the next section we discuss generalizations. In this restricted case, the set $\{j \text{ s.t. } \exists i : \mathbf{x}_j \in \mathcal{D}(i)\}$ simply amounts to the set of negatively labelled instances and does not depend on the index i. For brevity we drop the dependency on i and simply denote it by $\bar{\mathcal{D}}$. We further simplify the learning setting and assume that $\Delta(i, j) = 0$. Again, we discuss relaxations of this assumption in the next section. The ranking function f is restricted to the class of linear functions, $f(\mathbf{x}) = \mathbf{w} \cdot \mathbf{x}$. In this base ranking setting the empirical loss with respect to $\mathbf{w}$ distils to $\sum_i \log \left(1 + \sum_{j \in \bar{\mathcal{D}}} e^{\mathbf{w} \cdot (\mathbf{x}_j - \mathbf{x}_i)} \right)$. For brevity let us focus on a single domination loss term, $\ell_{\mathcal{D}}(\mathbf{x}_i; \mathbf{w})$. Fixing the query q and performing simple algebraic manipulations we get

$$\ell_{\mathcal{D}}(\mathbf{x}_i; \mathbf{w}) = \log \left(\frac{\sum_{j \in \bar{\mathcal{D}}(i)} e^{\mathbf{w} \cdot \mathbf{x}_j}}{e^{\mathbf{w} \cdot \mathbf{x}_i}} \right) = \log \left(\sum_{j \in \bar{\mathcal{D}}(i)} e^{\mathbf{w} \cdot \mathbf{x}_j} \right) - \mathbf{w} \cdot \mathbf{x}_i,$$

where $\bar{\mathcal{D}}(i) = \mathcal{D} \cup \{i\}$ consists of all the negatively labelled instances and the ith relevant instance.

There is no closed form solution for the optimum even when we restrict ourselves to a single coordinate of $\mathbf{w}$. We therefore use a bound optimization technique [9] by constructing a quadratic upper bound on the domination loss. This technique was used, for instance, in the context of boosting-style algorithms for regression. To construct the upper bound we need to calculate the gradient and the Hessian of the (simplified) domination loss. The gradient amounts to

$$\nabla_\mathbf{w} \ell_\mathcal{D}(\mathbf{x}_i; \mathbf{w}) = \frac{\sum_{j \in \bar{\mathcal{D}}(i)} e^{\mathbf{w} \cdot \mathbf{x}_j} \mathbf{x}_j}{\sum_{j \in \bar{\mathcal{D}}(i)} e^{\mathbf{w} \cdot \mathbf{x}_j}} - \mathbf{x}_i = \mathbb{E}_{\mathbf{p}_i}[\mathbf{x}] - \mathbf{x}_i ,$$

where $\mathbf{p}_i$ is the distribution induced by $\mathbf{w}$ whose rth component is $p_r = e^{\mathbf{w} \cdot \mathbf{x}_r} / \bar{Z}$ and $\bar{Z}$ is a normalization constant $\bar{Z} = \sum_{j \in \bar{\mathcal{D}}(i)} e^{\mathbf{w} \cdot \mathbf{x}_j}$. Using the above, the Hessian is

$$H_i(\mathbf{w}) = \mathbb{E}_{\mathbf{p}_i}[\mathbf{x}\mathbf{x}^\dagger] - \mathbb{E}_{\mathbf{p}_i}[\mathbf{x}] (\mathbb{E}_{\mathbf{p}_i}[\mathbf{x}])^\dagger .$$

Recalling the mean value theorem, the loss at $\mathbf{w} + \boldsymbol{\delta}$ can now be written as

$$\ell_\mathcal{D}(\mathbf{x}_i; \mathbf{w} + \boldsymbol{\delta}) = \ell_\mathcal{D}(\mathbf{x}_i; \mathbf{w}) + \nabla_\mathbf{w} \cdot \boldsymbol{\delta} + \frac{1}{2} \boldsymbol{\delta}^\dagger H_i(\mathbf{w} + \alpha\boldsymbol{\delta}) \boldsymbol{\delta}$$

$$= \ell_\mathcal{D}(\mathbf{x}_i; \mathbf{w}) + (\mathbb{E}_{\mathbf{p}_i}[\mathbf{x}] - \mathbf{x}_i) \cdot \boldsymbol{\delta} + \frac{1}{2} \mathbb{E}_{\tilde{\mathbf{p}}_i}\left[(\boldsymbol{\delta} \cdot \mathbf{x})^2\right] - \frac{1}{2}(\boldsymbol{\delta} \cdot \mathbb{E}_{\tilde{\mathbf{p}}_i}[\mathbf{x}])^2 ,$$

where $\tilde{\mathbf{p}}_i$ is the distribution induced at $\mathbf{w} + \alpha\boldsymbol{\delta}$ for an unknown $\alpha \in [0, 1]$. We now derive an update which minimizes the bound along a single coordinate, denoted r, of $\mathbf{w}$. Let $\mathbf{e}_r$ denote the vector whose components are 0 except for the rth component, which is 1; then,

$$\ell_\mathcal{D}(\mathbf{x}_i; \mathbf{w} + \delta\mathbf{e}_r) = \kappa(i) + \delta \left(\mathbb{E}_{\bar{\mathcal{D}}(i)}[\mathbf{p}, \mathbf{x}_r] - x_{i,r} \right)$$

$$+ \frac{1}{2} \delta^2 (\mathbb{E}_{\bar{\mathcal{D}}(i)}[\tilde{\mathbf{p}}, \mathbf{x}_r^2] - \mathbb{E}^2_{\bar{\mathcal{D}}(i)}[\tilde{\mathbf{p}}, \mathbf{x}_r]) ,$$

where $\mathbb{E}_{\bar{\mathcal{D}}(i)}[\mathbf{p}, \mathbf{x}_r] = \sum_{j \in \bar{\mathcal{D}}(i)} p_j x_{j,r}$. where $\kappa(i)$ is a constant that does not depend on δ. Since $\tilde{\mathbf{p}}_j$ is not known we need to bound the loss further. Let B_r denote the maximum value of the square of the rth feature over the instances retrieved, namely $B_r = \max_{j \in \bar{\mathcal{D}}} x_{j,r}^2$. We now can bound the multiplier of $\frac{1}{2}\delta^2$ by $\sum_{j \in \bar{\mathcal{D}}(i)} \tilde{p}_j x_{j,r}^2 \leq B_r$. Let m denote the number of relevant training elements for the current query. The bound of the domination loss restricted to coordinate r is $L(\mathbf{w} + \delta\mathbf{e}_r) \leq \kappa + \delta \sum_{i \notin \mathcal{D}} \left(\sum_{j \in \bar{\mathcal{D}}(i)} p_j x_{j,r} - x_{i,r} \right) + \frac{1}{2} B_r \delta^2 m$. The last term is a simple quadratic term in δ and thus the optimal step $\delta^\star$ along coordinate r can be trivially computed. Alas, our derivation of an efficient coordinate descent update does not end here, as the term multiplying δ in the bound above depends on the pairs

Fig. 22.1 Coordinate descent algorithm for ranking-learning with the domination loss

```
initialize:    Z = m,  ∀i : ρ_i = 0,  ∀r : B_r = max_j x²_{j,r}
while not converged do
   for r ∈ {1, ..., n} do
      ν_r = 0;  μ_r = 0
      for j ∈ D do
         μ_r ← μ_r + e^{ρ_j} x_{j,r}
      end for
      for each i ∉ D do
         ν_r ← ν_r + (Z x_{i,r} − μ_r)/(Z + e^{ρ_i})
      end for
      δ_r ← ν_r/(m B_r)  ;   w_r ← w_r + δ_r
      for each j ∈ D do
         Z ← Z − e^{ρ_j}  ;  ρ_j ← ρ_j + δ_r x_{j,r}  ;  Z ← Z + e^{ρ_j}
      end for
      for each i ∉ D do
         ρ_i ← ρ_i + δ_r x_{i,r}
      end for
   end for
end while
```

$i \notin \mathcal{D}$ and $j \in \bar{\mathcal{D}}(i)$. We now exploit the fact that the sole instance in $\bar{\mathcal{D}}(i)$ which depends on i is $\mathbf{x}_i$ itself, and rewrite the gradient, ν_r, with respect to r,

$$\nu_r + \sum_{i \notin \mathcal{D}} x_{i,r} = \sum_{i \notin \mathcal{D}} \sum_{j \in \bar{\mathcal{D}}(i)} p_j x_{j,r} = \sum_{i \notin \mathcal{D}} \left(\sum_{j \in \mathcal{D}} p_j x_{j,r} + p_i x_{i,r} \right)$$

$$= \sum_{i \notin \mathcal{D}} \sum_{j \in \mathcal{D}} \frac{e^{\mathbf{w} \cdot \mathbf{x}_j}}{Z + e^{\mathbf{w} \cdot \mathbf{x}_i}} x_{j,r} + \sum_{i \notin \mathcal{D}} \frac{e^{\mathbf{w} \cdot \mathbf{x}_i}}{Z + e^{\mathbf{w} \cdot \mathbf{x}_i}} x_{i,r}$$

$$= \left(\sum_{i \notin \mathcal{D}} \frac{1}{Z + e^{\mathbf{w} \cdot \mathbf{x}_i}} \right) \left(\sum_{j \in \mathcal{D}} e^{\mathbf{w} \cdot \mathbf{x}_j} x_{j,r} \right) + \sum_{i \notin \mathcal{D}} \frac{e^{\mathbf{w} \cdot \mathbf{x}_i} x_{i,r}}{Z + e^{\mathbf{w} \cdot \mathbf{x}_i}} ,$$

where $Z = \sum_{j \in \mathcal{D}} e^{\mathbf{w} \cdot \mathbf{x}_j}$. The latter expression can be computed in time linear in the size (number of instances) of each query, and *not* the number of comparable pairs in each query. Finally, to alleviate the dependency in the dimension of the instances, due to the products $\mathbf{w} \cdot \mathbf{x}_i$, on each iteration, we introduce the variables $\rho_i = \mathbf{w} \cdot \mathbf{x}_i$ and $\mu_r = \sum_{j \in \mathcal{D}} e^{\rho_j} x_{j,r}$. Then, the change to the rth coordinate of $\mathbf{w}$ is,

$$\delta^\star = \frac{1}{m B_r} \sum_{i \notin \mathcal{D}} \frac{Z x_{i,r} - \mu_r}{Z + e^{\rho_i}} .$$

To recap we provide the pseudo-code of the algorithm in Fig. 22.1.

22.4 Extensions

In this section we describe a few generalizations and extensions of the base coordinate descent algorithm described in the previous section. Concretely, we show a generalization of the algorithm to multi-valued feedback, we describe the addition of margin values over pairs of elements, and we define the incorporation of regularization throughout the course of the algorithm.

We start with the extension to multi-valued feedback. In the more general case, the feedback can take one of K predefined values. We can assume without loss of generality that the feedback set is $\{1, \ldots, K\}$. We thus can divide the instances retrieved for a query into k disjoint sets denoted by $G(1), \ldots, G(K)$. The set $G(K)$ consists of all the top-ranked instances and analogously $G(1)$ contains all the bottom-ranked elements. The set of instances dominated by any instance in $G(r)$ is denoted by $\mathcal{D}(r) = \{j \in G(l) | l < r\}$. The form of the bound for $\ell_{\mathcal{D}}(\mathbf{x}_i; \mathbf{w})$ with multi-valued feedback remains intact. When summing over all dominating x_i, however, we break down the summation by group k, yielding

$$L(\mathbf{w}+\delta\mathbf{e}_r) \leq \kappa + \sum_{k>1}\sum_{i\in G(k)}\left(\sum_{j\in\bar{\mathcal{D}}(i)}p_j x_{j,r} - x_{i,r}\right)\delta$$
$$+ \frac{1}{2}B_r\sum_{k>1}m(k)\delta^2, \quad (22.1)$$

where $m(k)$ is the number of elements in group k. The quadratic term can still be bounded by $mB_r/2$, where we redefine m to be the number of dominating elements, $\sum_{k>1}m(k)$. Again, efficient computation requires decomposing the linear multiplier. Using the same argument as earlier we get

$$\sum_{k>1}\sum_{i\in G(k)}\sum_{j\in\bar{\mathcal{D}}(i)}p_j x_{j,r}$$
$$= \sum_{k>1}\sum_{i\in G(k)}\frac{1}{Z_k+e^{\mathbf{w}\cdot\mathbf{x}_i}}\sum_{j\in\mathcal{D}(k)}e^{\mathbf{w}\cdot\mathbf{x}_j}x_{j,r} + \sum_{k>1}\sum_{i\in G(k)}\frac{e^{\mathbf{w}\cdot\mathbf{x}_i}x_{i,r}}{Z_k+e^{\mathbf{w}\cdot\mathbf{x}_i}},$$

where $Z_k = \sum_{j\in\mathcal{D}(k)}e^{\mathbf{w}\cdot\mathbf{x}_j}$. The gradient for the multi-valued feedback amounts to

$$\sum_{k>1}\sum_{i\in G(k)}\left(\sum_{j\in\bar{\mathcal{D}}(i)}p_j x_{j,r} - x_{i,r}\right) = \sum_{k>1}\sum_{i\in G(k)}\frac{\mu_{k,r} - Z_k x_{i,r}}{Z_k+e^{\mathbf{w}\cdot\mathbf{x}_i}},$$

and $\mu_{k,r} = \sum_{j\in\mathcal{D}(k)}e^{\mathbf{w}\cdot\mathbf{x}_j}x_{j,r}$. Note that Z and μ can be constructed recursively in linear time as follows: $Z_k = \sum_{j\in G(k-1)}e^{\mathbf{w}\cdot\mathbf{x}_j} + Z_{k-1}$ and

$\mu_{k,r} = \sum_{j \in G(k-1)} e^{\mathbf{w} \cdot \mathbf{x}_j} x_{j,r} + \mu_{k-1,r}$. To recap, each generalized update of δ is computed as follows,

$$\delta = \frac{1}{mB_r} \sum_{k>1} \sum_{i \in G(k)} \frac{Z_k x_{i,r} - \mu_{k,r}}{Z_k + e^{\mathbf{w} \cdot \mathbf{x}_i}}.$$

Next we discuss the infusion of margin requirements, which amounts to using the loss $\sum_i \log\left(1 + \sum_{j \in \tilde{\mathcal{D}}(i)} e^{\mathbf{w} \cdot (\mathbf{x}_j - \mathbf{x}_i) + \Delta(i,j)}\right)$. When Δ is of general form, the problem is no longer decomposable and we were not able to devise an efficient extension. However, when the margin requirement can be expressed as a sum of two functions, $\Delta(i, j) = s(i) - s(j)$, it is possible to extend the efficient coordinate descent procedure. Adding a separable margin, each term $e^{\mathbf{w} \cdot \mathbf{x}_j}$ is replaced with $e^{\mathbf{w} \cdot \mathbf{x}_j \pm s(j)}$. Concretely, we obtain the following gradient:

$$\nabla_{\mathbf{w}} \ell_{\mathcal{D}}(\mathbf{x}_i; \mathbf{w}) = \frac{\sum_{j \in \mathcal{D}} e^{\mathbf{w} \cdot \mathbf{x}_j - s(j)} \mathbf{x}_j + e^{\mathbf{w} \cdot \mathbf{x}_i + s(i)} \mathbf{x}_i}{\sum_{j \in \mathcal{D}} e^{\mathbf{w} \cdot \mathbf{x}_j - s(j)} + e^{\mathbf{w} \cdot \mathbf{x}_i + s(i)}} - \mathbf{x}_i.$$

The rest of the derivation is identical to that for the zero-margin case, and yields

$$\delta = \frac{1}{mB_r} \sum_i \frac{Z x_{i,r} - \mu_r}{Z + e^{\mathbf{w} \cdot \mathbf{x}_i + s(i)}}$$

where

$$\mu_r = \sum_{j \in \mathcal{D}(k)} e^{\mathbf{w} \cdot \mathbf{x}_j - s(j)} x_{j,r} \quad \text{and} \quad Z = \sum_{j \in \mathcal{D}(k)} e^{\mathbf{w} \cdot \mathbf{x}_j - s(j)}.$$

We conclude this section by showing how to incorporate a regularization term and perform feature selection. We use the ℓ_1 norm of the weight vector $\mathbf{w}$ as the means for regularizing the weights. We would like to note though that closed form extensions can be derived for other ℓ_p norms, in particular the ℓ_2^2 norm. We focus on the ℓ_1 norm since it promotes sparse solutions. Adding an ℓ_1 penalty to the quadratic bound and performing a coordinate descent step on the penalized bound amounts to the following (scalar) optimization problem:

$$\min_{\delta} g_r \delta + \frac{\beta_r}{2} \delta^2 + \lambda \|w_r + \delta\|_1, \tag{22.2}$$

where g_r and β_r are the expressions appearing in (22.1). To find the optimal solution, denoted by $\delta^\star$, of the above equation we need the following lemmas. The proofs of the lemmas employ routine analysis tools. For brevity we omit the subscript r in the following lemmas.

Lemma 22.1. *If $\beta w - g > 0$; then $w + \delta^\star \geq 0$ and if $\beta w - g < 0$ then $w + \delta^\star \leq 0$.*

Proof. Let us consider without loss of generality the case where $\beta w - g > 0$. Suppose that $w + \delta^\star < 0$, then, (22.2) reduces to, $0 = g + \beta \delta^\star - \lambda$. Combining the above equation with the bound on $\beta w - g$, and recalling that $\lambda \geq 0$ and $\beta > 0$, we obtain that $\beta(w + \delta^\star) > \lambda$ and thus $w + \delta^\star > 0$. We thus get a contradiction; hence $w + \delta^\star \geq 0$. The symmetric case follows similarly. □

Lemma 22.2. *The optimal solution $\delta^\star$ equals $-w$ if and only if $|g - \beta w| \leq \lambda$.*

Proof. Without loss of generality, let us consider the case where $\beta w - g > 0$. We can then use Lemma 22.1 to simplify (22.2). Substituting $w + \delta$ for $\|w + \delta\|_1$ and adding the constraint $w + \delta \geq 0$ we get $\delta^\star = \mathrm{argmin}_\delta \; g\delta + \frac{\beta}{2}\delta^2 + \lambda(w+\delta)$ s.t. $w + \delta \geq 0$. If the inequality constraint holds with a strict inequality, then $\delta^\star = -(\lambda+g)/\beta > -w$. If, however, the minimum $(\lambda+g)/\beta \geq w$ then the optimum must be at the constraint boundary, namely $\delta^\star = -w$. Therefore, if $\lambda \geq \beta w - g \geq 0$ then $\delta^\star = -w$, whereas if $\beta w - g > \lambda$, then $\delta^\star = -(\lambda + g)/\beta$. The symmetric case where $g - \beta w \leq 0$ follows similarly. □

Equipped with the above lemmas, the update amounts to the following two-step procedure. Given the current value of w_r and g_r we check the condition stated in Lemma 22.2. If it is satisfied we set the new value of w_r to 0. Otherwise, we set $\delta^\star = -(g + \lambda)/\beta$ if $\beta w_r - g_r > 0$ and $\delta^\star = (-g + \lambda)/\beta$ if $\beta w_r - g_r < 0$.

As we discuss in the experiments, the combination of the robust loss, the coordinate descent procedure, and the sparsity inducing regularization often yields compact models consisting of about 700 non-zero weights out of 10,000 features for representing images, and even sparser models for documents.

22.5 Experiments

We evaluated and compared the algorithm and its extension on various datasets. We first evaluated the algorithm on the Microsoft's LETOR collection, whose are of modest size. On these datasets we compared the algorithm to RankSVM [6], AdaRank [16], and RankBoost [4]. These algorithms take different approaches to the ranking problem. To compare all algorithms we used three evaluation criteria: NDCG5, NDCG10, and Precision at 10. The results are provided in Table 22.1. Note that RankSVM is not presented in the table, since there are published results for RankSVM on only a subset of the datasets, on which its performance was average. We tested our algorithm with and without margin requirements using three-tier feedback. While the performance of our algorithm was often better than that of AdaRank and RankBoost, the results were not conclusive and all the versions we tested exhibited similar performance. We believe that the lack of ability to discriminate between the algorithms is largely due to the modest size of the LETOR collection and the tacit overfitting of the test sets due to repeated experiments that have been conducted on the LETOR collection. We therefore focused on experiments with larger datasets and compared our approach to a fast

Table 22.1 Results for LETOR dataset. Average rank (lower is better) over eight datasets is shown

Algorithm measure	Domination rank			AdaRank		RankBoost
	Multi-valued	Base	Margin	NDCG	MAP	
NDCG 10	3.62	3.38	2.75	4.12	3.62	3.38
Precision 10	3.25	3.12	2.38	4.00	4.38	3.12
NDCG 5	3.12	3.38	3.62	4.12	3.38	3.38

Table 22.2 Results for the large image dataset

Algorithm	PAMIR	Domination rank			
		Base-ℓ_2^2	Base-ℓ_1	Multi-valued	Margin
Avg. precision	0.051	0.050	0.050	0.050	0.050
Precision @ 10	0.080	0.073	0.073	0.073	0.073
Domination error	0.964	0.964	0.963	0.964	0.964
All pairs error	0.234	0.213	0.237	0.235	0.236
% zero weights	3.4	1.4	94.3	93.4	93.2

online algorithm called PAMIR [5] which can handle large ranking problems. One particular aspect that we tested is the ability of our algorithm to yield compact yet accurate models.

Image Ranking Experiments. The first *large* ranking experiment we conducted was an image ranking task. This image dataset consists of about 2.3 million training images and 0.4 million test images. The dataset is a subset of Google's image search repository and is not publicly available. The evaluation included 1,000 different queries and the results represent the average over these queries. The dataset is similar to the Corel image dataset used in [5], albeit it is much larger. We used the feature extraction scheme described in [5], which yielded a 10,000-dimension vector representation for each image with an average density (non-zero features) of 2 %. The results are given in Table 22.2. We used both ℓ_2^2 and ℓ_1 regularization in order to check the algorithm's performance with compact models and handle the extremely noisy labels. For each image in our dataset we have real-valued feedback which is based on the number of times the image was selected by a user when returned as a result for its associated query. As reported in Table 22.2 we used the user feedback information in two ways: first to construct three-valued feedback (relevant, somewhat relevant, and irrelevant), and second as the means to impose margin constraints. We defined the margin to be the scaled difference between the user counts of the two images. The scaling factor was chosen using cross-validation. It is apparent from the table that all the variants attain comparable performance. However, the ℓ_1 version of the domination loss yielded vastly sparse models, which renders the algorithm usable for very large ranking tasks. Disappointingly, despite our careful design, neither the multi-valued feedback nor the margin requirements resulted in improved performance on the image dataset.

Table 22.3 Results for the RCV1 collection

	PAMIR	Domination rank
Avg. precision	0.705	0.670
Precision @ 10	0.915	0.918
Domination error	0.974	0.976
All pairs error	0.014	0.024
% zero weights	78.1	98.8

Document Ranking Experiments. The second large dataset we experimented with is the Reuters RCV1 dataset. This dataset consists of roughly 800,000 news articles spanning a 1-year period. Each document is associated with one or more of 103 topics. Most documents are labelled with at least two topics and many with three or more. We view each topic as a ranking task across the entire collection. From each article, we extracted the raw text as input. We used unigram features as performed in [11]. We randomly partitioned the dataset, placing half of the documents in the training set and splitting the remainder evenly between a validation set and a test set. Of the 103 topics in the dataset, two had too few topics to provide meaningful results. We thus excluded these topics and reported results averaged over the remaining 101 topics. For each of the 101 topics, we learned a ranking function and used it to score all the test instances. The results were averaged across topics and are provided in Table 22.3. Here again we see that the ℓ_2 penalized PAMIR, which is a pure dual algorithm, and the ℓ_1 penalized coordinate descent algorithm achieve similar performance with the exception of the number of misordered pairs, which is closely related to the loss PAMIR employs. The main advantage again is the sparsity of the resulting models. Our approach uses fewer than 2 % of the original features whereas PAMIR uses a large portion. (PAMIR does not use all of the features despite the ℓ_2 regularization since some topics consist of a small number of relevant documents.)

22.6 Conclusions

We derived in this chapter an efficient coordinate descent algorithm for the task of learning to rank. Our construction is tightly based on the domination loss first proposed in [3]. We described a convex relaxation of that loss with an associated update that scales linearly with the number of training instances. We also derived several extensions of the basic algorithm, including the ability to handle multiple valued feedback, margin requirements over pairs of instances, and ℓ_1 regularization. Furthermore, the algorithm's efficiency is retained for these extensions. Experiments with several datasets show that by using ℓ_1 regularization, the resulting ranking models are trained considerably faster and yield significantly more compact models, yet attain performance competitive with some of the state-of-the-art learning to rank approaches.

References

1. Cao, Z., Qin, T., Liu, T.Y., Tsai, M.F., Li, H.: Learning to rank: from pairwise approach to listwise approach. In: ICML '07: Proceedings of the 24th International Conference on Machine Learning, Corvalis, pp. 129–136 (2007)
2. Cohen, W.W., Schapire, R.E., Singer, Y.: Learning to order things. J. Artif. Intell. Res. **10**, 243–270 (1999)
3. Dekel, O., Manning, C., Singer, Y.: Log-Linear Models for Label Ranking. Advances in Neural Information Processing Systems, vol. 14, Vancouver. MIT Press, Cambridge (2004)
4. Freund, Y., Iyer, R., Schapire, R.E., Singer, Y.: An efficient boosting algorithm for combining preferences. J. Mach. Learn. Res. **4**, 933–969 (2003)
5. Grangier, D., Bengio, S.: A discriminative kernel-based model to rank images from text queries. IEEE Trans. Pattern Anal. Mach. Intell. **30**(8), 1371–1384 (2008)
6. Joachims, T.: Optimizing search engines using clickthrough data. In: Proceedings of the ACM Conference on Knowledge Discovery and Data Mining (KDD), Edmonton (2002)
7. Joachims, T.: A support vector method for multivariate performance measures. In: Proceedings of the International Conference on Machine Learning (ICML), Bonn (2005)
8. Luo1, Z., Tseng, P.: On the convergence of the coordinate descent method for convex differentiable minimization. J. Optim. Theory Appl. **72**(1), 7–35 (1992)
9. Roweis, S.T., Salakhutdinov, R.: Adaptive overrelaxed bound optimization methods. In: Proceedings of the International Conference on Machine Learning (ICML), Washington, DC, pp. 664–671 (2003)
10. Salton, G.: Automatic Text Processing: The Transformation, Analysis and Retrieval of Information by Computer. Addison-Wesley, Boston (1989)
11. Singhal, A., Buckley, C., Mitra, M.: Pivoted document length normalization. In: Research and Development in Information Retrieval, Zurich, pp. 21–29 (1996)
12. Tseng, P., Yun, S.: A coordinate gradient descent method for nonsmooth separable minimization. Math. Program. B **117**, 387–423 (2007)
13. Vapnik, V.N.: Estimation of Dependences Based on Empirical Data. Springer, New York (1982)
14. Vapnik, V.N.: The Nature of Statistical Learning Theory. Springer, New York (1995)
15. Vapnik, V.N.: Statistical Learning Theory. Wiley, New York (1998)
16. Xu, J., Li, H.: Adarank: a boosting algorithm for information retrieval. In: SIGIR '07: Proceedings of the 30th Annual international ACM SIGIR Conference on Research and Development in Information Retrieval, Amsterdam, pp. 391–398 (2007)

Chapter 23
Direct Approximation of Divergences Between Probability Distributions

Masashi Sugiyama

Abstract Approximating a divergence between two probability distributions from their samples is a fundamental challenge in the statistics, information theory, and machine learning communities, because a divergence estimator can be used for various purposes such as two-sample homogeneity testing, change-point detection, and class-balance estimation. Furthermore, an approximator of a divergence between the joint distribution and the product of marginals can be used for independence testing, which has a wide range of applications including feature selection and extraction, clustering, object matching, independent component analysis, and causality learning. In this chapter, we review recent advances in direct divergence approximation that follow the general inference principle advocated by Vladimir Vapnik—one should not solve a more general problem as an intermediate step. More specifically, direct divergence approximation avoids separately estimating two probability distributions when approximating a divergence. We cover direct approximators of the Kullback–Leibler (KL) divergence, the Pearson (PE) divergence, the relative PE (rPE) divergence, and the L^2-distance. Despite the overwhelming popularity of the KL divergence, we argue that the latter approximators are more useful in practice due to their computational efficiency, high numerical stability, and superior robustness against outliers.

23.1 Introduction

Let us consider the problem of approximating a divergence D between two probability distributions P and P' on $\mathbb{R}^d$ from two sets of independent and identically distributed samples $\mathcal{X} := \{x_i\}_{i=1}^n$ and $\mathcal{X}' := \{x'_{i'}\}_{i'=1}^{n'}$ following P and P'.

M. Sugiyama (✉)
Tokyo Institute of Technology, 2-12-1 O-okayama, Meguro-ku, Tokyo 152-8552, Japan
e-mail: sugi@cs.titech.ac.jp

A divergence approximator can be used for various purposes such as two-sample testing [12, 26], change detection in time-series [14], class-prior estimation under class-balance change [7], salient object detection in images [49], and event detection from movies [48] and Twitter [18]. Furthermore, an approximator of the divergence between the joint distribution and the product of marginal distributions can be used for solving a wide range of machine learning problems [22], including independence testing [23], feature selection [10, 34], feature extraction [33, 46], canonical dependency analysis [13], object matching [43], independent component analysis [32], clustering [16, 28], and causality learning [42]. For this reason, accurately approximating a divergence between two probability distributions from their samples has been an important challenge in the statistics, information theory, and machine learning communities.

A naive way to approximate the divergence from P to P', denoted by $D(P\|P')$, is to first obtain estimators $\hat{P}_\mathcal{X}$ and $\hat{P}'_{\mathcal{X}'}$ of the distributions P and P' separately from their samples $\mathcal{X}$ and $\mathcal{X}'$, and then compute a plug-in approximator $D(\hat{P}_\mathcal{X}\|\hat{P}'_{\mathcal{X}'})$. However, this naive approach violates *Vapnik's principle* [39]:

> If you possess a restricted amount of information for solving some problem, try to solve the problem directly and never solve a more general problem as an intermediate step. It is possible that the available information is sufficient for a direct solution but is insufficient for solving a more general intermediate problem.

More specifically, if we know the distributions P and P', we can immediately know their divergence $D(P\|P')$. However, knowing the divergence $D(P\|P')$ does not necessarily imply knowing the distributions P and P', because different pairs of distributions can yield the same divergence values. Thus, estimating the distributions P and P' is more general than estimating the divergence $D(P\|P')$. Following Vapnik's principle, direct divergence approximators $\widetilde{D}(\mathcal{X}, \mathcal{X}')$ that do not involve the estimation of distributions P and P' have been developed recently [11, 19, 24, 31, 47].

The purpose of this chapter is to give an overview of such direct divergence approximators.

23.2 Divergence Measures

In this section, we introduce useful divergence measures.

Kullback–Leibler (KL) Divergence: The most popular divergence measure in statistics and machine learning would be the KL divergence [17], defined as

$$\mathrm{KL}(p\|p') := \int p(\boldsymbol{x}) \log \frac{p(\boldsymbol{x})}{p'(\boldsymbol{x})} \mathrm{d}\boldsymbol{x},$$

where $p(\boldsymbol{x})$ and $p'(\boldsymbol{x})$ are probability density functions of P and P', respectively.

Advantages of the KL divergence are that it is compatible with maximum likelihood estimation, it is invariant under input metric change, its Riemannian geometric structure is well studied [2], and it can be approximated accurately via *direct density-ratio estimation* [19, 24, 29]. However, it is not symmetric, it does not satisfy the triangle inequality, its approximation is computationally expensive due to the log function, and it is sensitive to outliers and numerically unstable because of the strong non-linearity of the log function and possible unboundedness of the density-ratio function p/p' [4, 47].

Pearson (PE) Divergence: The PE divergence [20] is a squared-loss variant of the KL divergence defined as

$$\mathrm{PE}(p\|p') := \int p'(x) \left(\frac{p(x)}{p'(x)} - 1 \right)^2 \mathrm{d}x.$$

Because both the PE and KL divergences belong to the class of Ali–Silvey–Csiszár divergences (which is also known as f-divergences) [1, 6], they share similar theoretical properties such as invariance under input metric change. The quadratic function the PE divergence adopts is compatible with least-squares estimation.

The PE divergence can also be accurately approximated via direct density-ratio estimation in the same way as the KL divergence [11, 29], but its approximator can be obtained *analytically* in a computationally much more efficient manner than that of the KL divergence. Furthermore, the PE divergence tends to be more robust against outliers than the KL divergence [30]. However, other weaknesses of the KL divergence such as asymmetry, violation of the triangle inequality, and possible unboundedness of the density-ratio function p/p' remain unsolved in the PE divergence.

Relative Pearson (rPE) Divergence: To overcome the possible unboundedness of the density-ratio function p/p', the rPE divergence was introduced recently [47], which is defined as

$$\mathrm{rPE}(p\|p') := \mathrm{PE}(p\|q_\alpha) = \int q_\alpha(x) \left(\frac{p(x)}{q_\alpha(x)} - 1 \right)^2 \mathrm{d}x,$$

where $q_\alpha = \alpha p + (1-\alpha) p'$ for $0 \le \alpha < 1$. When $\alpha = 0$, the rPE divergence is reduced to the plain PE divergence. The quantity p/q_α is called the *relative density ratio*, which is always upper-bounded by $1/\alpha$ for $\alpha > 0$. Thus, it can overcome the unboundedness problem of the PE divergence, while the invariance under input metric change is still maintained.

The rPE divergence is still compatible with least-squares estimation, and it can be approximated in almost the same way as the PE divergence via *direct relative density-ratio estimation*. Indeed, an rPE divergence approximator can still be obtained analytically in an accurate and computationally efficient manner. However, it still violates symmetry and the triangle inequality in the same way as the KL and PE divergence, and the choice of α is not straightforward in practice.

L^2-Distance: The L^2-distance is another standard distance measure between probability distributions defined as

$$L^2(p, p') := \int \Big(p(x) - p'(x)\Big)^2 dx.$$

The L^2-distance is a proper distance measure, and thus it is symmetric and satisfies the triangle inequality. Furthermore, the density difference $p(x) - p'(x)$ is always bounded as long as each density is bounded. Therefore, the L^2-distance is stable, without the need of tuning any control parameter such as α in the rPE divergence.

The L^2-distance is also compatible with least-squares estimation, and it can be accurately and analytically approximated in a computationally efficient and numerically stable manner via *direct density-difference estimation* [31]. However, the L^2-distance is not invariant under input metric change, which is a unique property inherent in ratio-based divergences.

23.3 Direct Divergence Approximation

In this section, we review recent advances in direct divergence approximation.

KL Divergence Approximation [24]: The key idea is to estimate the density ratio p/p' without estimating the densities p and p'. More specifically, a density ratio approximator $\hat{r}$ is obtained by minimizing the empirical KL divergence from p to $r \cdot p'$ with respect to a density-ratio model r:

$$\hat{r} := \underset{r}{\arg\min} \; \frac{1}{n} \sum_{i=1}^{n} \log r(x_i) \quad \text{subject to } r \geq 0 \text{ and } \frac{1}{n'} \sum_{i'=1}^{n'} r(x'_{i'}) = 1.$$

For a linear-in-parameter density-ratio model defined by

$$r(x) = \sum_{i=1}^{n} \theta_i \exp\left(-\frac{\|x - x_i\|^2}{2\sigma^2}\right), \tag{23.1}$$

the above optimization problem is convex and thus the global optimal solution can be obtained easily, e.g., by a gradient-projection iteration. The Gaussian width σ can be tuned by cross-validation with respect to the objective function. Given the density ratio estimator $\hat{r}$, a KL divergence estimator $\widehat{\text{KL}}(\mathcal{X}\|\mathcal{X}')$ can be constructed as

$$\widehat{\text{KL}}(\mathcal{X}\|\mathcal{X}') := \frac{1}{n} \sum_{i=1}^{n} \log \hat{r}(x_i).$$

A MATLAB implementation of the above KL divergence approximator (called the *KL importance estimation procedure*, KLIEP) is available from http://sugiyama-www.cs.titech.ac.jp/~sugi/software/KLIEP/.

Variations of this procedure for various density ratio models have been developed, including the log-linear model [38], the Gaussian mixture model [41], and the mixture of probabilistic principal component analyzers [45]. Also, an unconstrained variant, which corresponds to approximately maximizing the *Legendre–Fenchel lower bound* of the KL divergence [15], was also proposed [19]:

$$\widehat{\mathrm{KL}}'(\mathcal{X}\|\mathcal{X}') := \max_r \left[\frac{1}{n} \sum_{i=1}^{n} \log r(\boldsymbol{x}_i) - \frac{1}{n'} \sum_{i'=1}^{n'} r(\boldsymbol{x}'_{i'}) + 1 \right].$$

PE Divergence Approximation [11]: The PE divergence can also be directly approximated without estimating the densities p and p' via direct estimation of the density ratio p/p'. More specifically, a density ratio approximator $\hat{r}$ is obtained by minimizing the empirical p'-weighted squared difference between a density ratio model r and the true density ratio p/p':

$$\hat{r} := \underset{r}{\operatorname{argmin}} \left[\frac{1}{n'} \sum_{i'=1}^{n'} r^2(\boldsymbol{x}'_{i'}) - \frac{2}{n} \sum_{i=1}^{n} r(\boldsymbol{x}_i) \right].$$

For the linear-in-parameter density-ratio model (23.1), possibly together with ℓ_2-regularization [9], the density ratio estimator $\hat{r}$ can be obtained analytically, with a closed-form leave-one-out cross-validation score [40]. Moreover, together with the ℓ_1-regularization [35], the coefficients $\{\theta_i\}_{i=1}^n$ tend to be sparse and can be learned in a computationally efficient way [36], further equipped with a regularization path tracking algorithm [8].

A MATLAB implementation with the ℓ_2-regularizer (called *unconstrained least-squares importance fitting*, uLSIF) is available from http://sugiyama-www.cs.titech.ac.jp/~sugi/software/uLSIF/.

rPE Divergence Approximation [47]: The rPE divergence can be estimated in the same way as the PE divergence as

$$\hat{r} := \underset{r}{\operatorname{argmin}} \left[\frac{\alpha}{n} \sum_{i=1}^{n} r^2(\boldsymbol{x}_i) + \frac{1-\alpha}{n'} \sum_{i'=1}^{n'} r^2(\boldsymbol{x}'_{i'}) - \frac{2}{n} \sum_{i=1}^{n} r(\boldsymbol{x}_i) \right].$$

Thus, all the computational advantages of PE divergence approximation mentioned above are inherited by rPE divergence approximation.

A MATLAB implementation of this algorithm (called *relative uLSIF*; RuLSIF) is available from http://sugiyama-www.cs.titech.ac.jp/~yamada/RuLSIF.html.

L^2-Distance Approximation [31]: The key idea is to directly estimate the density difference $p - p'$ without estimating each density. More specifically, a density difference approximator $\hat{f}$ is obtained by minimizing the empirical squared difference between a density difference model f and the true density difference $p - p'$:

$$\hat{f} := \underset{f}{\mathrm{argmin}} \left[\int f(x)^2 dx - \left(\frac{2}{n} \sum_{i=1}^{n} f(x_i) - \frac{2}{n'} \sum_{i'=1}^{n'} f(x'_{i'}) \right) \right].$$

In practice, the use of the Gaussian kernel model,

$$f(x) = \sum_{i=1}^{n} \theta_i \exp\left(-\frac{\|x - x_i\|^2}{2\sigma^2}\right) + \sum_{i'=1}^{n'} \theta_{n+i'} \exp\left(-\frac{\|x - x'_{i'}\|^2}{2\sigma^2}\right),$$

is advantageous because the first term $\int f(x)^2 dx$ in the objective function can be computed analytically for this model. The above optimization problem is essentially of the same form as least-squares density-ratio approximation for the PE divergence, and therefore least-squares density-difference approximation can enjoy all the computational properties of least-squares density-ratio approximation.

A MATLAB implementation of the above algorithm (called *least-squares density difference*; LSDD) is available from http://sugiyama-www.cs.titech.ac.jp/~sugi/software/LSDD/.

Convergence Issues: All the direct divergence approximators reviewed above were proved to achieve the $\sqrt{n}$-consistency in the parametric case (suppose $n' = n$) [11, 24, 31, 47], which is the optimal convergence rate. Furthermore, they were also proved to achieve the minimax optimal convergence rate in the non-parametric case [11, 19, 24, 31, 47]. Also, experimentally, direct divergence approximators were shown to outperform the naive approaches based on density estimation [11, 24, 31, 47].

23.4 Usage of Divergence Estimators in Machine Learning

In this section, we show applications of divergence estimators in machine learning.

Change-Detection in Time-Series: The goal is to discover abrupt property changes behind time-series data. Let $y(t) \in \mathbb{R}^m$ be an m-dimensional time-series sample at time t, and let $Y(t) := [y(t)^\top, y(t+1)^\top, \ldots, y(t+k-1)^\top]^\top \in \mathbb{R}^{km}$ be a subsequence of time series at time t with length k. Instead of a single point $y(t)$, the subsequence $Y(t)$ is treated as a sample here, because time-dependent information can be naturally incorporated by this trick [14]. Let $\mathcal{Y}(t) := \{Y(t), Y(t+1), \ldots, Y(t+r-1)\}$ be a set of r retrospective subsequence

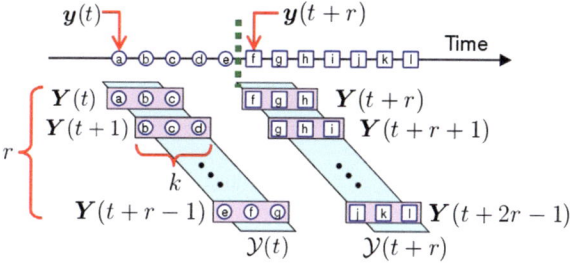

Fig. 23.1 Change-point detection in time-series

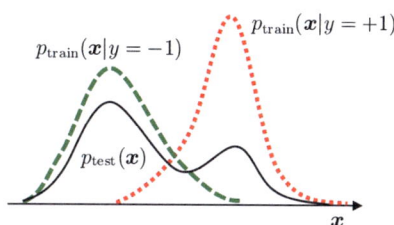

Fig. 23.2 Class-prior estimation

samples starting at time t. Then a divergence between the probability distributions of $\mathcal{Y}(t)$ and $\mathcal{Y}(t+r)$ may be used as the plausibility of change points (see Fig. 23.1).

The change-detection methods based on the rPE divergence [18] and the L^2-distance [31] were shown to be promising through experiments. In particular, the method based on the rPE divergence was successfully applied to event detection from movies [48] and Twitter [18].

Class-Prior Estimation Under Class-Balance Change: In real-world pattern recognition tasks, changes in class balance are often observed between the training and test phases. In such cases, naive classifier training produces significant estimation bias because the class balance in the training dataset does not properly reflect that of the test dataset. Here, let us consider a binary pattern recognition task of classifying pattern $x \in \mathbb{R}^d$ into class $y \in \{+1, -1\}$. The goal is to learn the class balance of a test dataset in a semi-supervised learning setup where unlabelled test samples are provided in addition to labelled training samples [3]. The class balance in the test set can be estimated by matching a mixture of class-wise training input densities,

$$q_{\text{test}}(x) := \pi p_{\text{train}}(x|y = +1) + (1 - \pi) p_{\text{train}}(x|y = -1),$$

to the test input density $p_{\text{test}}(x)$ under some divergence measure [7]. Here, $\pi \in [0, 1]$ is a mixing coefficient to be learned to minimize the divergence (see Fig. 23.2).

The class-balance estimation methods based on the PE divergence [7] and the L^2-distance [31] were shown to be promising through experiments.

Salient Object Detection in an Image: The goal is to find salient objects in an image. This can be achieved by computing a divergence between the probability

Fig. 23.3 Object detection in an image

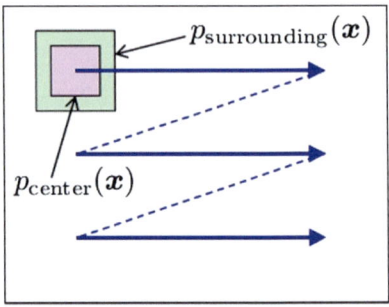

distributions of image features (such as brightness, edges, and colours) in the center window and its surroundings [49]. This divergence computation is swept over the entire image, possibly with changing scale (Fig. 23.3).

The object detection method based on the rPE divergence was demonstrated to be promising in experiments [49].

Measuring Statistical Independence: The goal is to measure how strongly two random variables U and V are statistically dependent by using paired samples $\{(\boldsymbol{u}_i, \boldsymbol{v}_i)\}_{i=1}^n$ drawn independently from the joint distribution with density $p_{U,V}(\boldsymbol{u}, \boldsymbol{v})$. Let us consider a divergence between the joint density $p_{U,V}$ and the product of marginal densities $p_U \cdot p_V$. This actually serves as a measure of statistical independence, because U and V are independent if and only if the divergence is zero (i.e., $p_{U,V} = p_U \cdot p_V$), and the dependence between U and V is stronger if the divergence is larger.

Such a dependence measure can be approximated in the same way as ordinary divergences by using the two datasets formed as $\mathcal{X} = \{(\boldsymbol{u}_i, \boldsymbol{v}_i)\}_{i=1}^n$ and $\mathcal{X}' = \{(\boldsymbol{u}_i, \boldsymbol{v}_j)\}_{i,j=1}^n$. The dependence measure based on the KL divergence is called *mutual information* [21], which plays a central role in information theory [5]. On the other hand, its PE divergence variant is called the *squared-loss mutual information*, which was shown to be useful for solving various machine learning tasks [22] such as independence testing [23], feature selection [10, 34], feature extraction [33, 46], canonical dependency analysis [13], object matching [43], independent component analysis [32], clustering [16, 28], and causality learning [42]. An L^2-distance variant of the dependence measure is called *quadratic mutual information* [37].

23.5 Conclusions

In this chapter, we reviewed recent advances in direct divergence approximation. Direct divergence approximators theoretically achieve optimal convergence rates both in parametric and non-parametric cases and experimentally compare favourably with their naive density estimation counterparts. However, direct divergence approximators still suffer from the curse of dimensionality. A possible cure for this problem is to combine them with dimension reduction, with the hope that two probability distributions share some commonality [25, 27, 44]. Further investigating this line would be a promising future direction.

Acknowledgements The author acknowledges support from the JST PRESTO program, KAKENHI 25700022, the FIRST program, and AOARD.

References

1. Ali, S., Silvey, S.: A general class of coefficients of divergence of one distribution from another. J. R. Stat. Soc. B **28**(1), 131–142 (1966)
2. Amari, S., Nagaoka, H.: Methods of Information Geometry. Oxford University Press, Providence (2000)
3. Chapelle, O., Schölkopf, B., Zien, A. (eds.): Semi-Supervised Learning. MIT Press, Cambridge (2006)
4. Cortes, C., Mansour, Y., Mohri, M.: Learning bounds for importance weighting. In: Lafferty, J., Williams, C., Zemel, R., Shawe-Taylor, J., Culotta, A. (eds.) Advances in Neural Information Processing Systems, Vancouver, vol. 23, pp. 442–450 (2010)
5. Cover, T., Thomas, J.: Elements of Information Theory, 2nd edn. Wiley, Hoboken (2006)
6. Csiszár, I.: Information-type measures of difference of probability distributions and indirect observation. Stud. Sci. Math. Hungarica **2**, 229–318 (1967)
7. du Plessis, M., Sugiyama, M.: Semi-supervised learning of class balance under class-prior change by distribution matching. In: Proceedings of 29th International Conference on Machine Learning (ICML'12), Edinburgh, pp. 823–830 (2012)
8. Efron, B., Hastie, T., Johnstone, I., Tibshirani, R.: Least angle regression. Ann. Stat. **32**(2), 407–499 (2004)
9. Hoerl, A., Kennard, R.: Ridge regression: biased estimation for nonorthogonal problems. Technometrics **12**(3), 55–67 (1970)
10. Jitkrittum, W., Hachiya, H., Sugiyama, M.: Feature selection via ℓ_1-penalized squared-loss mutual information. IEICE Trans. Inf. Syst. **E96-D**(7), 1513–1524 (2013)
11. Kanamori, T., Hido, S., Sugiyama, M.: A least-squares approach to direct importance estimation. J. Mach. Learn. Res. **10**, 1391–1445 (2009)
12. Kanamori, T., Suzuki, T., Sugiyama, M.: f-divergence estimation and two-sample homogeneity test under semiparametric density-ratio models. IEEE Trans. Inf. Theory **58**(2), 708–720 (2012)
13. Karasuyama, M., Sugiyama, M.: Canonical dependency analysis based on squared-loss mutual information. Neural Netw. **34**, 46–55 (2012)
14. Kawahara, Y., Sugiyama, M.: Sequential change-point detection based on direct density-ratio estimation. Stat. Anal. Data Min. **5**(2), 114–127 (2012)
15. Keziou, A.: Dual representation of ϕ-divergences and applications. C. R. Math. **336**(10), 857–862 (2003)
16. Kimura, M., Sugiyama, M.: Dependence-maximization clustering with least-squares mutual information. J. Adv. Comput. Intell. Intell. Inform. **15**(7), 800–805 (2011)
17. Kullback, S., Leibler, R.: On information and sufficiency. Ann. Math. Stat. **22**, 79–86 (1951)
18. Liu, S., Yamada, M., Collier, N., Sugiyama, M.: Change-point detection in time-series data by relative density-ratio estimations. Neural Netw. **43**, 72–83 (2013)
19. Nguyen, X., Wainwright, M., Jordan, M.: Estimating divergence functionals and the likelihood ratio by convex risk minimization. IEEE Trans. Inf. Theory **56**(11), 5847–5861 (2010)
20. Pearson, K.: On the criterion that a given system of deviations from the probable in the case of a correlated system of variables is such that it can be reasonably supposed to have arisen from random sampling. Philos. Mag. Ser. 5 **50**(302), 157–175 (1900)

21. Shannon, C.: A mathematical theory of communication. Bell Syst. Tech. J. **27**, 379–423 (1948)
22. Sugiyama, M.: Machine learning with squared-loss mutual information. Entropy **15**(1), 80–112 (2013)
23. Sugiyama, M., Suzuki, T.: Least-squares independence test. IEICE Trans. Inf. Syst. **E94-D**(6), 1333–1336 (2011)
24. Sugiyama, M., Suzuki, T., Nakajima, S., Kashima, H., von Bünau, P., Kawanabe, M.: Direct importance estimation for covariate shift adaptation. Ann. Inst. Stat. Math. **60**(4), 699–746 (2008)
25. Sugiyama, M., Kawanabe, M., Chui, P.: Dimensionality reduction for density ratio estimation in high-dimensional spaces. Neural Netw. **23**, 44–59 (2010)
26. Sugiyama, M., Suzuki, T., Itoh, Y., Kanamori, T., Kimura, M.: Least-squares two-sample test. Neural Netw. **24**(7), 735–751 (2011)
27. Sugiyama, M., Yamada, M., von Bünau, P., Suzuki, T., Kanamori, T., Kawanabe, M.: Direct density-ratio estimation with dimensionality reduction via least-squares hetero-distributional subspace search. Neural Netw. **24**(2), 183–198 (2011)
28. Sugiyama, M., Yamada, M., Kimura, M., Hachiya, H.: On information-maximization clustering: Tuning parameter selection and analytic solution. In: Proceedings of 28th International Conference on Machine Learning (ICML'11), Bellevue, pp. 65–72 (2011)
29. Sugiyama, M., Suzuki, T., Kanamori, T.: Density Ratio Estimation in Machine Learning. Cambridge University Press, Cambridge (2012)
30. Sugiyama, M., Suzuki, T., Kanamori, T.: Density ratio matching under the Bregman divergence: a unified framework of density ratio estimation. Ann. Inst. Stat. Math. **64**(5), 1009–1044 (2012)
31. Sugiyama, M., Suzuki, T., Kanamori, T., du Plessis, M., Liu, S., Takeuchi, I.: Density-difference estimation. Neural Comput. **25**(10), 2734–2775 (2013)
32. Suzuki, T., Sugiyama, M.: Least-squares independent component analysis. Neural Comput. **23**(1), 284–301 (2011)
33. Suzuki, T., Sugiyama, M.: Sufficient dimension reduction via squared-loss mutual information estimation. Neural Comput. **3**(25), 725–758 (2013)
34. Suzuki, T., Sugiyama, M., Kanamori, T., Sese, J.: Mutual information estimation reveals global associations between stimuli and biological processes. BMC Bioinform. **10**(1) (2009). S52, 12p
35. Tibshirani, R.: Regression shrinkage and subset selection with the Lasso. J. R. Stat. Soc. B **58**(1), 267–288 (1996)
36. Tomioka, R., Suzuki, T., Sugiyama, M.: Super-linear convergence of dual augmented Lagrangian algorithm for sparsity regularized estimation. J. Mach. Learn. Res. **12**, 1537–1586 (2011)
37. Torkkola, K.: Feature extraction by non-parametric mutual information maximization. J. Mach. Learn. Res. **3**, 1415–1438 (2003)
38. Tsuboi, Y., Kashima, H., Hido, S., Bickel, S., Sugiyama, M.: Direct density ratio estimation for large-scale covariate shift adaptation. J. Inf. Process. **17**, 138–155 (2009)
39. Vapnik, V.: Statistical Learning Theory. Wiley, New York (1998)
40. Wahba, G.: Spline Models for Observational Data. Society for Industrial and Applied Mathematics, Philadelphia (1990)
41. Yamada, M., Sugiyama, M.: Direct importance estimation with Gaussian mixture models. IEICE Trans. Inf. Syst. **E92-D**(10), 2159–2162 (2009)
42. Yamada, M., Sugiyama, M.: Dependence minimizing regression with model selection for non-linear causal inference under non-Gaussian noise. In: Proceedings of the Twenty-Fourth AAAI Conference on Artificial Intelligence (AAAI'10), Atlanta, pp. 643–648. AAAI (2010)
43. Yamada, M., Sugiyama, M.: Cross-domain object matching with model selection. In: Gordon, G., Dunson, D., Dudík, M. (eds.) Proceedings of the Fourteenth International Conference on Artificial Intelligence and Statistics (AISTATS2011), Ft. Lauderdale, JMLR Workshop and Conference Proceedings, vol. 15, pp. 807–815 (2011)

44. Yamada, M., Sugiyama, M.: Direct density-ratio estimation with dimensionality reduction via hetero-distributional subspace analysis. In: Proceedings of the Twenty-Fifth AAAI Conference on Artificial Intelligence (AAAI11), San Francisco, pp. 549–554. AAAI (2011)
45. Yamada, M., Sugiyama, M., Wichern, G., Simm, J.: Direct importance estimation with a mixture of probabilistic principal component analyzers. IEICE Trans. Inf. Syst. **E93-D**(10), 2846–2849 (2010)
46. Yamada, M., Niu, G., Takagi, J., Sugiyama, M.: Computationally efficient sufficient dimension reduction via squared-loss mutual information. In: Proceedings of the Third Asian Conference on Machine Learning (ACML'11), Taoyuan. JMLR Workshop and Conference Proceedings, vol. 20, pp. 247–262 (2011)
47. Yamada, M., Suzuki, T., Kanamori, T., Hachiya, H., Sugiyama, M.: Relative density-ratio estimation for robust distribution comparison. Neural Comput. **25**(5), 1324–1370 (2013)
48. Yamanaka, M., Matsugu, M., Sugiyama, M.: Detection of activities and events without explicit categorization. IPSJ Trans. Math. Model. Appl. **6**(2), 86–92 (2013)
49. Yamanaka, M., Matsugu, M., Sugiyama, M.: Salient object detection based on direct density-ratio estimation. IPSJ Trans. Math. Model. Appl. **6**(2), 78–85 (2013)

Index

Absolute loss, 63–65, 178–180, 191
AdaBoost, 37–39, 44, 45
AEF. *See* Approximation error function (AEF)
AERM. *See* Asymptotic empirical risk minimization (AERM)
Anticausal prediction, 133
Approximation error function (AEF), 28, 31, 33
Assumption of randomness, 105
Asymptotic empirical risk minimization (AERM), 67, 68

Bayes error, 48
Bayes risk, 25, 72, 74–77
  conditional, 73–75
Binary classification, 25–27, 29, 32, 34, 60, 163, 174, 219, 233
Boost-by-majority, 49
Boosting, 37–50, 264
Bootstrap, 231–237, 240, 243
Bregman divergence, 72, 74
BrownBoost, 49

C4.5, 40
Catoni's localization theorem, 100
Causality, 129–140, 273, 274, 280
Causal Markov condition, 130
Clustering, 84, 90, 133, 273, 274, 280
Collaborative filtering, 188
Conditional Bayes risk, 73–75
Conditional validity, 114, 115
Conformal predictor, 113–115
Consistency, 7–11, 25–31, 33, 48, 49, 59–63, 65–68, 122, 143–159, 162, 231–243, 278

Dantzig selector, 196
Decision tree, 40, 48, 83
Density functional theory (DFT), 255
Domination loss, 261–263, 265

Embedding model, 85
Empirical risk function, 82
Empirical risk minimization (ERM), 60–68, 82, 179, 180, 217, 261
Exclusion dimension, 53–57
Exponential loss, 43–47, 49, 219

Falsification of a scientific theory, 54
f-divergence, 72–75, 275
Feature selection, 267, 273, 274, 280

General learning, 59, 60, 66–68
Gradient descent, 44, 206, 214, 248, 251, 253, 255
Growth function, 4, 9–11, 20

Hellinger divergence, 73
Hinge loss, 26, 28–30, 32, 33, 83, 84, 213, 214, 233, 235
Hypothesis depth, 163, 165, 166, 168

Independent component analysis (ICA), 273, 274, 280
Index, 9

Inductive setting, 135
Information retrieval (IR), 262

Kernel methods, 87–88, 105, 245–247, 257
Kernel principal component analysis (kPCA), 245–248, 253
Kernel ridge regression (KRR), 105–115, 249
KL divergence. *See* Kullback–Leibler divergence
k nearest neighbours, 86–87, 134
kPCA. *See* Kernel principal component analysis (kPCA)
Kriging, 107
KRR. *See* Kernel ridge regression (KRR)
Kullback–Leibler divergence, 73, 96, 97, 274–277

Lasso, 46, 49, 107, 196, 205, 206, 211, 212, 221
L^2-Distance, 275–278
Learnability, 59–68, 78, 178, 179
Learning rate, 25–29, 32–34, 62
Least squares method, 221
Least squares regression, 25–27, 29, 31–33, 43, 214
  co-regularized, 137
Linear model, 84, 246
Linear regression, 49, 195–204, 214
Log loss, 71, 73, 77
Logistic function, 164
Logistic loss, 32, 43, 84, 214, 233, 235
Logistic regression, 49, 119, 164, 165, 214
0-1 Loss, 25, 29, 60, 61, 65, 71–73
Loss function, 43–46, 49, 60, 61, 63–66, 71–79, 82–84, 96, 162, 178, 179, 182, 185, 186, 188–190, 214, 217, 231–233, 235, 262, 263
ℓ_1-Regularization, 46, 269, 270, 277
ℓ_2-Regularization, 214, 269, 270, 277
ℓ_1-Regularized exponential loss, 47

MA. *See* Median approximation (MA)
MAP. *See* Maximum a posteriori (MAP)
MATLAB, 277, 278
Maximum a posteriori (MAP), 162
Mean squared error, 143–159
Median approximation (MA), 171–173
Minimax forecaster, 178
Minimum mean squared error, 143–159
MTL. *See* Multi-task learning (MTL)

MT-MKL. *See* Multi-task multiple kernel learning (MT-MKL)
Multiple kernel learning (MKL), 123
Multi-task learning (MTL), 117–126, 205, 212, 215
Multi-task multiple kernel learning (MT-MKL), 121–123

Nemitski loss function, 235, 239, 241
Neural network, 83, 88–89, 257
Nonlinear gradient denoising (NLGD), 254, 256, 257

Occam's razor, 39
Online learning, 177, 212
Overfitting, 39–43, 46, 47, 53, 82, 86, 233, 251, 268

PAC. *See* Probably approximately correct (PAC)
PAC-Bayesian, 95–102, 161, 162, 167
PC. *See* Principal component (PC)
PCA. *See* Principal component analysis (PCA)
Pearson (PE) divergence, 275, 277
Penalized empirical risk minimization, 217, 218
Popper dimension, 53–57
Prediction with expert advice, 105, 179, 188, 189
Pre-image, 248–256
Principal component (PC), 247
Principal component analysis (PCA), 247, 277
Probably approximately correct (PAC), 65, 95
Proper loss, 72–77

Quadratic loss. *See* Squared loss

R, 114
R^2. *See* Randomized rounding
Rademacher complexity, 178, 180, 188, 193
Radial basis function (RBF), 87, 113, 231, 235
Randomized rounding (R^2), 178, 182
Ranking function, 261, 263
Rate of convergence, 8, 72, 76–77, 143–145, 154–159, 166, 203, 231
RBF. *See* Radial basis function (RBF)
Regularization, 45–50, 60, 64, 82, 84, 117–126, 162, 178, 188, 189, 205–215, 218,

Index 287

233, 238, 240, 249, 250, 261, 262, 266–270, 277
Regularization-based multi-task learning, 119–121
Relative Pearson (rPE) divergence, 275, 277
Renormalized Bayes, 115
Reproducing kernel Hilbert space (RKHS), 26–28, 111, 233, 235, 240
Ridge regression (RR), 49, 107
Ridge Regression Confidence Machine (RRCM), 114
RKHS. *See* Reproducing kernel Hilbert space (RKHS)
Root mean square errors (RMSEs), 139
rPE divergence. *See* Relative Pearson (rPE) divergence
RR. *See* Ridge regression (RR)
RRCM. *See* Ridge Regression Confidence Machine (RRCM)

Self-tuned Dantzig estimator, 196, 197
Semi-supervised learning (SSL), 84, 89, 129–140, 162, 173, 279
Semi-supervised regression (SSR), 137, 139, 140
SLT. *See* Statistical learning theory (SLT)
Sparsity, 46, 84, 87, 106, 107, 123, 195, 196, 205–215, 253, 256, 261–270, 277
Squared loss, 25, 27–30, 34, 60, 63–65, 71, 73, 77, 214, 219, 232, 275, 280
SRM. *See* Structural risk minimization (SRM)
SSL. *See* Semi-supervised learning (SSL)
SSR. *See* Semi-supervised regression (SSR)
Stability, 59–68, 161, 163, 169, 273

Statistical learning theory (SLT), 3, 60, 61
Structural risk minimization (SRM), 33, 82, 217
Supervised learning, 25, 53, 54, 59, 60, 64, 66, 68, 89, 118, 119, 137, 140, 162, 206
Supervised regression, 140
Support vector machine (SVM), 25–34, 49, 78, 84, 87–88, 95, 105–107, 114, 118, 134, 205, 212, 231–243, 246
Support vector regression (SVR), 105–107
Surrogate loss, 74
SVM. *See* Support vector machine (SVM)
SVR. *See* Support vector regression (SVR)

Transductive online learning, 177–193
Transductive setting, 89, 134, 177–193
Transfer learning, 119, 130
Tukey depth, 163–166, 168, 174
Tukey median, 161, 163, 165, 174

Uniform convergence, 3–11, 20, 95, 170
Uniform learnability, 62, 66–68
Universal consistency, 29–31, 145–154
Unsupervised learning, 162

Variational divergence, 73
VC dimension, 4, 10, 20, 21, 39, 53–57, 66, 67, 82, 95, 96, 170, 172, 179, 187, 188

Weak learner, 38, 40, 44
Weak learning condition, 39, 40, 48, 49

If you have any concerns about our products,
you can contact us on
ProductSafety@springernature.com

In case Publisher is established outside the EU,
the EU authorized representative is:
**Springer Nature Customer Service Center GmbH
Europaplatz 3, 69115 Heidelberg, Germany**

Printed by Libri Plureos GmbH
in Hamburg, Germany